BIOMACROMOLECULAR MASS SPECTROMETRY YEARBOOK SERIES

BIOMACROMOLECULAR MASS SPECTROMETRY YEARBOOK. VOLUME 1

Biomacromolecular Mass Spectrometry Yearbook Series

Biomacromolecular Mass Spectrometry Yearbook, Volume 1
Simone Koenig (Editor)
2010. ISBN: 978-1-60876-167-8

BIOMACROMOLECULAR MASS SPECTROMETRY YEARBOOK SERIES

BIOMACROMOLECULAR MASS SPECTROMETRY YEARBOOK. VOLUME 1

SIMONE KOENIG
EDITOR

Nova Science Publishers, Inc.
New York

For permission to use material from this book please contact us:
Telephone 631-231-7269; Fax 631-231-8175
Web Site: http://www.novapublishers.com

LIBRARY OF CONGRESS CATALOGING-IN-PUBLICATION DATA

Available upon request.

ISBN: 978-1-60876-167-8

Published by Nova Science Publishers, Inc. ✦ New York

CONTENTS

PREFACE

Chapter 1 - Structural variability resulting from protein post-translational glycosylation vastly increases the range of biological functions produced from the genome. The chemistry of glycans differs from that of proteins and dictates the use of specialized mass spectrometric techniques for successful analysis. Moreover, the unique chemistry of each glycoconjugate class drives the need for tailored analytical techniques. Glycosaminoglycans (GAGs) are linear glycans that coat the surfaces of all animal cells, bound to syndecan and glypican proteoglycan core proteins. There, they play developmentally critical roles as co-receptors for many classes of growth factors and growth factor receptors. Animal cells grow in organized extracellular matrix environments consisting of proteins, glycoproteins and proteoglycans.

GAGs serve to create and support immobilized growth factor gradients and play structural roles in these extracellular matrices. Although, due to their acidity, fragility and polydispersity, GAGs have traditionally proven difficult to analyze using mass spectrometry, effective new approaches have emerged over the past few years. This review summarizes structural biochemistry of GAGs and the uses of mass spectrometry in their analysis, with an emphasis on recent developments that have proven effective.

Chapter 2 - FTICR (Fourier transform ion cyclotron resonance) mass spectrometry offers several advantages for the analysis of biological samples, including excellent mass resolution, ultra-high mass accuracy, and a wide mass range. This paper gives a short overview of recent applications and technological developments of FTICRMS, which are relevant to the field of protein analysis. Tandem mass spectrometry, "bottom up" and "top down" approaches for protein analysis as well as methods for studying protein conformations are described and illustrated with examples.

Chapter 3 - Matrix assisted laser desorption ionization time-of-flight mass spectrometry (MALDI-TOF MS) and isoelectric focusing (IEF) of pure recombinant human MIA (melanoma inhibitory activity) protein, as expressed in CHO cells, indicated that part of the protein was post-translationally modified.

Whereas standard positive ion mode MALDI analysis of intact MIA revealed an additional peak 80 Da higher than the theoretical mass, IEF showed three major protein bands which could all unequivocally be identified as MIA isoforms by peptide mass fingerprinting.

Negative ionization MALDI mass spectrometry on a quadrupole TOF instrument indicated that the differences between these isoforms were due to one or two 80 Da modifications confined to tryptic fragment 59-76. The identity of the 80 Da post-translational modification (PTM) was determined to be a Tyr-O-sulfation through enzymatic removal of the 80 Da moiety from MIA by arylsulfatase treatment, and through the absence of the 80 Da modification(s) in MIA harvested from cells cultured in the presence of chlorate, which is known to inhibit Tyr-O-sulfation. The PTM being sulfation was further corroborated by its metastability upon MALDI-TOF MS analysis. It is concluded that the CHO produced MIA is partially post-translationally modified by Tyr-O-sulfation at positions 69 and/or 70 (^{59}LFWGGSVQGD**YY**GDLAAR^{76}).

Protein Tyr-O-sulfation, has been reported to serve various functions, including modulation of protein-protein interactions. Combination of this notion with the observation that the eukaryotic MIA protein shows only a weak binding towards laminin and fibronectin leads to the speculation that the putative *in vivo* metastasis promoting activity of MIA may be based on the interaction of MIA with an as yet unknown factor via Tyr-(O)-sulfation enhanced binding kinetics.

Chapter 4 - Many prokaryotes use polyhydroxyalkanoates (PHA) as storage compounds for carbon and/or energy in the form of water-insoluble inclusion PHA granules in the cytoplasm. The surface of such granules is covered by proteins which also affect their form and structure. Here, phasin PhaP1 protein was identified in high abundance on the grana surface of *R. eutropha* employing gel electrophoretic separation and mass spectrometry. Homologous phasins PhaP3 and Pha4 were visualized using 2D-PAGE in the PhaP1 knock-out mutant, but no evidence for the presence of PhaP2 could be found. All detected phasins exhibit isoforms whose structural differences are still a matter of investigation, although C-terminally truncated isoforms of PhaP1 were detected with FTICR-MS.

Chapter 5 – This chapter features speaker abstracts from the Münster Conference in 2004.

Chapter 6 - Some of the best times of my five year postdoc at the National Institutes of Health in Bethesda, Maryland, I had during lunch with my boss, Henry M. Fales (Hank). About seventy at the time, he had fantastic stories to share about his life with organic chemistry and mass spectrometry. They are too good to be forgotten and Hank kindly agreed to put them on paper. I collected some information shortly after I left his lab in summer of 1999 and worked it up for this publication in summer of 2005. It might interest readers among the mass spectrometrists who find the history of their technique as fascinating as I do. A bibliography is attached as a timeline.

Chapter 7 - Myeloid-related protein MRP14 is a member of the S100 family of calcium-binding proteins and is known to be phosphorylated *in vivo*. With the combination of enzymatic dephosphorylation, reversed phase and metal ion affinity chromatography, and mass spectrometry methods the phosphorylation site T113 of human MRP14 was unambiguously determined.

Chapter 8 - The dynamic range of protein expression has been recognized as a main difficulty for analyses decades ago with the advent of two-dimensional gel electrophoresis (2D-PAGE) and later in particular with the option of mass spectrometric protein identification. The general observation is that the abundant proteins are seldom of interest and that the proteins which matter are hard to catch.

Chapter 9 - This chapter features speaker abstracts from the Münster Conference in 2005.

Chapter 10 - Mass spectrometry has found a growing number of applications in many disciplines of the life sciences. This review will focus on the field of molecular and clinical virology. Examples for established as well as potential applications for mass spectrometry techniques will be given from different fields of virology, among them receptor identification,

identification of cellular interaction partners, post translational modification analysis, virus particle structure and clinical applications.

Chapter 11 - *Nocardia africana* is a pulmonary pathogen which has been recently isolated from human in Sudan. There is great need for proteomic investigations but they are sparse at present. First attempts on *N. africana SD 769* (DSM44491) visualized 74 protein of molecular weights between 10 -250 kDa and isoelectric points between 4–7. Due to the limited database information, identified proteins were assigned by homology to sequences from the other bacterial species such as *Staphylococcus aureus, Oceanobacillus iheyensis, N. farcinica* and *Yersinia pestis.*

Chapter 12 - Protein kinase B (PKB) plays an important role in various signal transduction pathways. PKB in its active state is phosphorylated on Thr-308 and Ser-473 and phosphorylates various target proteins. Mass spectrometric data suggests that PKB also has a high potential for autophosphorylation at random sites on expense of regulatory sites affecting protein activity.

Chapter 13 - A method is described which allows the generation of bioactive endothelial cell membrane preparations for subsequent processing of proopiomelanocortin peptides. Semi-quantitative mass spectrometry is used as a detector for monitoring of the products of membrane protease (neutral endopeptidase and angiotensin-converting enzyme) action.

Chapter 14 – This chapter features speaker abstracts from the Münster Conference in 2006.

Chapter 15 - Time-of-Flight Secondary Ion Mass Spectrometry (ToF-SIMS) is an analytical tool for surface analysis. Applying ToF-SIMS the impact of primary ions leads, among others, to the desorption of secondary ions that are characteristic for the composition of the sample's surface. Recent progress in instrumental development now facilitates a considerable use of this technique for the analysis of biological samples.

Several aspects make ToF-SIMS a beneficial tool for bioanalysis: It just probes the uppermost three monolayers of the sample, offers imaging capabilities with a lateral resolution as good as 300 nm and, recently, a 3D imaging option for biological samples was established. The latter option allows a vertical resolution of 30 nm.

This paper reviews the state-of-the-art technology and results on spectrometry, imaging, depth profiling and 3D analysis on biological samples such as cells, tissue sections and Langmuir-Blodgett films in order to give an overview on the chances and limits of this technique.

Chapter 16 - The recent discovery of a polycistronic mRNA in the red flour beetle *Tribolium castaneum* qualifying as a gap gene albeit encompassing only four very short open reading frames offers a great potential for mass spectrometric studies specifically targeting at small peptides. Addressing the question if more putative candidate peptides are indeed directly translated into their functional form, those studies might greatly help in investigating this postulated new class of coding RNAs, determining whether they are a rather rare and special occurrence, or much more common than previously assumed and just as yet hidden from the routinely applied gene prediction methods due to their unexpected properties.

Chapter 17 - Nowadays surface plasmon resonance and mass spectrometry are fully developed technologies for biomolecule characterization. Considerable efforts have been made in recent years to combine these techniques, since SPR and MS provide complementary data – quantitative and qualitative, respectively. An extremely important aspect of SPR experiments is the correct choice of chip surface and surface modification. These factors need to be carefully considered especially when planning the SPR-MS analyses. There are two basic experimental setups for SPR-MS experiments: one of these uses the microrecovery technique, whilst the other is an on-chip-MS based approach. For high-throughput arrays only the second setup has significant value. Both SPR and MS are technologies with developed array solutions. Therefore these techniques may be coupled to create the SPR-MS biomacromolecule array platform.

Chapter 18 - Adduct ions attributed to copper were observed at low abundance in peptide maps measured with MALDI mass spectrometry on some re-usable stainless steel targets. These ions were enhanced after mechanical treatment of the target using sandpaper. Although the copper source was not be satisfactorily elucidated the contaminant adduct ions seemed to be more present with target plates often cleaned. Such ions generated unwanted chemical noise at low intensity. In the examples shown here, peptide adduct ions of considerable abundance were observed. Their appearance can be cause for misinterpretation of the data in protein identification experiments and it should be prevented.

Chapter 19 - This chapter features speaker abstracts from the Münster Conference in 2007.

In: Biomacromolecular Mass Spectrometry Yearbook. Volume 1 ISBN: 978-1-60876-167-8
Ed: Simone Koenig

Chapter 1

PRINCIPLES OF MASS SPECTROMETRY OF GLYCOSAMINOGLYCANS*

Joseph Zaia†
Boston University, Department of Biochemistry,
Mass Spectrometry Resource

1. ABSTRACT

Structural variability resulting from protein post-translational glycosylation vastly increases the range of biological functions produced from the genome. The chemistry of glycans differs from that of proteins and dictates the use of specialized mass spectrometric techniques for successful analysis. Moreover, the unique chemistry of each glycoconjugate class drives the need for tailored analytical techniques. Glycosaminoglycans (GAGs) are linear glycans that coat the surfaces of all animal cells, bound to syndecan and glypican proteoglycan core proteins. There, they play developmentally critical roles as co-receptors for many classes of growth factors and growth factor receptors. Animal cells grow in organized extracellular matrix environments consisting of proteins, glycoproteins and proteoglycans.

GAGs serve to create and support immobilized growth factor gradients and play structural roles in these extracellular matrices. Although, due to their acidity, fragility and polydispersity, GAGs have traditionally proven difficult to analyze using mass spectrometry, effective new approaches have emerged over the past few years. This review summarizes structural biochemistry of GAGs and the uses of mass spectrometry in their analysis, with an emphasis on recent developments that have proven effective.

2. ABBREVIATIONS

CE	capillary electrophoresis

* A version of this chapter was also published in *Biomacromolecular Mass Spectrometry Volume 1, Number 1*, published by Nova Science Publishers, Inc. It was submitted for appropriate modifications in an effort to encourage wider dissemination of research.

† Dr. Joseph Zaia, Boston University School of Medicine, 715 Albany St., R-806, Boston, MA 02118. e-mail: jzaia@bu.edu. ph: 617-638-6762. Fax: -6761

CID	collisional-induced dissociation
CS	chondroitin sulfate
DHB	2,5-dihydroxybenzoic acid
DS	dermatan sulfate
ESI	electrospray ionization
FAB	fast atom bombardment
GAG	glycosaminoglycan
Gal	galactose
GalNAc	*N*-acetylgalactosamine
GlcA	glucuronic acid
GlcN	glucosamine
GlcNAc	*N*-acetylglucosamine
HexA	glucuronic acid or iduronic acid
ΔHexA	4,5-unsaturated uronic acid
HS	heparan sulfate
IdoA	iduronic acid
KS	keratan sulfate
MALDI	matrix-assisted laser desorption/ionization
TOF	time-of-flight
Xyl	xylose

3. INTRODUCTION

As the field of proteomics matures, researchers are coming to appreciate the complexity of mature gene products [1] that arises from post-translational modifications. Even the simplest of enzymatic post-translational modifications, Ser/Thr/Tyr phosphorylation, results in a several-fold increase in the number of forms of a given protein expressed at a given time. Post-translational modifications are dynamic in nature and provide a mechanism whereby cells meet changing growth and signaling needs with a surprisingly small number of genes.

Glycans are added to specific sites on the protein backbone through the actions of glycosyltransferase enzymes under complex control mechanisms (for details see [2]). Unlike the nucleic acids and proteins, glycan biosynthesis occurs without a template. Enzyme and substrate availabilities, among other factors, control the expressed structures of glycoconjugate glycans. As a result, glycans consist of distributions of variations on a core structure, attached to a given aglycon (amino acid residue or lipid core). Viewed in this way, glycosylation vastly increases the complexity of gene products relative to the number of genes in a given eukaryotic genome [1].

Glycoproteins are proteins carrying glycosylation, resulting from enzymatic post-translational modifications. Proteoglycans are a sub-class of glycoproteins in which linear sulfated carbohydrates, known as a glycosaminoglycans (GAGs), are attached to specific Ser residues on the protein backbone through a β-linked Xylose-containing tetrasaccharide. Glycoprotein glycans include the *N*-linked and *O*-linked classes. *N*-Linked glycans modify, through a β-linked GlcNAc, Asn residues in amino acid sequences of NXT or NXS, where X = any residue other than proline. The *N*-linked structure consists of a common chitobiose core $[(GlcNAc)_2(Man)_3]$ and variable branched elaborations forming an antenna structure.

Additions of units of lactosamine [Galβ(1-4)GlcNAc] or neo-lactosamine [Galβ(1-3)GlcNAc] capped on the non-reducing end with NeuAc are common to mammalian *N*-linked glycans. The carbohydrate consists of repeating disaccharide units, the biosynthesis of which occurs through the action of a series of glycosylas enzymes. The major classes of *O*-linked glycans are linked through β-*O*-GlcNAc or α-*O*-GalNAc to Ser or Thr residues of the protein backbone. The enzymatic modification of Ser/Thr residues by a single β-*O*-GlcNAc monosaccharide occurs to many nuclear and cytosolic proteins [3-5]. Many of the same Ser and Thr residues that are phosphorylated as part of signal transduction events are also targets for β-*O*-GlcNAc modification. It appears that β-*O*-GlcNAc is a reciprocal switch to phosphorylation and plays important rolse in many intracellular mechanisms. *O*-linked glycans are often branched, sialylated, and typically smaller than *N*-linked glycans. They are bound to Ser or Thr residues, commonly through an α-linked GalNAc residue. *O*-linked glycans are often found in clusters, known as mucin-like domains, along the protein backbone, giving rise to a bottle brush-like structure.

The addition of glycans to the scope of a mass spectrometric analysis scheme entails planning based on clear-eyed understanding of their chemical properties. Native glycans are considerably more fragile than are amino acid chains, and care must be taken to avoid in-source fragmentation during MALDI or ESI MS. Many glycans are capped with sialic acid residues, the glycosidic bonds of which readily dissociate during the MALDI process and during collisional induced dissociation. This phenomenon limits the ability to analyze sialylated glycans using traditional vacuum MALDI with either reflectron TOF or ICR mass analyzers. Any of a number of measures will minimize this problem, including methyl esterification of sialic acids [6, 7], use of anionic MALDI matrix dopants [8, 9], and the use of cooled MALDI sources to diminish the extent to which in-source fragmentation of sialylated glycans occur [10, 11]. Fucose residues also modify many classes of glycans and will fragment readily. Rearrangements of fucose residues have been observed during CID of protonated glycans [12, 13].

Enzymatic addition of sulfate or phosphate groups to specific sites on glycans considerably increases their acidity and fragility during MS ionization and CID. Due to analytical difficulties, the roles of these groups in modulating glycan biological activities have not been fully appreciated. Although FAB mass spectrometry was successfully applied to peptide analysis in the early 1980s, it resulted in abundant fragmentation to glycan sulfate groups, the extent of which increased as the number of sulfations per molecule increased [14].

GAGs are linear oligosaccharides consisting of repeating disaccharide units, as shown in Figure 1. Hyaluronan, an unsulfated GAG not attached to a core protein, consists of repeating units of [GlcAβ(1-3)GlcNAcβ(1-4)], is found in extracellular matrices throughout the animal body, and serves as a tether for glycoproteins and proteoglycans that contain a lectin domain. The size range of hyaluronan is 10^5 to 10^7 Da [2], although smaller fragments are also present in many tissues. Keratan sulfates (KS), unique among GAG classes in that they have Gal instead of HexA residues, are found in many connective tissues. Type I KS, originally described in cornea, is an extension to the type 2 lactosamine [Galβ(1-4)GlcNAc]-containing antennae of *N*-linked glycans. Type II KS, found in skeletal tissue, is an *O*-linked glycan bound to Ser or Thr residues through a αGalNAc residue. KS chains are usually sulfated on the 6-position of GlcNAc and may be sulfated on the 6-position of Gal residues. KS chains often have specialized capping sequence on the non-reducing chain ends, the complexities of which vary among different tissues [15-18]. For example, while neuraminic acids in either the α(2-3) or α(2-6) linkage cap KS in skeletal tissue, considerably more variable capping structures have been observed for corneal KS. In addition, α(1-3) fucose substituted onto GlcNAc residues has been observed for articular cartilage KS.

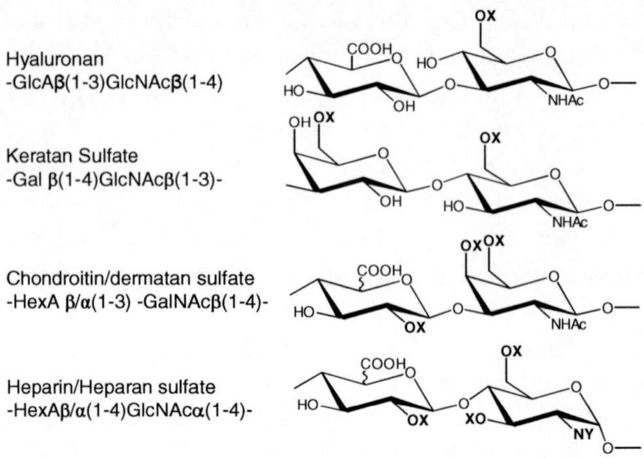

Figure 1. Disaccharide Repeating Structures for the Glycosaminoglycans. The possible positions of sulfation are indicated for each GAG class. X = H or SO_3H, Y = H or Ac or SO_3H.

Chondroitin sulfate (CS) consists of repeating units of [GlcAβ(1-3)GalNAcβ(1-4)] enzymatically polymerized on a protein-bound linker oligosaccharide of the sequence GlcAβ(1,3)Galβ(1,3)Galβ(1,4)-Xylβ(1-*O*)-Ser. Once polymerized, sulfotransferase and uronic acid epimerases modify the nascent chains to form mature CS with domains of high content of GlcA alternating with those containing high content of IdoA residues. Dermatan sulfate (DS), a form of CS, contains a high content of IdoA residues and a given chain will have CS-like domains alternating with DS-like domains. CS sequences are commonly sulfated on the 4- and/or 6-positions of GalNAc residues and less commonly at the 2-position of GlcA residues. DS sequences are commonly sulfated at the 4-position of GalNAc residues and less commonly at the 2-position of IdoA residues [19]. DS sequences are of biochemical interest because the high IdoA content confers conformational flexibility that facilitates interactions with protein binding partners (20).

The heparan sulfates (HS) are the most highly sulfated, complex, and information-rich of the GAGs [for a reviews see [21, 22]]. Biosynthetic enzymes polymerize repeats of [GlcAβ(1-4)GlcNAcα(1-4)] onto a linker of the same structure as CS/DS GAGs, and the same Ser residue may be modified with CS or HS chains under different conditions [23]. After biosynthesis, *N*-deacetylase/*N*-sulfotransferase enzymes act on the nascent chains to create alternating domains with high content of *N*-sulfated GlcN residues and those with high content of GlcNAc residues. Epimerases then convert many of the GlcA residues to IdoA and specific sulfotransferases modify the 3- and/or 6-positions of GlcN-sulfate residues and/or the 2-position of IdoA residues. These enzymatic modifications result in alternating *N*-acetylated domains with high GlcNAc and low sulfation and *N*-sulfated domains with high sulfation and epimerization. The enzymatic reactions do not go to completion, therefore, no two HS chains have exactly the same pattern of sulfation and epimerization. Heparin, a highly sulfated form of HS found in mast cell granules, contains a comparatively low content of *N*-acetylated domains. HS chains bound to syndecan and glypican core proteins coat the surfaces of all adherent animal cells, and bound to perlecan, agrin and other molecules play roles in extracellular matrices [for reviews see [22, 24, 25]]. The considerable ongoing biomedical interest in HS stems from the many classes of growth factors and growth factor receptors that bind to these molecules [26]. The HS sequences that bind to proteins of interest have been determined in only a few cases due to analytical difficulties. HS chains appear to serve as cell surface co-receptors that catalyze interactions of growth factors and growth factor receptors by virtue of their ability to bind both partners [27]. HS bound to extracellular matrix molecules appears to act by binding growth factors and setting up gradients for these molecules that are necessary for cell migration [28, 29].

Polysulfation confers chemical properties on the GAGs that dictate modification of the mass spectrometric techniques that have become standard for analysis of peptides and *N*- and *O*-linked oligosaccharides. Sulfate group fragility drives the need for gentle ionization conditions to prevent in-source fragmentation of these molecules. The value of tandem mass spectra generated from polysulfated oligosaccharides depends on the charge state of the precursor ion to a greater extent than observed for less acidic biological macromolecules (peptides and other glycan classes). The sections below summarize mass spectrometric approaches for analysis of sulfated GAGs. Although space limitations do not permit an exhaustive review of the literature, emphasis is given to the techniques that have proven the most practical for GAG analysis.

4. Techniques for GAG Analysis

a. Molecular Weight Measurement of Intact GAG Chains

Hyaluronan is a non-sulfated GAG that plays important roles in extracellular matrices through tethering glycoproteins and proteoglycans to cell surfaces. Oligosaccharides produced from hyaluronan have been used to characterize interactions with binding proteins. In some cases, low molecular weight fractions of hyaluronan have biological activities not associated with the parent molecule (reviewed in [30]). These observations have catalyzed efforts to generate methods for production of hyaluronan olgosaccharides in high purity. One such method combines SEC and high performance anion exchange chromatography [30]. ESI has been used effectively in the characterization of hyaluronan oligomers [30-32]. Negative ESI is a natural choice for hyaluronan fragments due the high uronic acid content.

Sulfated GAGs are polydisperse with respect to chain length and sulfate content, resulting in a distribution of glycoforms for a chain bound to a given Ser residue. Although it is possible to produce abundant ions from intact GAG chains using negative ESI, the polydispersity and high mass results in a complex pattern of overlapping charge states. This effect may be seen clearly with a quadrupole mass analyzer. Fourier transform MS has sufficient resolving power to separate isotopes and identify charge states for ions produced from molecules as large as GAGs (up to ~50 kDa) but is limited by space charge effects. Thus, an intact crude GAG sample will produce so many different *m/z* values that none is above the noise level. Determination of intact GAG chains using FTMS will require some preliminary chromatographic separation or ion filtering to limit the number of *m/z* values produced in a single spectrum.

The use of capillary and slab electrophoresis for the analysis of GAGs has recently been reviewed [33, 34]. Gradient polyacrylamide gel electrophoresis has been used to characterize GAGs in urine of mucopolysaccharidosis patients [35]. The value of slab gel electrophoresis is that it provides an estimation of the molecular weight distribution for these very complex samples. GAGs may be blotted onto cationic nitrocellulose membranes, a technique that potentially allows subsequent manipulation of the samples [36]. GAGs are similar to nucleic acids in that they are linear and anionic. Like nucleic acids, GAG chains may be separated using polyacrylamide gel electrophoresis with laser induced fluorescence detection. Unlike nucleic acids, sulfated GAGs are extremely polydisperse, and clear bands are not observed.

SEC has been widely used to determine the average size distributions in GAG preparations (for examples see [37-43]). GAG behavior during SEC separation has been shown to be distinct from that of proteins [44]. A method for determination of the average molecular weight of heparin preparations using SEC with universal calibration has been described [45].

b. GAG Degrading Enzymes

GAG degrading enzymes and their activities have been reviewed [46]. To give a brief background, GAGs are commonly depolymerized through the use of commercially available enzymes of two varieties. GAG lyase enzymes act through an eliminative mechanism to cleave a HexNAc-HexA bond, leaving a (4,5)-unsaturated (termed Δ-unsaturated) HexA residue at the new non-reducing terminus. GAG endohydrolase enzymes act by a hydrolytic mechanism to produce saturated residues at both the new reducing and non-reducing ends. For hyaluronan, testicular hyaluronidase is an effective endohydrolase enzyme for depolymerization.that cleaves the GlcNAc-GlcA bond (for a brief review see [30]). Testicular hyaluronidase also cleaves CS, albeit slowly. This enzyme is reported to have transglycosylase activity [47-50] and for this reason is not used for structural analysis of CS. For keratan sulfate, keratanase (from *Pseudomonas*), and keratanase II (from *Bacillus sp. Ks36*) enzymes are available. Keratanase is an endo-β-1,4-galactosidase that requires an unsulfated Gal residue flanked by 6-*O*-sulfated GlcNAc residues. Keratanase II is an endo-β-1,3-glucosaminidase that requires sulfation at the 6-*O* position of GlcN, but acts independently of Gal sulfation. Thus, it cleaves at a majority of 6-*O*-sulfated GlcNAc residues. For an overview of the use of keratanases for structural profiling of KS, see [18]. Each of the five commercially available chondroitin lyase enzymes cleave the GalNAc-HexA bond by an eliminative mechanism. Chondroitinase ABC, from *Proteus vulgaris*, is non-specific and cleaves all forms of CS. Chondroitinase ACI, purified from *Flavobacterium heparinum*, cleaves GalNAc-GlcA bonds in which GalNAc is 4- or 6-*O*-sulfated but is inactive against GalNAc-IdoA bonds. Chondroitinase ACI is active against the GlcA containing regions of CS/DS co-polymers. Chondroitinase ACII is purified from the culture broth of *Arthrobacter aurescens* and has similar activity but does not act on CS/DS co-polymers. Chondroitinase ACII is active against hyaluronan. Chondroitinase B cleaves GalNAc-IdoA bonds. Chondroitinase C is specific for GalNAc-GlcA bonds in which the GalNAc residue is 4-*O*-sulfated.

Enzymatic depolymerization of HS and heparin is discussed in detail in [26]. Three commercially available heparin lyases are available, each of which cleaves the GlcN-HexA bond in HS/heparin. Heparin lyase I, from *Flavobacterium heparinum*, cleaves highly sulfated regions, known as *N*-sulfated regions, of HS/heparin. Heparin lyase III, from *Flavobacterium heparinum*, cleaves regions with high content of GlcNAc residues. Heparin lyase II, from *Flavobacterium heparinum*, has some activity for *N*-acetylated regions and some for *N*-sulfated regions. HS/heparin may also be chemically cleaved using nitrous acid treatment. Nitrous acid at pH <2.5 cleaves *N*-sulfated GlcN residues, forming anhydromannose and leaving the uronic acid residue intact [51]. In this regard, nitrous acid is complementary to lyase digestion. Details on the use and mechanism of nitrous acid cleavage are given in [26].

d. Determination of GAG Depolymerization Products

The average chain length of GAG chains may be determined by quantifying the non-reducing or reducing terminal saccharides relative to those in the interior of the sequence. Lyase depolymerization results in the formation of a disaccharide containing a saturated HexA residue, corresponding to the non-reducing terminus of the intact chain. An ion exchange HPLC method for quantification of the non-reducing terminal saccharides relative to interior disaccharides resulting from lyase digestion has been used to estimate the size of CS chains [52, 53]. More recently, fluorophore-assisted carbohydrate electrophoresis has been used to characterize GAG depolymerization products [30, 54-56]. The method has the advantage of high sensitivity, due to the fluorescence of the 2-aminoacridone label used. This technique is similar in capability to capillary electrophoresis with laser induced fluorescence detection used for GAG disaccharide analysis [57-60]. More details on GAG disaccharide analysis are given in section 0.

e. GAG Sequencing

Analytical techniques for sequencing of HS/heparin oligosaccharides have been reviewed [26, 61, 62]. Generally, techniques involve isolation of GAG chains from a biological source followed by partial depolymerization and purification of an oligosaccharide of interest using affinity chromatography. Labeling of the oligosaccharides with a radioisotope, chromophore or fluorophore allows sensitive detection of the GAG oligosaccharides. Alternatively, cultured cells can be metabolically labeled with $^{35}SO_3$. The purified, labeled oligosaccharides are then subjected to further enzymatic and/or chemical cleavage and detection of the products using HPLC or electrophoresis [61, 63, 64]. These approaches carry the advantage that they have proven successful at sequencing selected GAG oligosaccharides from cellular sources.

5. Mass Spectrometric Ionization of GAG Oligosaccharides

The energetic nature of the desolvation process in early ESI sources resulted in similar problems to FAB MS, with intact ions from polysulfated glycans observed only in low abundances. Once optimized sources became available, however, the usefulness of ESI for the analysis of sulfated glycans became clear and the development of tandem mass spectrometric methods for determination of the structures of sulfated glycans became possible [65-71]. Traditional vacuum MALDI results in the deposition of a considerable amount of internal energy into analyte ions, and this causes very weak molecular ion signals for polysulfated glycans. The addition of basic peptides to pair with polysulfated glycans overcomes this problem and results in the detection of a peptide-oligosaccharide complex [72, 73].

The pKa of a typical sulfate group is ~3, rendering the sulfated GAGs the most acidic biomolecules. Although this property suggests the use of negative mode ionization, the engineering of mass spectrometers and orientation of mass spectrometrists tends toward the positive mode. This inclination goes a long way towards explaining the history of mass spectrometric analysis of sulfated GAGs in which the use of positive ionization resulted in very limited success toward the production of intact molecular ions. Fast atom bombardment (FAB), also known as liquid secondary ionization, was applied to the analysis of HS [74-79] and CS [75, 80] chains during the same timeframe as to peptides, in the early 1980s. In general, FAB ionization resulted in fragmentation to the sulfate groups and the observation of intact molecular ions in low abundances. At the same time, permethylated glycans [81] derived from a variety of glycoconjugates were observed to produce significantly stronger signals than their native counterparts [82-84].

a. MALDI of GAG Oligosaccharides

In the early days of MALDI, researchers observed that acidic biomolecules produce very weak intact molecular ions in either the positive or negative ionization modes [73]. After several years of work, it became clear that all biomolecules tend to undergo fragmentation in the MALDI source under vacuum conditions [85, 86], and that native oligosaccharides undergo this process to a significantly greater extent than do peptides [87]. Sialylated oligosaccharides pose a particular problem using vacuum MALDI, as shown by the fact that intact molecular ions cannot be transmitted through the ion reflector, indicating that metastable decay occurs in the first field free region [7]. Sulfate groups confer an even greater fragility when present on an oligosaccharide structure [72]. Figure 2 illustrates the tendency for protonated sulfate groups to dissociate as the internal energy increases during ionization or a dissociation experiment. During CID, deprotonated sulfate groups or those paired with a metal cation tend not to fragment to produce losses of SO_3 from the precursor ion and require significantly higher energies to undergo fragmentation than do protonated sulfate groups [14].

Hyaluronan lacks sulfation, and MALD mass spectrometric approaches are similar to those used for other acidic polysaccharide classes. MALDI approaches for carbohydrates have been reviewed [88, 89]. Use of negative MALDI-TOF MS in linear [90] or reflectron mode [91, 92] from 2,5-dihydroxybenzoic acid (DHB) matrix to profile hyaluronidase digestion products of hyaluronan resulted in detection of $[M-H]^-$ ions with no evidence of fragmentation of the ions. In the negative reflectron mode, DP4, 6 and 8 were detected, while in the positive mode only the DP4 was detected. The same MALDI conditions resulted in fragmentation to the sulfate groups of CS oligosaccharides [91]. Linear negative MALDI-TOF MS from DHB matrix was used to profile complex hyaluronan digest mixtures after separation by SEC [93]. Other researchers prefer 2,4,6-trihydroxyacetophenone/ammonium citrate matrix for negative reflectron MALDI of HA oligosaccharides [30], and the choice of matrix may depend on the instrument used. Reductive amination using anthanilic acid was found to enhance the negative MALDI signal strength from DHB matrix of mixtures of hyaloronan oligosaccharides (94). The reflectron mode was found to be effective for profiling DP4-10 and the linear mode most appropriate for DP12-36. In summary, the use of negative MALDI for profiling hyaluronan oligosaccharides is a well-established method.

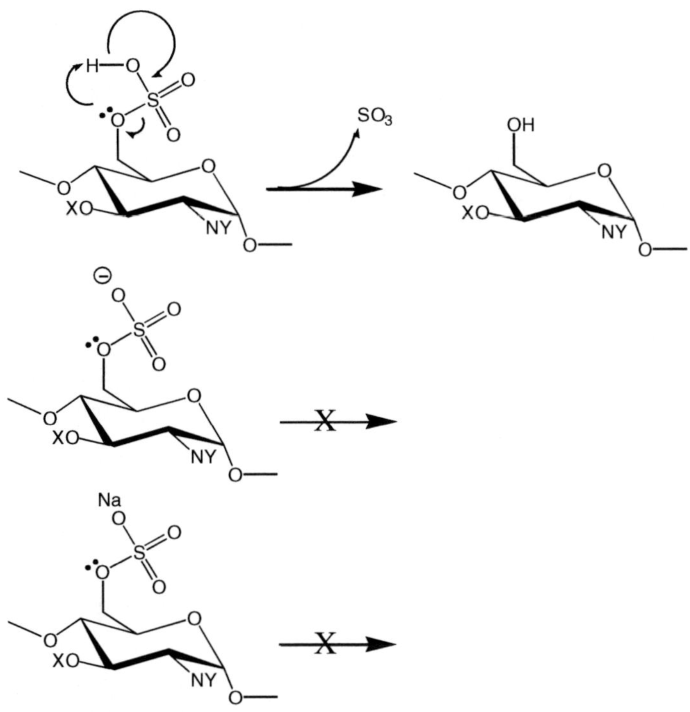

Figure 2. Production of abundant ions from losses of SO_3 in sulfated oligosaccharides during CID. X = SO_3H or H, Y = Ac or SO_3H or H.

Direct MALDI-TOF MS analysis of CS oligosaccharides using DHB matrix has shown results similar to those using FAB ionization [91] in that fragmentation to the sulfate groups was evident in the mass spectra. Chondroitin lyase degradation products analyzed as both positive and negative ions from DHB matrix showed fragmentation to the sulfate groups through losses of 102 u ($NaSO_3$) [91]. The use of positive linear mode MALDI-TOF MS from DHB matrix has been reported for CS oligosaccharides [95]. Multiply sodiated ions were detected, and it is likely that sulfate groups were lost during the ionization process. The basic peptide $(RG)_{15}$ has been used for positive MALDI-TOF MS analysis of CS oligosaccharides from caffeic acid matrix [96]. The use of basic peptides for the MALDI-TOF MS analysis of polysulfated GAG oligosaccharides is strongly recommended (details on this method are given below).

The addition of basic peptides or proteins to the matrix mixture containing polysulfated oligosaccharides results in the observation of a complex in positive mode MALDI MS [72, 73]. These complexes evidently arise from pairing of negatively charged sulfate groups with positively charged arginine or lysine residues, and the *m/z* value of the protein-oligosaccharide complex provides a means of molecular weight determination given the known protein mass. Measuring the masses of HS oligosaccharides using this approach before and after digestion with enzymes of known specificities provides an effective means of sequencing these molecules [97-100]. The observed *m/z* values for protein-carbohydrate complexes provides a means of determining the ability of a given HS oligosaccharide to bind a basic protein (growth factor) molecule using positive mode MALDI-TOF MS [101, 102]. Alternatively, as shown in Figure 2, pairing a sodium or other alkali cation with a sulfate group also stabilizes it during the ionization or CID process.

The use of MALDI-TOF MS to characterize GAGs and their interactions with proteins has recently been reviewed [103]. The most effective uses of MALDI-TOF MS for analysis of sulfated GAGs involve pairing the oligosaccharide to a basic peptide or protein, thereby minimizing the extent of in-source fragmentation. Measurement of HS oligosaccharide mass typically involves the use of the synthetic peptide $(RG)_{19}R$, isolated as the free base [100]. Caffeic acid matrix has proven effective for the observation of GAG-peptide complexes in the positive mode. Measurement of oligosaccharide mass before and after enzymatic or chemical degradation is a useful means of sequence determination [97-100, 104]. This approach has been used to study the binding of HS sequences to fibroblastic growth factors [104, 105]. The detection of intact proteins bound to a GAG oligosaccharide has been used to study growth factor oligomerization [102]. Mass spectrometric analysis of protein-GAG complexes has been reviewed in detail elsewhere [14].

b. ESI of GAG Oligosaccharides

Comparisons of the use of negative ESI and negative MALDI for analysis of hyaluronan oligosaccharides points out the importance of mass spectrometric operating parameters for success of the experiment [106, 107]. The use of ESI desolvation conditions that are appropriate for peptides will result in fragmentation to the HA oligosaccharides, with the result that ions corresponding to the loss of a monosaccharide residue from the oligosaccharides are detected. This in-source fragmentation may be eliminated by reducing the magnitude to the electrospray desolvation voltages. It is the author's experience that sources producing slow heating of the electrospray droplets (heated capillary sources or Z-spray) result in the most gentle desolvation. This is particularly important for analysis of polysulfated oligosaccharides. An LC/MS of system for analysis of hyaluronan oligosaccharides has been described that employs a C2 column for separation [107].

Observation of intact ions from sulfated GAG oligosaccharides using ESI requires special care despite the gentle nature of this ionization technique relative to vacuum MALDI. Desolvation conditions optimized for the analysis of peptides in the positive mode will surely cause in-source fragmentation of sulfated oligosaccharides. Greater success results from switching to the negative mode and using the lowest (in absolute terms) desolvation energies possible. In addition, the use of a solution containing ammonium sulfate [108], ammonium acetate [109], or ammonium hydroxide [110] is recommended. Negative ESI results in the observation of intact molecular ions, enabling tandem mass spectrometric experiments using this ionization mode. The production of positive ions by pairing polysulfated oligosaccharides with cationic counter ions has been demonstrated using ESI [111] and applied to on-line

LC/MS using reversed phase ion pairing chromatography [112, 113], as discussed in detail in section 0. While cationic ion paring is useful for conducting LC-MS analyses in positive ion mode, it is not useful in MS/MS analysis, due to the absence of oligosaccharide backbone cleavage products.

6. MASS SPECTROMETRIC GAG DISACCHARIDE PROFILING

Analysis of an unknown GAG sample entails determining the composition of the products of complete enzymatic degradation, often achieved using chondroitin or heparin lyase enzymes. Non-specific chondroitin lyase digestion of a CS/DS proteoglycan or released GAG sample results in the formation of the eight Δ-unsaturated structures shown in Table 1. Complete lyase digestion of a HS/heparin sample results in the formation of 12 Δ-unsaturated disaccharides, see Table 2. Note that in producing a 4,5-unsaturated uronic acid residue, the action of the lyase enzyme destroys the epimeric center in position 5. Several techniques will separate and quantify Δ-unsaturated lyase mixtures, including ion exchange, normal phase and reversed phase ion pairing chromatography [114]. The most sensitive methods of disaccharide analysis entail detection of fluorescent disaccharide derivatives. Capillary electrophoresis with laser-induced fluorescence detection (60, 115) or fluorescence assisted carbohydrate electrophoresis [54-56] of reductively aminated disaccharides with the 2-aminoacridone label have both proven effective for quantitative analysis. Separation of disaccharides using reversed phase ion pairing chromatography followed by post-column derivatization with 2-cyanoacetamide and fluorescence detection [116, 117] provides sufficient sensitivity for the analysis of CS/DS and HS disaccharides released from *Drosophila* and *C. elegans* preparations [118, 119] and has been used in a number of other biochemical studies [120-122].

Mass spectrometric determination of Δ-unsaturated disaccharides entails the use of a combination of MS and MS^n scans on an ion trap instrument. This approach hinges on the determination of the ionization response of each disaccharide relative to an internal standard to enable calculation of a correction factor for the raw ion abundances. Quantification of disaccharides with unique molecular masses entails determination of the corrected ion abundances relative to that of the internal standard. Tandem MS is necessary for isomeric disaccharides, and product ions uniquely abundant to a given structure must be identified using pure standards. The abundances of these ions in isomeric mixtures are used to quantify the composition of a mixture of isomers using a system of equations [123, 124]. This approach has been adapted for the analysis of CS [108, 125] and HS [126] disaccharides. Consider for example the Δ-unsaturated disaccharides derived from CS/DS, shown in Table 1. Distinct product ions that distinguish each of the three isomeric singly sulfated disaccharides ΔDi-UA2S, ΔDi-4S and ΔDi-6S [127] are listed in the table. The abundances of the three diagnostic product ions (*m/z* 237, 282, and 300 as shown in Table 1) for pure standards are used to mathematically determine the composition of the three disaccharides in a mixture based on the abundances of the diagnostic ions using a system of equations [108]. Although the published work has been limited to the singly sulfated CS/DS Δ-unsaturated disaccharides, the same principle will likely determine the compositions of mixtures containing the isomeric di-sulfated disaccharides shown in Table 1.

Lyase digestion of HS/heparin results in 12 possible Δ-unsaturated disaccharides, as shown in Table 2. Those disaccharides with the –A suffix are *N*-acetylated, –S are *N*-sulfated, and –H are unmodified GlcN residues [26]. The –A and –S disaccharides compose the bulk of lyase digestion products, with –H structures in very low abundances [128]. Mass spectrometric quantification of the isomeric III-A, II-A pair and the III-S, II-S pair [126] involves an approach similar to that used for the CS oligosaccharides, described above. Tandem mass spectrometry of pure II-A shows an abundant $^{0,2}A_2$ ion while that of III-A an abundant B_1 ion (see Table 2), useful for determining mixture compositions. Although the II-S disaccharide ion produces an abundant Y_1 as a diagnostic, the III-S disaccharide ion does not produce a useful ion in the MS^2 stage. Subsequent fragmentation of the $^{0,2}A_2$ ion (247.5→357.0) results in the formation of the diagnostic $^{2,4}A_2$ ion. Thus, tandem mass spectrometry using an ion trap based on ion abundances in the MS, MS^2 or MS^3 stages provides a means of quantification of HS/heparin disaccharides relative to a single internal standard. Presumably, similar methods will distinguish the –H disaccharides shown in Table 2.

Table 1. Structures of Δ-unsaturated CS/DS disaccharides, and diagnostic mass spectrometric ions useful for quantitation [108]. ND = not determined

Disaccharide	Structure	Mass	MS mode	Diagnostic ion	m/z
ΔDi-OS	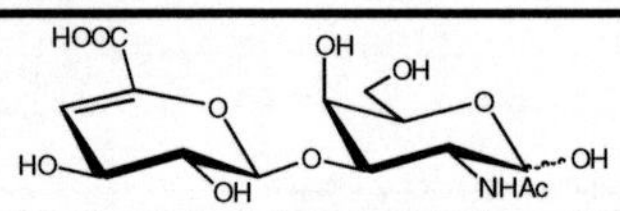	379.11	MS	$[M-H]^-$	378.1
ΔDi-4S	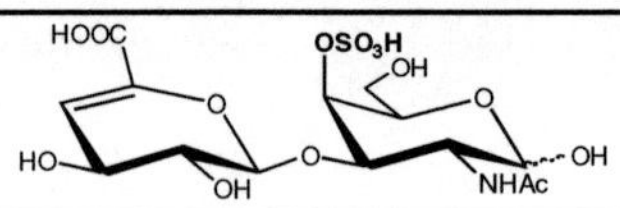	459.07	MS^2	Y_1	300.0
ΔDi-6S	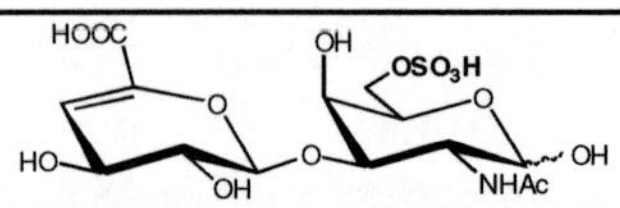	459.07	MS^2	Z_1	282.0
ΔDi-UA2S	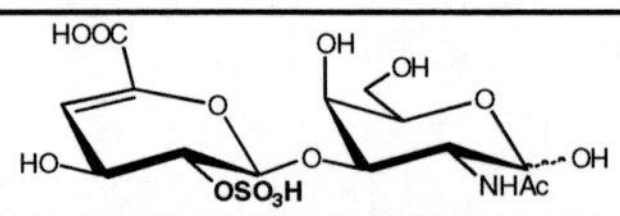	459.07	MS^2	B_1	237.0
ΔDi-S_B	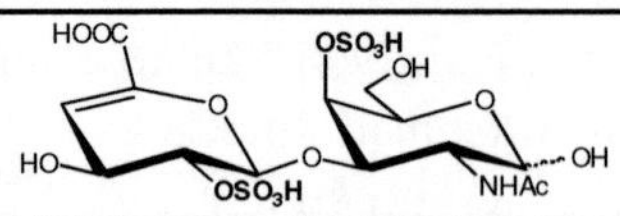	539.03	ND		
ΔDi-S_D	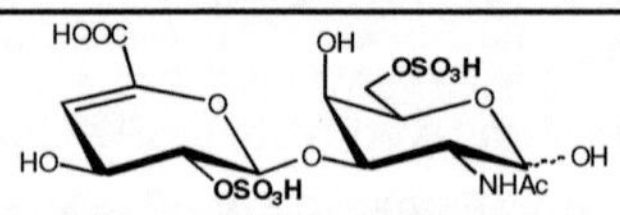	539.03	ND		
ΔDi-S_E	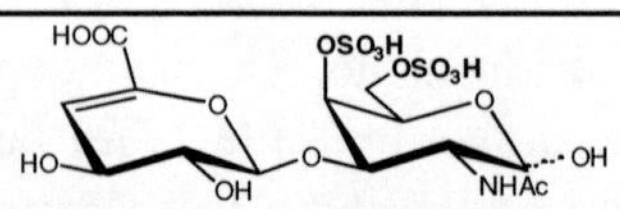	539.03	ND		
ΔDi-TriS	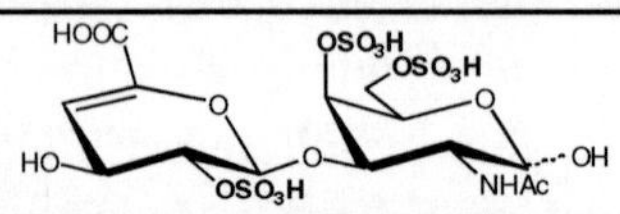	618.98	ND		

Table 2. Structures of Δ-unsaturated HS/heparin disaccharides, and diagnostic mass spectrometric ions useful for quantitation [126]. ND = not determined

Disaccharide	Structure	Mass	MS mode	Diagnostic ion	m/z
HS IV-A	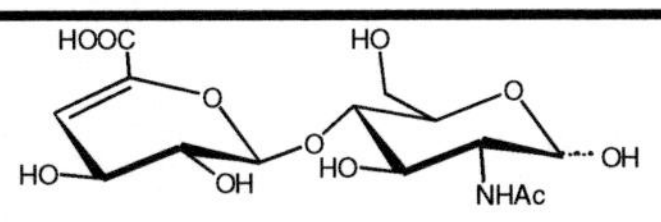	379.11	MS	$[M-H]^-$	378.1
HS III-A	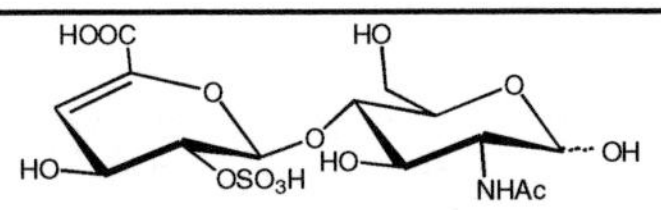	459.07	MS^2	B_1	237.0
HS II-A	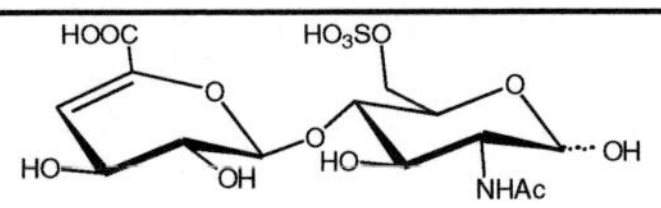	459.07	MS^2	$^{0,2}A_2$	357.1
HS I-A	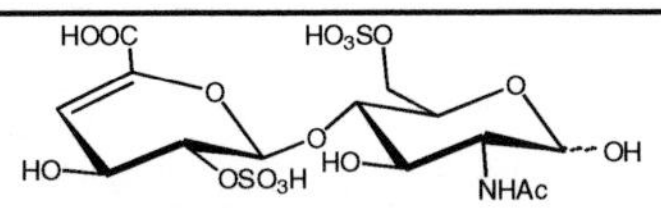	539.03	MS	$[M-2H]^{2-}$	268.5
HS IV-S	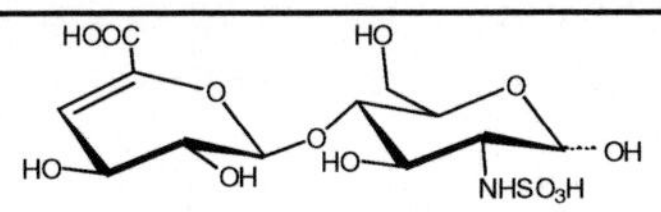	417.06	MS	$[M-H]^-$	416.0
HS III-S	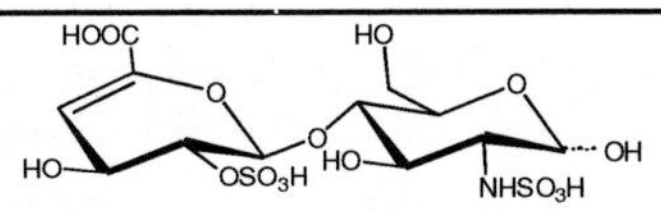	497.01	MS^3	$^{2,4}A_2$	247.5^{2-} → 357.0
HS II-S	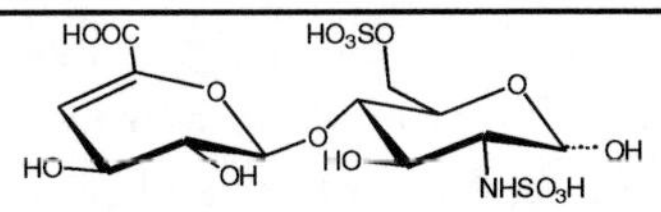	497.01	MS^2	Y_1	338.0
HS I-S	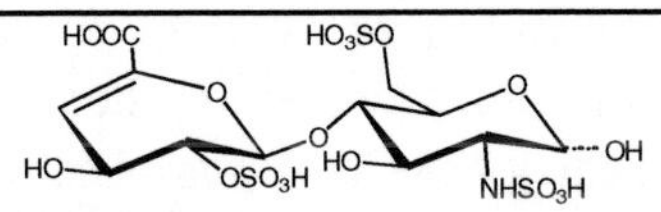	576.97	MS	$[M-H]^-$	576.0
HS IV-H	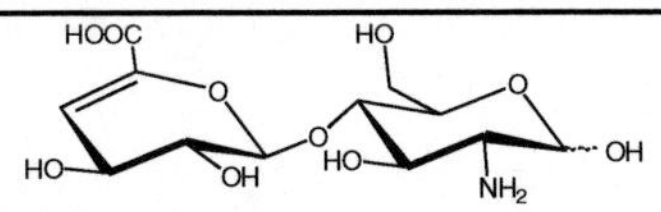	337.10	ND		
HS III-H	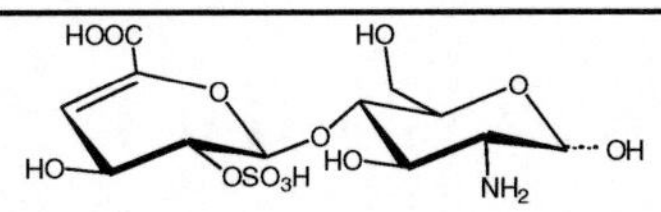	417.06	ND		
HS II-H	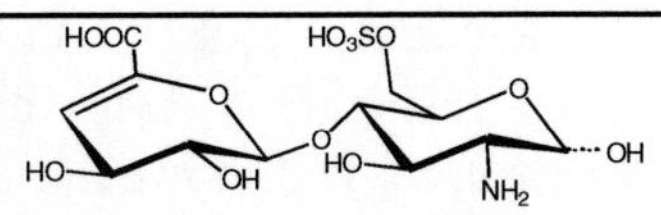	417.06	ND		
HS I-H	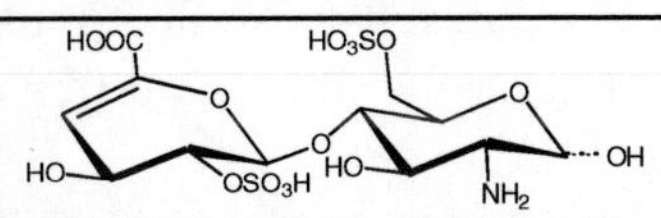	497.01	ND		

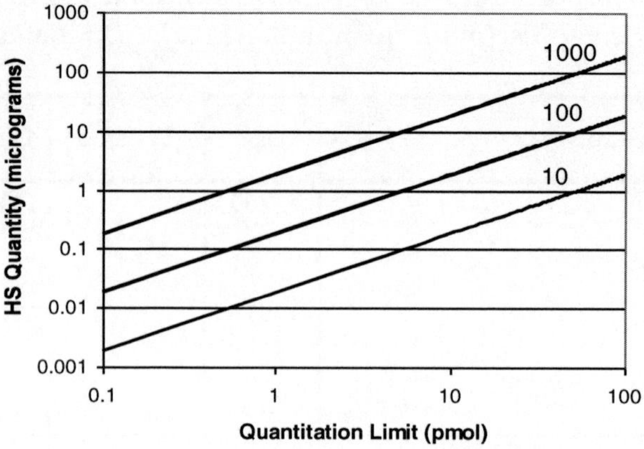

Figure 3. Total quantity of HS required for disaccharide analysis as a function of analytical detection limit and range of disaccharide concentrations. The three curves show 1000:1, 100:1, and 10:1 ranges in the concentrations of the most:least abundant disaccharides

The amount of HS required for disaccharide analysis depends on the limit of quantification of the analytical detection technique and on the range of concentration between the most abundant disaccharide and the least. As shown in Figure 3, a quantity of 2 μg of HS is required with an analytical detection limit of 1 pmol of disaccharide if the range of the constituent disaccharide concentrations varies over a factor of 1000. CE with laser-induced detection generates effective detection limits in this range for reductively aminated disaccharides [60, 115]. Note that the sample quantity limitations arise due to sample handling issues and not to the absolute sensitivity of the detector. For analysis of GAGs from small organisms such as *Drosophila*, sub-microgram quantities may be available and only the HPLC method with post-column 2-cyanoacetamide derivatization and fluorescence detection has demonstrated the sensitivity required for these samples [129]. Although mass spectrometric quantification requires no derivatization or workup of the enzymatically produced disaccharide mixture (108), the detection limits remain untested and the method has not yet been used for analysis of small quantities of GAG samples. Given the known performance of modern electrospray mass spectrometers, however, detection limits of 1 pmol per disaccharide or better seem reasonable. If so, then mass spectrometric quantification will compete with CE/LIF in terms of sensitivity.

7. Tandem MS of Sulfated GAGs

Although FAB ionization produced the first tandem mass spectra of sulfated GAGs [127, 130, 131], the singly charged ions fragmented by losing SO_3 from the precursor ions, diminishing the diagnostic value of the product ion pattern. ESI, however, results in multiple charging and the formation of glycosidic bond cleavages. The product ion patterns resulting from CID of multiply charged negative GAG oligosaccharide ions differ depending on the GAG oligosaccharide class, allowing for their analysis.

Chondroitin/dermatan sulfate. Commercially available chondroitin lyase enzymes act to cleave CS/DS chains to the reducing side of HexNAc residues by an eliminative mechanism to generate cleavage products with Δ-unsaturated uronic acid at the non-reducing terminus. The unsaturated bond absorbs at 232 nm, allowing precise control of the extent of enzymatic digestion. Tandem mass spectrometric fragmentation of Δ-unsaturated CS disaccharides results in an abundant $^{0,2}X_1$ ion that arises from rearrangement of the unsaturated uronic acid ring. CS/DS oligosaccharides also produce an analogous ion albeit at lower relative abundance.

The product ion patterns of CS/DS oligosaccharide ions change dramatically with charge state, as shown in Figure 4 [132]. With only one charge, two sulfate groups remain protonated for the $[M-H]^-$ ion, and fragmentation to the sulfate groups dominates the product ion profile. Glycosidic bond cleavage ions become significantly more abundant for the $[M-2H]^{2-}$ ion, and are most abundant to the reducing side of the GalNAc residues, similar to those observed adjacent to HexNAc residues in other classes of oligosaccharides [133]. This pattern changes dramatically during fragmentation of the $[M-3H]^{3-}$ charge state in that glycosidic cleavage is produced to the non-reducing side of the GalNAc residue. Evidently, the most labile glycosidic bond is no longer to the reducing side of the GalNAc residue, as is typical for glycans. Interestingly, the abundances of the glycosidic bond cleavages depend on the positions of GalNAc sulfation and uronic acid epimerization [134]. The usefulness of tandem MS for structural analysis of CS oligosaccharides has now been shown in a number of applications [135-137]. CS/DS oligosaccharides tend to form negative electrospray ions in which the most abundant corresponds to one charge per sulfate group, resulting in the formation of abundant ions from glycosidic bond cleavages with low abundances for ions from fragmentation to the sulfate groups [132, 138].

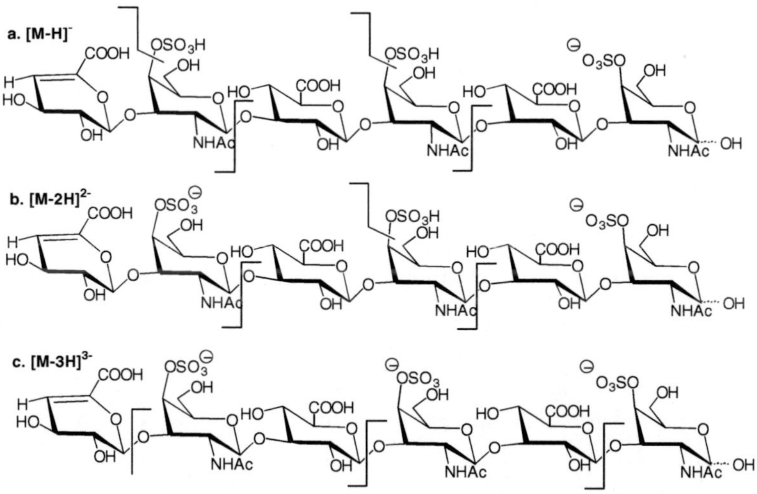

Figure 4. The effect of charge state on product ion pattern for CS/DS oligosaccharides. (a) $[M-H]^-$, (b) $[M-2H]^{2-}$, (c) $[M-3H]^{3-}$. The structures shown correspond to CSA-type (GlcA, GalNAc4S) and Similar trends are observed for oligosaccharides of different lengths and those derived from CSB/DS (IdoA, GalNAc4S) and CSC (GlcA, GalNAc6S). The most abundant bond cleavage ions are shown

Figure 5 illustrates the general trends for the dependence of negative ESI product ion pattern on deprotonated precursor ion charge in the tandem mass spectra of CS/DS oligosaccharides, combining data from the published literature [14, 108, 125, 132, 134, 136, 139, 140]. For the $[M-H]^-$ ion generated from a tetra-sulfated CS octamer, three of the four sulfate groups remain protonated, and the ion carries the equivalent of 0.125 charge per monosaccharide residue. Under these conditions, the most abundant product ion results from loss of SO_3 from the precursor. For the same octamers, $[M\text{-}4H]^{4-}$ ions have one charge per sulfate group, corresponding to 0.5 charge per monosaccharide residue and produce abundant glycosidic bond fragmentation with very low abundances for losses of SO_3. Note that losses of SO_3 are more abundant for GalNAc-6-sulfated CS than for GalNAc-4-sulfated CS/DS [134]. The 5- charge state for a CS/DS tetrasulfated octamer corresponds to 0.625 charge per monosaccharide residue and results in a useful product ion pattern with slightly elevated losses of the equivalent of $[H_2SO_4]^-$. As a general trend, the abundances of such losses increasee with increasing absolute value of negative charge state.

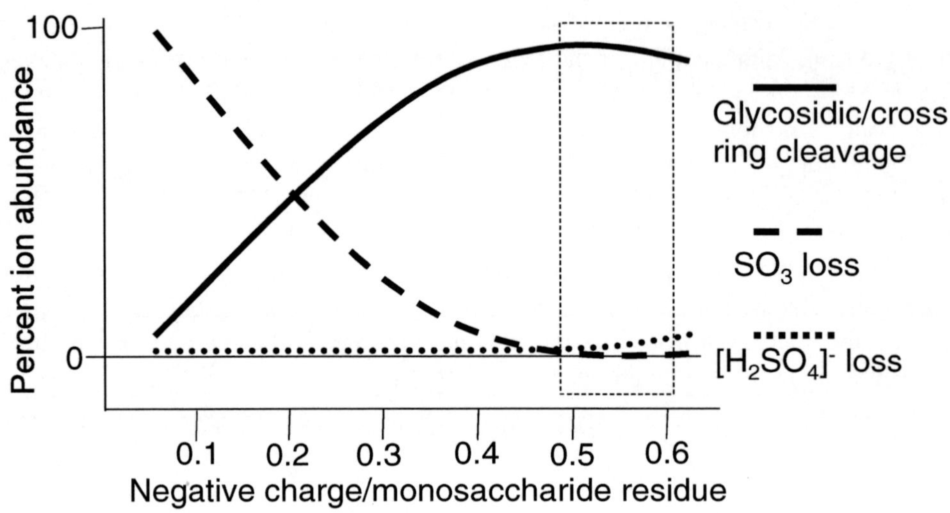

Figure 5. Trends in fragmentation types observed for CS oligosaccharides as a function of charge state. The plots are constructed from published CS/DS tandem mass spectrometric data and represent general trends rather than exact product ion abundances. The rectangle highlights the region of precursor ion negative charge/residue observed to produce the most useful product ion patterns

Keratan sulfates. For the analysis of KS chains there exist two commercially available glycosidase enzymes, keratanase and keratanase II. Keratanase is an endo-β-galactosidase, cleaving to the reducing side of unsulfated Gal residues and keratanase II is an endo-β-*N*-acetylglucosaminidase, cleaving to the reducing side of 6-*O*-sulfated GlcNAc residues, acting independently of sulfation to Gal residues. Both enzymes act by a hydrolytic mechanism to produce fully saturated oligosaccharides. Use of similar negative ion electrospray mass spectrometric conditions for KS oligosaccharides as used for CS/DS results in the production of ions in which all sulfate groups are charged. While, relative to CS/DS, KS oligosaccharides contain a similar number of sulfate groups per disaccharide repeat, they produce markedly different product ion patterns. Most oligosaccharides containing a reducing terminal 4-linked HexNAc residue produce an abundant $^{0,2}A_n$ cross-ring cleavage ion [141, 142], and KS oligosaccharides follow this rule. The cross-ring cleavage mechanism entails anomeric ring opening followed by a retro-aldol rearrangement that requires a free 3-hydroxyl group [141, 143], therefore, the 3-linked GalNAc residue on the reducing terminus of lyase-generated CS/DS oligosaccharides does not produce this $^{0,2}A_n$ ion [144]. In addition, KS oligosaccharides fragment to produce an abundant ion from loss of water from the precursor ion [145, 146]. Isolation and MS^3 fragmentation of the [M-H_2O] ion results in abundant and useful glycosidic bond cleavage ions [147]. Thus, the use of trapped ion mass spectrometers will achieve best results for analysis of KS oligosaccharides.

Heparan sulfates. Commercially available heparin lyases cleave HS chains by an eliminative mechanism to the reducing side of GlcN residues, creating a Δ-unsaturated HexA residue at the new non-reducing end. The absorbance of the unsaturated bond provides a means of controlling the extent of reaction precisely. Tandem mass spectrometric analysis of HS Δ-unsaturated disaccharides demonstrates that abundant $^{0,2}X_1$ ions are formed in addition to glycosidic bond and $^{0,2}A_2$ ions from cleavage of the 4-linked reducing terminal HexN residue [126]. The $^{0,2}X_1$ ions result from a retro-Diels-Alder rearrangement and the $^{0,2}A_2$ from anomeric ring opening followed by retro-aldol rearrangement [144].

Studies on synthetic heparin oligosaccharides have shown that the product ion pattern depends strongly on the precursor ion charge state. In a manner similar to that observed for CS/DS oligosaccharides, HS oligosaccharide ions with low negative charge states (in absolute value terms) contain several protonated sulfate groups and the most abundant product ions result from losses of SO_3 from the precursor ion [148]. Although the relative abundances of glycosidic bond cleavage product ions increases with higher negative charge states (absolute value terms), the abundances of other sulfate group fragmentations also increase. For oligosaccharides with triply sulfated GlcN residues, the abundances of ions produced from losses of the equivalent of $[H_2SO_4]^-$ increase as the precursor ion charge increases [149]. Fragmentations to the sulfate groups thus compete with glycosidic bond cleavages during the CID process for deprotonated HS oligosaccharides, entailing special care in interpreting the mass spectral profiles. The only way to insure that an observed product ion has not undergone secondary fragmentation to a sulfate group is through the detection of a complementary ion. The sum of the neutral mass equivalents of an ion and its complement total the mass of the neutral molecule. Alternatively, the pairing of alkali or alkaline earth metals with heparin oligosaccharides results in sulfate group stabilization, preventing both losses of SO_3 and the equivalent of $[H_2SO_4]^-$ to a large extent [148].Figure **6** illustrates the general trend for the effects of charge state on the observed product ion profiles. Precursor ions with fewer than 0.5 charge per monosaccharide residue have several sulfate groups that remain protonated, and losses of SO_3 dominate the product ion profiles. The formation of ions with approximately one charge per monosaccharide residue maximizes the abundances of useful glycosidic bond and cross-ring cleavage ions, as highlighted by the rectangle on the figure. With more than one charge per residue, the abundances of ions produced from losses of the equivalent of $[H_2SO_4]^-$ occur with increasing abundances relative to glycosidic bond cleavages. For this process, the loss of a charged group evidently reduces excessive like-charge repulsions due to excessive density of charge. For synthetic heparins bearing four sulfate groups per GlcN residue, it is possible to produce an ion with 1.5 charge per monosaccharide residue that produces an abundant ion the *m/z* value of which is consistent with loss of $[H]^-$ from the precursor [149]. The fact that excessive like-charge repulsions favor fragmentation that are uninformative dictates that the number of charges per residue for HS oligosaccharide ions should be limited to between 0.75 and 1.25.

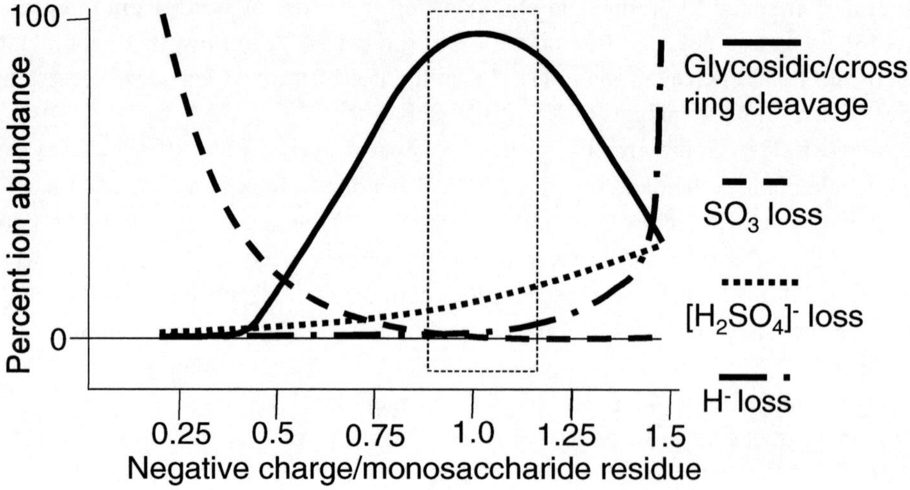

Figure 6. Trends in fragmentation types observed for HS oligosaccharides as a function of charge state. The plots are constructed from published HS/heparin tandem mass spectrometric data and represent general trends rather than exact product ion abundances. The rectangle highlights the region of precursor ion negative charge/residue observed to produce the most useful product ion patterns

8. LC/MS of GAG Oligosaccharides

The complexity of mixtures of oligosaccharides generated by enzymatic or chemical depolymerization of sulfated GAG chains precludes direct MS analysis without prior separation. Separation of CS/DS chains using high performance SEC provides sufficient separation to allow for direct MS analysis [139]. This mode of chromatography requires approximately 100 mM salt concentration to achieve adequate resolution and the use of ammonium acetate fulfills this requirement while not inhibiting mass spectrometric ion formation. SEC results in dilution of the GAG mixture, however, and the ammonium acetate salts will reduce signal strength to some degree. SEC may not, therefore, provide sufficient ion signal for the analysis of CS released from biological samples available in small supply (*i.e.* less than a few micrograms). On-line SEC/MS for separation and analysis of low molecular weight heparins using an ESI time-of-flight instrument has been used to determine the compositions of over 60 components in the GAG mixture [150]. In this work, the formation of heparin oligosaccharides using negative ESI resulted in ambiguous mass determinations, necessitating the removal of cations from the methanol/water/ammonium bicarbonate chromatographic solvent. The authors demonstrate that a commercially available self regenerating cation exchange column results in a 100-fold improvement in signal intensity and an increase in the negative ion charge state (in absolute terms). A chemical reaction mass spectrometric detector has been used to determine the elemental and isotopic composition of heparin preparations eluting from an SEC column [151, 152].

The use of basic peptides to pair polysulfated oligosaccharides has been used in ESI MS analysis in the positive mode. Cationic dyes have long been used in the analysis of sulfated GAGs both as a means of compound-class specific precipitation and as a means of spectrophotometric quantification. Taking advantage of this phenomenon, researchers have paired alkylammonium ions with polysulfated carbohydrates to enhance their hydrophobicity and thus facilitate reversed phase separations [153-155]. Tetrabutylammonium salts effectively pair with HS disaccharide digestion products, allowing reversed phase analysis [117, 119]. More recent research has identified dibutyl amine as an ion pairing agent compatible with on-line positive ion ESI LC/MS of heparin oligosaccharides [112]. Other researchers have used tributylammonium salts for this purpose [156]. Sulfated amines such as aminonaphthalenetrisulfonic acid, when used as reductive amination reagents, confer acidity on neutral glycans, thereby facilitating their analysis by electrophoresis [157] or capillary electrophoresis-MS [158] and triethylammonium salts have been used to pair such derivatized oligosaccharides for on-line LC/MS studies [159]. The use of ion pairs carries the advantage that it combines a high resolution reversed phase chromatographic separation with on-line mass spectrometric detection. It carries the disadvantage that the amines must be continuously pumped into the ion source.

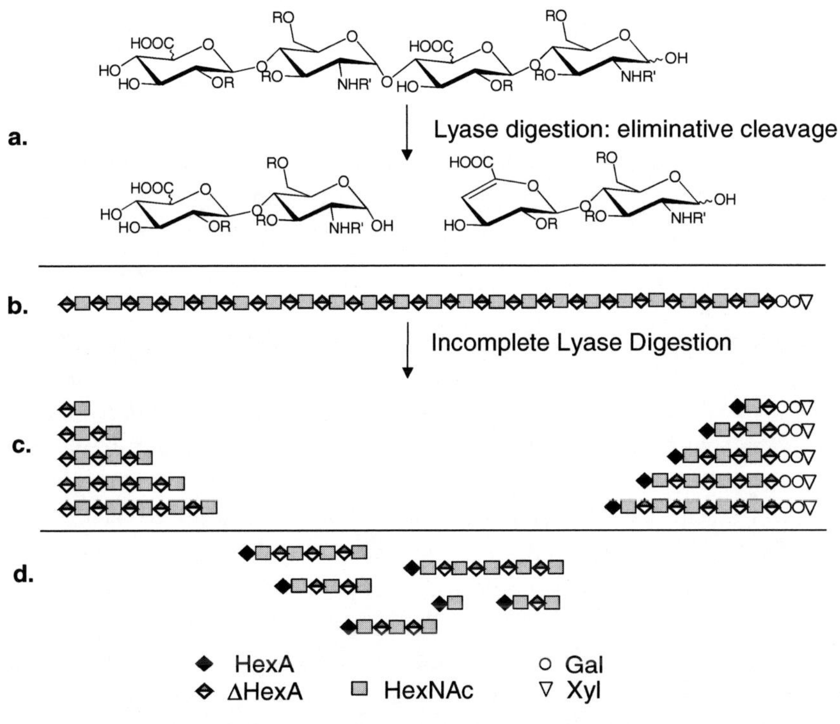

Figure 7. Lyase based sequence ladders of CS/DS and HS/heparin GAGs, (a) eliminative cleavage of a heparin tetramer, R = H or SO_3H, **R' =** H **or** SO_3H or Ac, resulting in the formation of a 4,5-unsaturated (Δ-unsaturated) uronic acid residue on the new non-reducing terminus, (b) a GAG chain (CS/DS or HS/heparin) is digested with a lyase enzyme using conditions to generate inclomplete extent of digestion to generate (c) products containing either the original reducing end (that include the linker structure) or the non-reducing end (that include a saturated uronic acid residue at their non-reducing ends) and (d) products from the interior of the original GAG chain with Δ-unsaturated HexA on the non-reducing ends and HexN on the reducing ends

Determination of the masses of HS oligosaccharides defines the compositions with respect to the numbers of HexA, HexN, sulfate and acetate groups. In addition, lyase digestion products containing the original reducing or non-reducing ends of the HS/heparin chain have unique mass values relative to other oligosaccharides in the mixture as shown in

Figure **7**. Under partial lyase digestion conditions, a series of oligosaccharides will be detected using the on-line reversed phase ion pairing LC/MS approach. Some of these oligosaccharides will have mass values consistent with a saturated HexA residue at their non-reducing terminus. Figure **7**c) that derives from the non-reducing end of the parent chain. Ions that contain the reducing end of the parent GAG will contain part of the linker oligosaccharide, depending on the method used to release the GAG from the core protein, and also have unique masses. The mass differences among the terminal series define the composition of each disaccharide in the sequence [156]. Other oligosaccharides will have mass values consistent with ΔHexA at the non-reducing terminus that result from lyase digestion to the interior of the GAG chainFigure 7d). Therefore, an on-line ion pairing LC/MS data set will contain information on the structure of the entire GAG chain, provided that an extended series of ions containing the reducing and non-reducing ends of the intact chain can be detected.

9. Capillary Electrophoresis/MS of GAG Oligosaccharides

Sulfated GAG di- and oligosaccharides produced by lyase digestion, by virtue of their acidity, migrate towards the anode in reversed polarity capillary electrophoresis (CE) and may be detected using the unique absorbance at 232 nm afforded by the Δ-unsaturated bond in the non-reducing terminal HexA residue [160]. The most highly sulfated oligosaccharides exhibit the fastest migration in the reversed polarity mode of CE, and the unsulfated molecules migrate slowly. In forward polarity CE, the electroosmotic flow suffices to carry sulfated GAG oligosaccharides toward the cathode despite their anionic nature and the most highly sulfated GAG oligosaccharides migrate the slowest [161, 162]. Because of the comparatively small path length in the capillary and the low volume of injection, the CE analysis requires at least 250 pmol of each disaccharide in a five microliter volume for detection of the Δ-unsaturated HexA residue at 232 nm. The addition of a chromophore or fluorophore via reductive amination dramatically reduces the detection limits for CE of GAG disaccharides, and derivatives appropriate for laser induced fluorescence detection produce impressive results [59, 60, 115].

The use of a sheathless interface for on-line CE/MS, first shown for separation of complex *O*-glycosylated peptide mixtures using ammonium acetate/ammonium hydroxide salts [163], allows screening of a several glycan classes [164]. Separation of CS/DS oligosaccharide mixtures using forward polarity CE with ammonium acetate buffer at pH 12.0 results in the detection of 10 peaks in a 10 minute electropherogram [140]. Subsequent negative ion nano ESI MS on a Q-TOF instrument reveals several different CS/DS compositions in a collected CE fraction and sufficient ion signal for tandem mass spectrometric analysis [136]. A similar approach has been used for off-line CE MS of ganglioside mixtures [137]. The extension of this work to on-line sheathless CE/MS of GAG oligosaccharides yielded impressive results in which CS/DS oligosaccharides with more sulfate groups than GalNAc residues, termed oversulfated oligosaccharides, were identified [135].

Mass spectrometric analysis of GAG oligosaccharides released from small tissue samples or from model organisms such as *Drosophila* or *C. elegans* requires miniaturization of the sample interface. As shown from the published work, CE provides a powerful means for both resolving and on-line introduction of the GAG oligosaccharide mixture into the mass spectrometer. It is unclear at this time, however, if CE/MS will be sensitive enough to enable analysis of sub-microgram quantities of GAGs released from small tissue samples.

10. Applications to Mucopolysaccharidosis

Catabolism of proteoglycans occurs by an ordered process, culminating in the formation of monosaccharides in the lysosome. Proteolytic enzymes first degrade the core protein, followed by endoglycosidic cleavage of the GAG chains into large pieces. A series of exoglycosidases and sulfatases then remove residues from the non-reducing terminus of each GAG oligosaccharide until only monosaccharides remain. Because of the exolytic nature of the final steps of GAG degradation, the lack or misfunction of one exolytic enzyme causes the accumulation of partially degraded material in the lysosome, with deleterious health effects of varying severity. Mucopolysaccharidosis (MPS) is a class of lysosomal storage disorder [for a review see [165]] in which the lack of normal GAG-degrading enzyme activity results in accumulation of partially degraded oligosaccharides in the lysosome and elevated levels of GAG excretion into the urine relative to normal subjects. In general, subjects with the MPS diseases suffer from neurological and skeletal abnormalities and often do not live past young adulthood. Because therapies must commence as early in life as possible to minimize developmental abnormalities, effective analytical diagnostic and treatment-monitoring methods are needed. Treatment measures include enzyme replacement therapy [166, 167] and gene therapy [168]. A number of diagnostic approaches exist for characterizing MPS phenotypes, including electrophoretic patterning [35]. Mass spectrometric approaches to diagnosis of MPS phenotypes and for monitoring treatment have been developed [167, 169, 170].

The four types of Sanfilippo syndrome (mucopolysaccharidosis type III) arise from the lack of a HS/heparin degrading enzyme activity (type A, heparan *N*-sulfatase; type B, *N*-acetylglucosaminidase; type C, acetyl CoA: α-glucosaminideacetyltransferase; and type D, *N*-acetylglucosamine 6-sulfatase), each leading to similar clinical symptoms. Effective clinical treatment depends on identification of the deficient enzyme, presently carried out using radiometry [171], spectrophotometry [172], or fluorometry [173]. In an effort to measure all four enzyme activities on a single platform, an affinity mass spectrometry scheme has been developed to measure the four heparin degrading enzyme activities in cell lysates derived from forearm skin biopsies [174, 175]. The researchers synthesized enzymatic substrates conjugated to biotin and deuterated versions of the digestion products also conjugated to biotin. By adding both compounds to the lysates and allowing enzymatic digestion to occur, the biotin-conjugated substrate is converted to product of the same structure as the deuterated biotin-conjugated internal standard. Thus, mass spectra from normal level of enzymatic activity display deuterated and non-deuterated forms of the product and abnormal enzymatic activity results in the absence of the non-deuterated ion. The biotin handle carries the advantage that the activity measurement occurs in a complex lysate not directly compatible with mass spectrometry. The MS approach described combines measurement of all four enzymatic activities in a single platform.

Analysis of urine samples spiked with radiolabeled monosaccharides by ion exchange chromatography followed by paper chromatography established that sulfated hexosamine levels correlate with the types and severity of MPS diseases [176-178]. A recently described ESI mass spectrometric method quantifies mono- and disaccharides in urine samples of MPS patients with minimal sample workup, improving on existing methods [179]. Urine (or blood plasma) was lyophilized and derivatized using 1-phenyl-3-methyl pyrazolone (PMP) [180] and the excess reagent removed by solvent extraction. The aqueous layer was subjected to reversed phase chromatography and analyzed off-line using negative ESI MS. A deuterated, PMP-derivatized *N*-acetylglucosamine-6-sulfate standard was used as an internal control to quantitate the total monosulfated HexNAc levels, a combination of GalNAc4S, GalNAc6S, and GlcNAc6S. In addition, disulfated HexNAc and monosulfated HexNAc-HexA were quantified relative to the internal standard. The levels of these three ions (HexNAcS, HexNAcS$_2$ and HexNAcS-HexA) in urine effectively differentiated MPS patients from controls. This methodology has been used to monitor dose response of enzyme replacement therapy in feline MPS VI (*N*-acetylgalactosamine-4-sulfatase deficiency) [167]. In this work, the HexNAcS/HexNAcS-HexA ratio was significantly higher in urine of untreated MPS VI than in normal or treated animals.

Negative ESI MS of GAGs isolated from urine by a combination of anion exchange and gel filtration chromatography with subsequent PMP derivatization resulted in the identification of oligosaccharides the levels of which correlate with MPS type I (α-L-iduronidase deficiency) [169]. Tandem MS was used to characterize a tetrasaccharide, the levels of which were elevated in MPS I patient urine.

11. Conclusions

Determination of the mass of GAG oligosaccharides serves to define the composition with respect to monosaccharides residues, sulfate and acetate groups. The published record on MS of GAG oligosaccharides demonstrates effective use of MALDI of these molecules paired with basic peptides in the positive mode, ESI paired with basic peptides or other ammonium cations, and direct negative ESI to generate deprotonated ions. Any of these approaches may be used to determine the masses of GAG oligosaccharides before and after enzymatic or chemical digestion to reveal structural information or partial sequences. Tandem MS of positively charged GAG oligosaccharide ions typically results in abundant ions produced from fragmentation to the sulfate groups and loss of valuable information. Such analysis of ions with multiple negative charges, however, produce ion profiles with considerable structural value. The product ion patterns depend strongly on the ion charge state, and it is generally possible to observe abundant ions from glycosidic bond cleavages. While sulfated GAG oligosaccharides may be analyzed using either MALDI or ESI, only the latter technique produces ions amenable to tandem mass spectrometry.

References

[1] Mann, M., and Jensen, O. N. Proteomic analysis of post-translational modifications (2003) *Nat Biotechnol* 21, 255-61.

[2] Varki, A., Cummings, R., Esko, J., Freeze, H., Hart, G., and Marth, G. (1999) Essentials of Glycobiology, Cold Spring Harbor Laboratory Press, Cold Spring Harbor, NY.

[3] Wells, L., Whelan, S. A., and Hart, G. W. O-GlcNAc: a regulatory post-translational modification (2003) *Biochem Biophys Res Commun* 302, 435-41.

[4] Zachara, N. E., and Hart, G. W. O-GlcNAc a sensor of cellular state: the role of nucleocytoplasmic glycosylation in modulating cellular function in response to nutrition and stress (2004) *Biochim Biophys Acta* 1673, 13-28.

[5] Wells, L., and Hart, G. W. O-GlcNAc turns twenty: functional implications for post-translational modification of nuclear and cytosolic proteins with a sugar (2003) *FEBS Lett* 546, 154-8.

[6] Handa, S., and Nakamura, K. Modification of sialic acid carboxyl group of ganglioside (1984) *J Biochem (Tokyo)* 95, 1323-29.

[7] Powell, A. K., and Harvey, D. J. Stabilization of sialic acids in *N*-linked oligosaccharides and gangliosides for analysis by positive ion matrix-assisted laser desorption/ionization mass spectrometry (1996) *Rapid Commun Mass Spectrom* 10, 1027-32.

[8] Wong, A. W., Wang, H., and Lebrilla, C. B. Selection of anionic dopant for quantifying desialylation reactions with MALDI-FTMS (2000) *Anal Chem* 72, 1419-25.

[9] Wong, A. W., Cancilla, M. T., Voss, L. R., and Lebrilla, C. B. Anion dopant for oligosaccharides in matrix-assisted laser desorption/ionization mass spectrometry (1999) *Anal Chem* 71, 205-11.

[10] O'Connor, P. B., Mirgorodskaya, E., and Costello, C. E. High pressure matrix-assisted laser desorption/ionization Fourier transform mass spectrometry for minimization of ganglioside fragmentation (2002) *J Am Soc Mass Spectrom* 13, 402-07.

[11] O'Connor, P. B., and Costello, C. E. A high pressure matrix-assisted laser desorption/ionization Fourier transform mass spectrometry ion source for thermal stabilization of labile biomolecules (2001) *Rapid Commun Mass Spectrom* 15, 1862-68.

[12] Ernst, B., Müller, D. R., and Richter, W. J. False sugar sequence ions in electrospray tandem mass spectrometry of underivatized sialyl-Lewis-type oligosaccharides (1997) *Int J Mass Spectrom Ion Proc* 160, 283-90.

[13] Harvey, D. J., Mattu, T. S., Wormald, M. R., Royle, L., Dwek, R. A., and Rudd, P. M. "Internal residue loss": rearrangements occurring during the fragmentation of carbohydrates derivatized at the reducing terminus (2002) *Anal Chem* 74, 734-40.

[14] Zaia, J. Mass spectrometry of oligosaccharides (2004) *Mass Spectrom Rev* 23, 161-227.

[15] Tai, G. H., Nieduszynski, I. A., Fullwood, N. J., and Huckerby, T. N. Human corneal keratan sulfates (1997) *J Biol Chem* 272, 28227-31.

[16] Lauder, R. M., Huckerby, T. N., and Nieduszynski, I. A. The structure of the keratan sulphate chains attached to fibromodulin from human articular cartilage (1997) *Glycoconj J* 14, 651-60.

[17] Tai, G. H., Huckerby, T. N., and Nieduszynski, I. A. Multiple non-reducing chain termini isolated from bovine corneal keratan sulfates (1996) *J Biol Chem* 271, 23535-46.

[18] Whitham, K. M., Hadley, J. L., Morris, H. G., Andrew, S. M., Nieduszynski, I. A., and Brown, G. M. An improved method for the structural profiling of keratan sulfates:

analysis of keratan sulfates from brain and ovarian tumors (1999) *Glycobiology* 9, 285-91.

[19] Cheng, F., Heinegard, D., Malmstrom, A., Schmidtchen, A., Yoshida, K., and Fransson, L. A. Patterns of uronosyl epimerization and 4-/6-*O*-sulphation in chondroitin/dermatan sulphate from decorin and biglycan of various bovine tissues (1994) *Glycobiology* 4, 685-96.

[20] Venkataraman, G., Sasisekharan, V., Cooney, C. L., Langer, R., and Sasisekharan, R. A stereochemical approach to pyranose ring flexibility: its implications for the conformation of dermatan sulfate (1994) *Proc Natl Acad Sci U S A* 91, 6171-75.

[21] Lindahl, U., Kusche-Gullberg, M., and Kjellen, L. Regulated diversity of heparan sulfate (1998) *J Biol Chem* 273, 24979-82.

[22] Bernfield, M., Gotte, M., Park, P. W., Reizes, O., Fitzgerald, M. L., Lincecum, J., and Zako, M. Functions of cell surface heparan sulfate proteoglycans (1999) *Annu Rev Biochem* 68, 729-77.

[23] Esko, J. D., and Zhang, L. Influence of core protein sequence on glycosaminoglycan assembly (1996) *Curr Opin Struct Biol* 6, 663-70.

[24] Perrimon, N., and Bernfield, M. Specificities of heparan sulphate proteoglycans in developmental processes (2000) *Nature* 404, 725-8.

[25] Esko, J. D., and Lindahl, U. Molecular diversity of heparan sulfate (2001) *J Clin Invest* 108, 169-73.

[26] Conrad, H. E. (1998) Heparin Binding Proteins, Academic Press, New York.

[27] Lander, A. D. Proteoglycans: master regulators of molecular encounter? (1998) *Matrix Biol* 17, 465-72.

[28] Nugent, M. A., Nugent, H. M., Iozzo, R. V., Sanchack, K., and Edelman, E. R. Perlecan is required to inhibit thrombosis after deep vascular injury and contributes to endothelial cell-mediated inhibition of intimal hyperplasia (2000) *Proc Natl Acad Sci U S A* 97, 6722-7.

[29] Nybakken, K., and Perrimon, N. Heparan sulfate proteoglycan modulation of developmental signaling in Drosophila (2002) *Biochim Biophys Acta* 1573, 280-91.

[30] Mahoney, D. J., Aplin, R. T., Calabro, A., Hascall, V. C., and Day, A. J. Novel methods for the preparation and characterization of hyaluronan oligosaccharides of defined length (2001) *Glycobiology* 11, 1025-33.

[31] Roboz, J., Deng, L., Ma, L., and Silides, D. Monitoring the Stepwise Decomposition of Hyaluronan by Hyaluronidase in Pleural/Peritoneal Effusions from Patients with Malignant Mesothelioma Using Negative Electrospray Mass Spectrometry. (2000) *Proc 48th ASMS Conf Mass Spectrom Allied Topics*.

[32] Price, K. N., Al, T., Baker, D. C., Chisena, C., and Cysyk, R. L. Isolation and characterization by electrospray-ionization mass spectrometry and high-performance anion-exchange chromatography of oligosaccharides derived from hyaluronic acid by hyaluronate lyase digestion: observation of some heretofore unobserved oligosaccharides that contain an odd number of units (1997) *Carbohydr Res* 303, 303-11.

[33] Koketsu, M., and Linhardt, R. J. Electrophoresis for the analysis of acidic oligosaccharides (2000) *Anal Biochem* 283, 136-45.

[34] Mao, W. J., Thanawiroon, C., and Linhardt, R. J. Capillary electrophoresis for the analysis of glycosaminoglycans and glycosaminoglycan-derived oligosaccharides (2002) *Biomed Chromatogr* 16, 77-94.

[35] Byers, S., Rozaklis, T., Brumfield, L. K., Ranieri, E., and Hopwood, J. J. Glycosaminoglycan accumulation and excretion in the mucopolysaccharidoses: characterization and basis of a diagnostic test for MPS (1998) *Mol Genet Metab* 65, 282-90.

[36] Maccari, F., and Volpi, N. Glycosaminoglycan blotting on nitrocellulose membranes treated with cetylpyridinium chloride after agarose-gel electrophoretic separation (2002) *Electrophoresis* 23, 3270-7.

[37] Horenberg, J., and De Vries, J. X. Characterization of heparinsby high-pressure size exclusion liquid chromatography (1983) *J Chrom* 261, 287-92.

[38] de Vries, J. X. Analysis of heparins by size-exclusion and reversed-phase high-performance liquid chromatography with photodiode-array detection (1989) *J Chromatogr* 465, 297-304.

[39] Kristensen, H. I., Tromborg, E. M., Nielsen, J. R., Nielsen, J. I., Johansen, K. B., and Ostergaard, P. B. Development and validation of a size exclusion chromatography method for determination of molecular masses and molecular mass distribution in low molecular weight heparin (1991) *Thromb Res* 64, 131-41.

[40] Maderich, A. B., and Sugita, E. T. Determination of polyanionic macromolecules by size-exclusion chromatography (1993) *J Chromatogr* 622, 278-83.

[41] Malsch, R., and Harenberg, J. High-performance size exclusion chromatography and polyacrylamide gel electrophoresis for characterization of unfractionated and low molecular mass glycosaminoglycans (1994) *Semin Thromb Hemost* 20, 135-43.

[42] Soltes, L., Mislovicova, D., and Sebille, B. Insight into the distribution of molecular weights and higher-order structure of hyaluronans and some beta-(1 --> 3)-glucans by size exclusion chromatography (1996) *Biomed Chromatogr* 10, 53-9.

[43] Knobloch, J. E., and Shaklee, P. N. Absolute molecular weight distribution of low-molecular-weight heparins by size-exclusion chromatography with multiangle laser light scattering detection (1997) *Anal Biochem* 245, 231-41.

[44] Volpi, N., and Bolognani, L. Glycosaminoglycans and proteins: different behaviours in high-performance size-exclusion chromatography (1993) *J Chromatogr* 630, 390-6.

[45] Guo, X., Condra, M., Kimura, K., Berth, G., Dautzenberg, H., and Dubin, P. L. Determination of molecular weight of heparin by size exclusion chromatography with universal calibration (2003) *Anal Biochem* 312, 33-9.

[46] Ernst, S., Langer, R., Cooney, C. L., and Sasisekharan, R. Enzymatic degradation of glycosaminoglycans (1995) *Crit Rev Biochem Mol Biol* 30, 387-444.

[47] Takagaki, K., Ishido, K., Kakizaki, I., Iwafune, M., and Endo, M. Carriers for enzymatic attachment of glycosaminoglycan chains to peptide (2002) *Biochem Biophys Res Commun* 293, 220-4.

[48] Takagaki, K., Munakata, H., Kakizaki, I., Iwafune, M., Itabashi, T., and Endo, M. Domain structure of chondroitin sulfate E octasaccharides binding to type V collagen (2002) *J Biol Chem* 277, 8882-9.

[49] Takagaki, K., Munakata, H., Majima, M., Kakizaki, I., and Endo, M. Chimeric glycosaminoglycan oligosaccharides synthesized by enzymatic reconstruction and their

use in substrate specificity determination of Streptococcus hyaluronidase (2000) *J Biochem (Tokyo)* 127, 695-702.

[50] Takagaki, K., Munakata, H., Majima, M., and Endo, M. Enzymatic reconstruction of a hybrid glycosaminoglycan containing 6- sulfated, 4-sulfated, and unsulfated *N*-acetylgalactosamine (1999) *Biochem Biophys Res Commun* 258, 741-4.

[51] Shively, J. E., and Conrad, H. E. Formation of anhydrosugars in the chemical depolymerization of heparin (1976) *Biochemistry* 15, 3932-42.

[52] Plaas, A. H., Wong-Palms, S., Roughley, P. J., Midura, R. J., and Hascall, V. C. Chemical and immunological assay of the nonreducing terminal residues of chondroitin sulfate from human aggrecan (1997) *J Biol Chem* 272, 20603-10.

[53] Plaas, A. H., Hascall, V. C., and Midura, R. J. Ion exchange HPLC microanalysis of chondroitin sulfate: quantitative derivatization of chondroitin lyase digestion products with 2- aminopyridine [published erratum appears in Glycobiology 1997 Sep;7(6):xiv] (1996) *Glycobiology* 6, 823-9.

[54] Calabro, A., Midura, R., Wang, A., West, L., Plaas, A., and Hascall, V. C. Fluorophore-assisted carbohydrate electrophoresis (FACE) of glycosaminoglycans (2001) *Osteoarthritis Cartilage* 9 Suppl A, S16-22.

[55] Calabro, A., Hascall, V. C., and Midura, R. J. Adaptation of FACE methodology for microanalysis of total hyaluronan and chondroitin sulfate composition from cartilage (2000) *Glycobiology* 10, 283-93.

[56] Calabro, A., Benavides, M., Tammi, M., Hascall, V. C., and Midura, R. J. Microanalysis of enzyme digests of hyaluronan and chondroitin/dermatan sulfate by fluorophore-assisted carbohydrate electrophoresis (FACE) (2000) *Glycobiology* 10, 273-81.

[57] Militsopoulou, M., Lamari, F., and Karamanos, N. K. Capillary electrophoresis: a tool for studying interactions of glycans/proteoglycans with growth factors (2003) *J Pharm Biomed Anal* 32, 823-28.

[58] Lamari, F. N., Kuhn, R., and Karamanos, N. K. Derivatization of carbohydrates for chromatographic, electrophoretic and mass spectrometric structure analysis (2003) *J Chromatogr B Analyt Technol Biomed Life Sci* 793, 15-36.

[59] Lamari, F. N., Militsopoulou, M., Mitropoulou, T. N., Hjerpe, A., and Karamanos, N. K. Analysis of glycosaminoglycan-derived disaccharides in biologic: samples by capillary electrophoresis and protocol for sequencing glycosaminoglycans (2002) *Biomed Chromatogr* 16, 95-102.

[60] Militsopoulou, M., Lamari, F. N., Hjerpe, A., and Karamanos, N. K. Determination of twelve heparin- and heparan sulfate-derived disaccharides as 2-aminoacridone derivatives by capillary zone electrophoresis using ultraviolet and laser-induced fluorescence detection (2002) *Electrophoresis* 23, 1104-9.

[61] Stringer, S. E., Kandola, B. S., Pye, D. A., and Gallagher, J. T. Heparin sequencing (2003) *Glycobiology* 13, 97-107.

[62] Turnbull, J., Powell, A., and Guimond, S. Heparan sulfate: decoding a dynamic multifunctional cell regulator (2001) *Trends Cell Biol* 11, 75-82.

[63] Drummond, K. J., Yates, E. A., and Turnbull, J. E. Electrophoretic sequencing of heparin/heparan sulfate oligosaccharides using a highly sensitive fluorescent end label (2001) *Proteomics* 1, 304-10.

[64] Turnbull, J. E., Hopwood, J. J., and Gallagher, J. T. A strategy for rapid sequencing of heparan sulfate and heparin saccharides (1999) *Proc Natl Acad Sci U S A* 96, 2698-703.

[65] Geyer, H., Bahr, U., Liedtke, S., Schachner, M., and Geyer, R. Core structures of polysialylated glycans present in neural cell adhesion molecule from newborn mouse brain (2001) *Eur J Biochem* 268, 6587-99.

[66] Wheeler, S. F., and Harvey, D. J. Extension of the In-Gel Release Method for Structural Analysis of Neutral and Sialylated *N*-Linked Glycans to the Analysis of Sulfated Glycans: Application to the Glycans from Bovine Thyroid-Stimulating Hormone (2001) *Anal Biochem* 296, 96-100.

[67] Karlsson, N. G., Johansson, M. E., Asker, N., Karlsson, H., Gendler, S. J., Carlstedt, I., and Hansson, G. C. Molecular characterization of the large heavily glycosylated domain glycopeptide from the rat small intestinal Muc2 mucin (1996) *Glycoconj J* 13, 823-31.

[68] Karlsson, N. G., Karlsson, H., and Hansson, G. C. Sulphated mucin oligosaccharides from porcine small intestine analysed by four-sector tandem mass spectrometry (1996) *J Mass Spectrom* 31, 560-72.

[69] Thomsson, K. A., Karlsson, N. G., and Hansson, G. C. Liquid chromatography-electrospray mass spectrometry as a tool for the analysis of sulfated oligosaccharides from mucin glycoproteins (1999) *J Chromatogr A* 854, 131-9.

[70] Thomsson, K. A., Karlsson, H., and Hansson, G. Sequencing sulfated oligosaccharides from mucins by liquid chromatography and electrospray ionization tandem mass spectrometry (2000) *Anal Chem* 72, 4543-49.

[71] Robbe, C., Capon, C., Coddeville, B., and Michalski, J. C. Diagnostic ions for the rapid analysis by nano-electrospray ionization quadrupole time-of-flight mass spectrometry of O-glycans from human mucins (2004) *Rapid Commun Mass Spectrom* 18, 412-20.

[72] Juhasz, P., and Biemann, K. Utility of non-covalent complexes in the matrix-assisted laser desorption ionization mass spectrometry of heparin-derived oligosaccharides (1995) *Carbohydr Res* 270, 131-47.

[73] Juhasz, P., and Biemann, K. Mass spectrometric molecular-weight determination of highly acidic compounds of biological significance via their complexes with basic polypeptides (1994) *Proc Natl Acad Sci U S A* 91, 4333-37.

[74] Reinhold, V. N., Carr, S. A., Green, B. N., Petitou, M., Choay, J., and Sinay, P. Structural characterization of sulfated glycosaminoglycans by fast atom bombardment mass spectrometry: application to heparin fragments prepared by chemical synthesis (1987) *Carbohydr Res* 161, 305-13.

[75] Ii, T., Okuda, S., Hirano, T., and Ohashi, M. Tandem mass spectrometry for characterization of unsaturated disaccharides from chondroitin sulfate, dermatan sulfate and hyaluronan (1994) *Glycoconj J* 11, 123-32.

[76] Ii, T., Kubota, M., Hirano, T., Ohashi, M., Yoshida, K., and Suzuki, S. FAB CID-MS/MS characterization of tetrasaccharide tri- and tetrasulfate derived from the antigenic determinant recognized by the anti- chondroitin sulfate monoclonal antibody MO-225 (1995) *Glycoconj J* 12, 282-89.

[77] Khoo, K. H., Morris, H. R., McDowell, R. A., Dell, A., Maccarana, M., and Lindahl, U. FABMS/derivatisation strategies for the analysis of heparin-derived oligosaccharides (1993) *Carbohydr Res* 244, 205-23.

[78] Chai, W., Rosankiewicz, J. R., and Lawson, A. M. TLC-LSIMS of neoglycolipids of glycosaminoglycan disaccharides and of oxymercuration cleavage products of heparin fragments that contain unsaturated uronic acid (1995) *Carbohydr Res* 269, 111-24.

[79] Sugahara, K., Takemura, Y., Sugiura, M., Kohno, Y., Yoshida, K., Takeda, K., Khoo, K. H., Morris, H. R., and Dell, A. Chondroitinase ABC-resistant sulfated trisaccharides isolated from digests of chondroitin/dermatan sulfate chains (1994) *Carbohydr Res* 255, 165-82.

[80] Carr, S. A., and Reinhold, V. N. Structural characterization of sulfated glycosaminoglycans by fast atom bombardment mass spectrometry: application to chondroitin sulfate (1984) *J Carbohydr Chem* 3, 381-401.

[81] Ciucanu, I., and Kerek, F. A simple and rapid method for the permethylation of carbohydrates (1984) *Carbohydr Res* 131, 209-17.

[82] McNeil, M., Darvill, A. G., Aman, P., Franzen, L. E., and Albersheim, P. Structural analysis of complex carbohydrates using high-performance liquid chromatography, gas chromatography, and mass spectrometry (1982) *Methods Enzymol* 83, 3-45.

[83] Dell, A., Oates, J. E., Morris, H. R., and Egge, H. Structure determination of carbohydrates and glycosphingolipids by fast atom bombardment mass spectrometry (1983) *Int J Mass Spectrom Ion Phys* 46, 415-18.

[84] Dell, A., Morris, H. R., Egge, H., von Nicolai, H., and Strecker, G. Fast-atom bombardment mass spectrometry for carbohydrate structure determination (1983) *Carbohydr Res* 115, 41-52.

[85] Spengler, B., Kirsch, D., and Kaufmann, R. Metastable decay of peptides and proteins in matrix-assisted laser-desorption mass-spectrometry (1991) *Rapid Commun Mass Spectrom* 5, 198-202.

[86] Yu, W., Vath, J. E., Huberty, M. C., and Martin, S. A. Identification of the facile gas-phase cleavage of the Asp-Pro and Asp- Xxx peptide bonds in matrix-assisted laser desorption time-of-flight mass spectrometry (1993) *Anal Chem* 65, 3015-23.

[87] Spengler, B., Kirsch, D., Kaufmann, R., and Lemoine, J. Structure-Analysis of Branched Oligosaccharides Using Post-Source Decay in Matrix-Assisted Laser-Desorption Ionization Mass-Spectrometry (1994) *Organic Mass Spectrometry* 29, 782-87.

[88] Harvey, D. J. Matrix-assisted laser desorption/ionization mass spectrometry of carbohydrates and glycoconjugates (2003) *Mass Spectrom Rev* 226, 1-35.

[89] Harvey, D. J. Matrix-assisted laser desorption/ionization mass spectrometry of carbohydrates and glycoconjugates (2003) *International Journal of Mass Spectrometry* 226, 1-35.

[90] Vercruysse, K. P., Ziebell, M. R., and Prestwich, G. D. Control of enzymatic degradation of hyaluronan by divalent cations (1999) *Carbohydr Res* 318, 26-37.

[91] Schiller, J., Arnhold, J., Benard, S., Reichl, S., and Arnold, K. Cartilage degradation by hyaluronate lyase and chondroitin ABC lyase: a MALDI-TOF mass spectrometric study (1999) *Carbohydr Res* 318, 116-22.

[92] Courel, M. N., Maingonnat, C., Tranchepain, F., Deschrevel, B., Vincent, J. C., Bertrand, P., and Delpech, B. Importance of hyaluronan length in a hyaladherin-based assay for hyaluronan (2002) *Anal Biochem* 302, 285-90.

[93] Yeung, B., and Marecak, D. Molecular weight determination of hyaluronic acid by gel filtration chromatography coupled to matrix-assisted laser desorption ionization mass spectrometry (1999) *J Chromatogr A* 852, 573-81.

[94] Seyfried, N. T., Blundell, C. D., Day, A. J., and Almond, A. Preparation and application of biologically active fluorescent hyaluronan oligosaccharides (2004) *Glycobiology*.

[95] Mikami, T., Mizumoto, S., Kago, N., Kitagawa, H., and Sugahara, K. Specificities of three distinct human chondroitin/dermatan N-acetylgalactosamine 4-O-sulfotransferases demonstrated using partially desulfated dermatan sulfate as an acceptor: implication of differential roles in dermatan sulfate biosynthesis (2003) *J Biol Chem* 278, 36115-27.

[96] Ueoka, C., Nadanaka, S., Seno, N., Khoo, K. H., and Sugahara, K. Structural determination of novel tetra- and hexasaccharide sequences isolated from chondroitin sulfate H (oversulfated dermatan sulfate) of hagfish notochord (1999) *Glycoconj J* 16, 291-305.

[97] Venkataraman, G., Shriver, Z., Raman, R., and Sasisekharan, R. Sequencing complex polysaccharides (1999) *Science* 286, 537-42.

[98] Ernst, S., Rhomberg, A. J., Biemann, K., and Sasisekharan, R. Direct evidence for a predominantly exolytic processive mechanism for depolymerization of heparin-like glycosaminoglycans by heparinase I (1998) *Proc Natl Acad Sci U S A* 95, 4182-87.

[99] Rhomberg, A. J., Ernst, S., Sasisekharan, R., and Biemann, K. Mass spectrometric and capillary electrophoretic investigation of the enzymatic degradation of heparin-like glycosaminoglycans (1998) *Proc Natl Acad Sci U S A* 95, 4176-81.

[100] Rhomberg, A. J., Shriver, Z., Biemann, K., and Sasisekharan, R. Mass spectrometric evidence for the enzymatic mechanism of the depolymerization of heparin-like glycosaminoglycans by heparinase II (1998) *Proc Natl Acad Sci U S A* 95, 12232-37.

[101] Fromm, J. R., Hileman, R. E., Weiler, J. M., and Linhardt, R. J. Interaction of fibroblast growth factor-1 and related peptides with heparan sulfate and its oligosaccharides (1997) *Arch Biochem Biophys* 346, 252-62.

[102] Venkataraman, G., Shriver, Z., Davis, J. C., and Sasisekharan, R. Fibroblast growth factors 1 and 2 are distinct in oligomerization in the presence of heparin-like glycosaminoglycans (1999) *Proc Natl Acad Sci U S A* 96, 1892-97.

[103] Sturiale, L., Naggi, A., and Torri, G. MALDI mass spectrometry as a tool for characterizing glycosaminoglycan oligosaccharides and their interaction with proteins (2001) *Semin Thromb Hemost* 27, 465-72.

[104] Kreuger, J., Salmivirta, M., Sturiale, L., Gimenez-Gallego, G., and Lindahl, U. Sequence analysis of heparan sulfate epitopes with graded affinities for fibroblast growth factors 1 and 2 (2001) *J Biol Chem* 276, 30744-52.

[105] Guerrini, M., Agulles, T., Bisio, A., Hricovini, M., Lay, L., Naggi, A., Poletti, L., Sturiale, L., Torri, G., and Casu, B. Minimal heparin/heparan sulfate sequences for binding to fibroblast growth factor-1 (2002) *Biochem Biophys Res Commun* 292, 222-30.

[106] Prebyl, B. S., Kaczmarek, C., Tuinman, A. A., and Baker, D. C. Characterizing the electrospray-ionization mass spectral fragmentation pattern of enzymatically derived hyaluronic acid oligomers (2003) *Carbohydr Res* 338, 1381-7.

[107] Kuhn, A. V., Raith, K., Sauerland, V., and Neubert, R. H. Quantification of hyaluronic acid fragments in pharmaceutical formulations using LC-ESI-MS (2003) *J Pharm Biomed Anal* 30, 1531-7.

[108] Desaire, H., and Leary, J. Detection and quantification of the sulfated disaccharides in chondoritin sulfate by electrospray tandem mass spectrometry (2000) *J Am Soc Mass Spectrom* 11, 916-20.

[109] Chai, W., Luo, J., Lim, C. K., and Lawson, A. M. Characterization of heparin oligosaccharide mixtures as ammonium salts using electrospray mass spectrometry (1998) *Anal Chem* 70, 2060-6.

[110] Kim, Y. S., Ahn, M. Y., Wu, S. J., Kim, D. H., Toida, T., Teesch, L. M., Park, Y., Yu, G., Lin, J., and Linhardt, R. J. Determination of the structure of oligosaccharides prepared from acharan sulfate (1998) *Glycobiology* 8, 869-77.

[111] Gunay, N. S., Tadano-Aritomi, K., Toida, T., Ishizuka, I., and Linhardt, R. Evaluation of Counterions for Electrospray Ionization Mass Spectral Analysis of A Highly Sulfated Carbohydrate, Sucrose Octasulfate (2003) *Anal Chem* 75, 3226-31.

[112] Kuberan, B., Lech, M., Zhang, L., Wu, Z. L., Beeler, D. L., and Rosenberg, R. Analysis of Heparan Sulfate Oligosaccharides with Ion Pair-Reverse Phase Capillary High Performance Liquid Chromatography-Microelectrospray Ionization Time-of-Flight Mass Spectrometry (2002) *J Am Chem Soc* 124, 8707-18.

[113] Thanawiroon, C., and Linhardt, R. J. Separation of a complex mixture of heparin-derived oligosaccharides using reversed-phase high-performance liquid chromatography (2003) *J Chromatogr A* 1014, 215-23.

[114] Imanari, T., Toida, T., Koshiishi, I., and Toyoda, H. High-performance liquid chromatographic analysis of glycosaminoglycan- derived oligosaccharides (1996) *J Chromatogr A* 720, 275-93.

[115] Militsopoulou, M., Lecomte, C., Bayle, C., Couderc, F., and Karamanos, N. K. Laser-induced fluorescence as a powerful detection tool for capillary electrophoretic analysis of heparin/heparan sulfate disaccharides (2003) *Biomed Chromatogr* 17, 39-41.

[116] Toyoda, H., Nagashima, T., Hirata, R., Toida, T., and Imanari, T. Sensitive high-performance liquid chromatographic method with fluorometric detection for the determination of heparin and heparan sulfate in biological samples: application to human urinary heparan sulfate (1997) *J Chromatogr B Biomed Sci Appl* 704, 19-24.

[117] Toyoda, H., Yamamoto, H., Ogino, N., Toida, T., and Imanari, T. Rapid and sensitive analysis of disaccharide composition in heparin and heparan sulfate by reversed-phase ion-pair chromatography on a 2 um porous silica gel column (1999) *Journal of Chromatography A* 830, 197-201.

[118] Toyoda, H., Kinoshita-Toyoda, A., Fox, B., and Selleck, S. B. Structural analysis of glycosaminoglycans in animals bearing mutations in sugarless, sulfateless, and tout-velu. Drosophila homologues of vertebrate genes encoding glycosaminoglycan biosynthetic enzymes (2000) *J Biol Chem* 275, 21856-61.

[119] Toyoda, H., Kinoshita-Toyoda, A., and Selleck, S. B. Structural analysis of glycosaminoglycans in *Drosophila* and *Caenorhabditis elegans* and demonstration that tout-velu, a Drosophila gene related to EXT tumor suppressors, affects heparan sulfate in vivo (2000) *J Biol Chem* 275, 2269-75.

[120] Sudo, M., Sato, K., Chaidedgumjorn, A., Toyoda, H., Toida, T., and Imanari, T. (1)H nuclear magnetic resonance spectroscopic analysis for determination of glucuronic and iduronic acids in dermatan sulfate, heparin, and heparan sulfate (2001) *Anal Biochem* 297, 42-51.

[121] Sakai, S., Onose, J., Nakamura, H., Toyoda, H., Toida, T., Imanari, T., and Linhardt, R. J. Pretreatment procedure for the microdetermination of chondroitin sulfate in plasma and urine (2002) *Anal Biochem* 302, 169-74.

[122] Barth, H., Schafer, C., Adah, M. I., Zhang, F., Linhardt, R. J., Toyoda, H., Kinoshita-Toyoda, A., Toida, T., Van Kuppevelt, T. H., Depla, E., Von Weizsacker, F., Blum, H. E., and Baumert, T. F. Cellular binding of hepatitis C virus envelope glycoprotein E2 requires cell surface heparan sulfate (2003) *J Biol Chem* 278, 41003-12.

[123] Roboz, J. (1968) Introduction to Mass Spectrometry: Instrumentation and Techniques, Interscience, New York.

[124] Desaire, H., and Leary, J. A. Multicomponent quantification of diastereomeric hexosamine monosaccharides using ion trap tandem mass spectrometry (1999) *Anal Chem* 71, 4142-47.

[125] Desaire, H., Sirich, T. L., and Leary, J. A. Evidence of block and randomly sequenced chondroitin polysaccharides: sequential enzymatic digestion and quantification using ion trap tandem mass spectrometry (2001) *Anal Chem* 73, 3513-20.

[126] Saad, O. M., and Leary, J. A. Compositional analysis and quantification of heparin and heparan sulfate by electrospray ionization ion trap mass spectrometry (2003) *Anal Chem* 75, 2985-95.

[127] Linhardt, R. J., Wang, H. M., Loganthan, D., Lamb, D. J., and Mallis, L. M. Analysis of glycosaminoglycan-derived oligosaccharides using fast atom bombardment mass spectrometry (1992) *Carbohydr Res* 225, 137-45.

[128] Lawrence, R., Kuberan, B., Lech, M., Beeler, D. L., and Rosenberg, R. D. Mapping critical biological motifs and biosynthetic pathways of heparan sulfate (2004) *Glycobiology* 14, 467-79.

[129] Staatz, W. D., Toyoda, H., Kinoshita-Toyoda, A., Chhor, K., and Selleck, S. B. Analysis of proteoglycans and glycosaminoglycans from Drosophila (2001) *Methods Mol Biol* 171, 41-52.

[130] Mallis, L. M., Wang, H. M., Loganathan, D., and Linhardt, R. J. Sequence analysis of highly sulfated, heparin-derived oligosaccharides using fast atom bombardment mass spectrometry (1989) *Anal Chem* 61, 1453-58.

[131] Lamb, D. J., Wang, H. M., Mallis, L. M., and Linhardt, R. J. Negative ion fast-atom bombardment tandem mass spectrometry to determine sulfate and linkage position in glycosaminoglycan derived disaccharides (1992) *J Am Soc Mass Spectrom* 3, 797-803.

[132] McClellan, J. M., Costello, C. E., O'Connor, P. B., and Zaia, J. Influence of charge state on product ion mass spectra and the determination of 4S/6S sulfation sequence of chondroitin sulfate oligosaccharides (2002) *Anal Chem* 74, 3760-71.

[133] Egge, H., Dell, A., and Von Nicolai, H. Fucose containing oligosaccharides from human milk. I. Separation and identification of new constituents (1983) *Arch Biochem Biophys* 224, 235-53.

[134] Zaia, J., Li, X.-Q., Chan, S.-Y., and Costello, C. E. Tandem mass spectrometric strategies for determination of sulfation positions and uronic acid epimerization in chondroitin sulfate oligosaccharides. (2003) *J Am Soc Mass Spectrom* 14, 1270-81.

[135] Zamfir, A., Seidler, D. G., Schonherr, E., Kresse, H., and Peter-Katalinic, J. On-line sheathless capillary electrophoresis/nanoelectrospray ionization-tandem mass spectrometry for the analysis of glycosaminoglycan oligosaccharides (2004) *Electrophoresis* 25, 2010-6.

[136] Zamfir, A., Seidler, D. G., Kresse, H., and Peter-Katalinic, J. Structural investigation of chondroitin/dermatan sulfate oligosaccharides from human skin fibroblast decorin (2003) *Glycobiology*.

[137] Zamfir, A., Vukelic, Z., and Peter-Katalinic, J. A capillary electrophoresis and off-line capillary electrophoresis/electrospray ionization-quadrupole time of flight-tandem mass spectrometry approach for ganglioside analysis (2002) *Electrophoresis* 23, 2894-903.

[138] Zaia, J., McClellan, J. E., and Costello, C. E. Tandem mass spectrometric determination of the 4S/6S sulfation sequence in chondroitin sulfate oligosaccharides (2001) *Anal Chem* 73, 6030-39.

[139] Zaia, J., and Costello, C. E. Compositional analysis of glycosaminoglycans by electrospray mass spectrometry (2001) *Anal Chem* 73, 233-39.

[140] Zamfir, A., Seidler, D. G., Kresse, H., and Peter-Katalinc, J. Structural characterization of chondroitin/dermatan sulfate oligosaccharides from bovine aorta by capillary electrophoresis and electrospray ionization quadrupole time-of-flight tandem mass spectrometry (2002) *Rapid Commun Mass Spectrom* 16, 2015-24.

[141] Spengler, B., Dolce, J. W., and Cotter, R. J. Infrared-Laser Desorption Mass-Spectrometry of Oligosaccharides - Fragmentation Mechanisms and Isomer Analysis (1990) *Anal Chem* 62, 1731-37.

[142] Harvey, D. J. Matrix-assisted laser desorption/ionization mass spectrometry of carbohydrates (1999) *Mass Spectrom Rev* 18, 349-450.

[143] Carroll, J., Willard, D., and Lebrilla, C. Energetics of cross-ring cleavages and their relevance to the linkage determination of oligosaccharides (1995) *Analytica Chimica Acta* 307, 431-47.

[144] Saad, O. M., and Leary, J. A. Delineating mechanisms of dissociation for isomeric heparin disaccharides using isotope labeling and ion trap tandem mass spectrometry (2004) *J Am Soc Mass Spectrom* 15, 1274-86.

[145] Oguma, T., Toyoda, H., Toida, T., and Imanari, T. Analytical Method for Keratan Sulfates by High-Performance Liquid Chromatography/Turbo-Ionspray Tandem Mass Spectrometry (2001) *Anal Biochem* 290, 68-73.

[146] Kubota, M., Yoshida, K., Tawada, A., and Ohashi, M. Characterization of oligosaccharides of the lactosamine series derived from keratan sulfates by tandem mass spectrometry (2000) *Eu J Mass Spectrom* 6, 193-203.

[147] Miller, M. J., Qian, J., Seward, R., Li, X., Costello, C. E., and Zaia, J. Evaluation of multistage MS for determination of individual sulfation sites on chondroitin sulfate and keratan sulfate oligosaccharides (2003) *Proc 51st ASMS Conf on Mass Spectrom Allied Topics*.

[148] Zaia, J., and Costello, C. E. Tandem mass spectrometry of sulfated heparin-like glycosaminoglycan oligosaccharides (2003) *Anal Chem* 75, 2445-55.

[149] Naggar, E., Costello, C. E., and Zaia, J. Competing fragmentation processes in tandem mass spectra of heparin-like glycosaminoglycans (2004) *J Am Soc Mass Spectrom* 15, 1534-44.

[150] Henriksen, J., Ringborg, L. H., and Roepstorrf, P. On-line size-exclusion chromatography/mass spectrometry of low molecular mass heparin (2004) *J Mass Spectrom* 39, 1305-12.

[151] Lecchi, P., and Abramson, F. P. Analysis of biopolymers by size-exclusion chromatography-mass spectrometry (1998) *J Chromatogr A* 828, 509-13.

[152] Lecchi, P., and Abramson, F. P. Size exclusion chromatography-chemical reaction interface mass spectrometry: "a perfect match" (1999) *Anal Chem* 71, 2951-5.
[153] Guo, Y. C., and Conrad, H. E. Analysis of oligosaccharides from heparin by reversed-phase ion-pairing high-performance liquid chromatography (1988) *Anal Biochem* 168, 54-62.
[154] Karamanos, N. K., Vanky, P., Tzanakakis, G. N., Tsegenidis, T., and Hjerpe, A. Ion-pair high-performance liquid chromatography for determining disaccharide composition in heparin and heparan sulphate (1997) *J Chromatogr A* 765, 169-79.
[155] Linhardt, R. J., Gu, K. N., Loganathan, D., and Carter, S. R. Analysis of glycosaminoglycan-derived oligosaccharides using reversed-phase ion-pairing and ion-exchange chromatography with suppressed conductivity detection (1989) *Anal Biochem* 181, 288-96.
[156] Thanawiroon, C., Rice, K. G., Toida, T., and Linhardt, R. J. LC/MS sequencing approach for highly sulfated heparin-derived oligosaccharides (2004) *J Biol Chem* 279, 2608-15.
[157] Jackson, P. The use of polyacrylamide-gel electrophoresis for the high-resolution separation of reducing saccharides labelled with the fluorophore 8-aminonaphthalene-1,3,6-trisulphonic acid. Detection of picomolar quantities by an imaging system based on a cooled charge-coupled device (1990) *Biochem J* 270, 705-13.
[158] Gennaro, L. A., Delaney, J., Vouros, P., Harvey, D. J., and Domon, B. Capillary electrophoresis/electrospray ion trap mass spectrometry for the analysis of negatively charged derivatized and underivatized glycans (2002) *Rapid Commun Mass Spectrom* 16, 192-200.
[159] Gennaro, L. A., Harvey, D. J., and Vouros, P. Reversed-phase ion-pairing liquid chromatography/ion trap mass spectrometry for the analysis of negatively charged, derivatized glycans (2003) *Rapid Commun Mass Spectrom* 17, 1528-34.
[160] Pervin, A., al-Hakim, A., and Linhardt, R. J. Separation of glycosaminoglycan-derived oligosaccharides by capillary electrophoresis using reverse polarity (1994) *Anal Biochem* 221, 182-8.
[161] Ampofo, S. A., Wang, H. M., and Linhardt, R. J. Disaccharide compositional analysis of heparin and heparan sulfate using capillary zone electrophoresis (1991) *Anal Biochem* 199, 249-55.
[162] al-Hakim, A., and Linhardt, R. J. Capillary electrophoresis for the analysis of chondroitin sulfate- and dermatan sulfate-derived disaccharides (1991) *Anal Biochem* 195, 68-73.
[163] Zamfir, A., and Peter-Katalinic, J. Glycoscreening by on-line sheathless capillary electrophoresis/electrospray ionization-quadrupole time of flight-tandem mass spectrometry (2001) *Electrophoresis* 22, 2448-57.
[164] Zamfir, A., and Peter-Katalinic, J. Capillary electrophoresis-mass spectrometry for glycoscreening in biomedical research (2004) *Electrophoresis* 25, 1949-63.
[165] Meikle, P. J., Hopwood, J. J., Clague, A. E., and Carey, W. F. Prevalence of lysosomal storage disorders (1999) *JAMA* 281, 249-54.
[166] Wraith, J. E. Enzyme replacement therapy in mucopolysaccharidosis type I: progress and emerging difficulties (2001) *J Inherit Metab Dis* 24, 245-50.

[167] Crawley, A., Ramsay, S. L., Byers, S., Hopwood, J., and Meikle, P. J. Monitoring dose response of enzyme replacement therapy in feline mucopolysaccharidosis type VI by tandem mass spectrometry (2004) *Pediatr Res* 55, 585-91.

[168] Bosch, A., and Heard, J. M. Gene therapy for mucopolysaccharidosis (2003) *Int Rev Neurobiol* 55, 271-96.

[169] Fuller, M., Meikle, P. J., and Hopwood, J. J. Glycosaminoglycan degradation fragments in mucopolysaccharidosis I (2004) *Glycobiology* 14, 443-50.

[170] Meikle, P. J., Fuller, M., and Hopwood, J. J. Mass spectrometry in the study of lysosomal storage disorders (2003) *Cell Mol Biol (Noisy-le-grand)* 49, 769-77.

[171] Reynertson, R., Campbell, P., Ford, J. D., Jacobsson, I., Roden, L., and Thompson, J. N. New oligosaccharides from heparin and heparan sulfate and their use as substrates for heparin-degrading enzymes (1983) *J Biol Chem* 258, 7449-59.

[172] Nowakowski, R. W., Thompson, J. N., and Taylor, K. B. Sanfilippo syndrome, type D: a spectrophotometric assay with prenatal diagnostic potential (1989) *Pediatr Res* 26, 462-6.

[173] Marsh, J., and Fensom, A. H. 4-Methylumbelliferyl alpha-N-acetylglucosaminidase activity for diagnosis of Sanfilippo B disease (1985) *Clin Genet* 27, 258-62.

[174] Gerber, S. A., Turecek, F., and Gelb, M. H. Design and synthesis of substrate and internal standard conjugates for profiling enzyme activity in the Sanfilippo syndrome by affinity chromatography/electrospray ionization mass spectrometry (2001) *Bioconjug Chem* 12, 603-15.

[175] Gerber, S. A., Scott, C. R., Turecek, F., and Gelb, M. H. Direct profiling of multiple enzyme activities in human cell lysates by affinity chromatography/electrospray ionization mass spectrometry: application to clinical enzymology (2001) *Anal Chem* 73, 1651-7.

[176] Hopwood, J. J., and Elliott, H. Urinary excretion of sulphated N-acetylhexosamines in patients with various mucopolysaccharidoses (1985) *Biochem J* 229, 579-86.

[177] Hopwood, J. J., and Elliott, H. Isolation and characterization of N-acetylglucosamine 6-sulfate from the urine of a patient with Sanfilippo type D syndrome and its occurrence in normal urine (1983) *Biochem Int* 6, 831-6.

[178] Hopwood, J. J., and Elliott, H. N-acetylglucosamine 6-sulfate residues in keratan sulfate and heparan sulfate are desulfated by the same enzyme (1983) *Biochem Int* 6, 141-8.

[179] Ramsay, S. L., Meikle, P. J., and Hopwood, J. J. Determination of monosaccharides and disaccharides in mucopolysaccharidoses patients by electrospray ionisation mass spectrometry (2003) *Mol Genet Metab* 78, 193-204.

[180] Honda, S., Akao, E., Suzuki, S., Okuda, M., Kakehi, K., and Nakamura, J. High-Performance Liquid-Chromatography of Reducing Carbohydrates as Strongly Ultraviolet-Absorbing and Electrochemically Sensitive 1-Phenyl-3-Methyl-5-Pyrazolone Derivatives (1989) *Anal Biochem* 180, 351-57.

In: Biomacromolecular Mass Spectrometry Yearbook. Volume 1 ISBN: 978-1-60876-167-8
Ed: Simone Koenig

Chapter 2

FOURIER TRANSFORM ION CYCLOTRON RESONANCE MASS SPECTROMETRY FOR PROTEIN STRUCTURE ANALYSIS*

Christian Ihling and Andrea Sinz*
Biotechnological-Biomedical Center, Faculty of Chemistry and Mineralogy,
University of Leipzig, D-04103 Leipzig, Germany

ABSTRACT

FTICR (Fourier transform ion cyclotron resonance) mass spectrometry offers several advantages for the analysis of biological samples, including excellent mass resolution, ultra-high mass accuracy, and a wide mass range. This paper gives a short overview of recent applications and technological developments of FTICRMS, which are relevant to the field of protein analysis. Tandem mass spectrometry, "bottom up" and "top down" approaches for protein analysis as well as methods for studying protein conformations are described and illustrated with examples.

1. INTRODUCTION

Structural characterization of proteins is currently delivering new insights into the behavior of individual molecules or large ensembles and recent technological advances have enabled researchers to examine smaller and smaller amounts of biological molecules. Mass spectrometric techniques are capable of analyzing small quantities (down to the level of single molecules)[1] of very large proteins (up to a molecular weight of several megadaltons)[2]. The advent of new ionization techniques such as electrospray ionization

* A version of this chapter was also published in *Biomacromolecular Mass Spectrometry Volume 1, Number 1*, published by Nova Science Publishers, Inc. It was submitted for appropriate modifications in an effort to encourage wider dissemination of research.

* Corresponding author: Dr. Andrea Sinz, Biotechnological-Biomedical Center, Faculty of Chemistry and Mineralogy, University of Leipzig, Linnéstr.3. D-04103 Leipzig, Germany. Phone: +49-341-9736078. Fax: +49-341-9736115. Email: sinz@chemie.uni-leipzig.de

(ESI)[3] and matrix-assisted laser desorption/ionization (MALDI)[4] is mainly responsible for these advances. Identification and quantification of proteins from complex mixtures pose great challenges for developing novel analytical strategies, as any human cell comprises roughly 10,000 proteins at any particular time, with concentrations that differ at least of a factor of 10[10]. Besides, many proteins are post-translationally modified and are therefore present in several variants. To date, more than 100 modification types are recorded and additional ones are yet to be discovered [5]. Currently, 2D gel electrophoresis (2-DE) in combination with MALDI-TOF mass spectrometry has emerged as the major technology for separation, identification, and quantification of different protein species [6]. However, closer examination of the protein classes, which are routinely identified by 2-DE/MS, suggests that this approach does not represent a truly global technique [7].

A mass spectrometric technique, which has become of increasing importance for protein analysis during the past few years, is Fourier transform ion cyclotron resonance mass spectrometry (FTICRMS)[8; 9]. There are already a number of publications and textbooks giving comprehensive descriptions on mass spectrometry[6;10-12] and especially FTICRMS[13-16] for protein analysis. Therefore, we will just give a short overview on the different aspects of FTICRMS that makes it a valuable tool for analyzing proteins. It is almost certain that the number of applications using this mass spectrometric technique for analyzing complex biological samples will greatly increase in the near future.

2. Basics of FTICR Mass Spectrometry

Among all mass spectrometric techniques, FTICR mass spectrometry offers the highest resolution and mass measurement accuracy [17]. Since its introduction in 1974[8; 9], there have been a number of studies demonstrating these capabilities. For example, Marshall and co-workers resolved more than 11,000 components of crude oil within onc broadband mass spectrum [18]. In another experiment, the two peptides RVMRGMR and RSHRGHR whose masses differ by only 0.00045 amu (an electron mass is 0.00055 amu) were baseline resolved [19]. Smith and colleages showed the measurement of 71 peptides from a tryptic digest of bovine serum albumin (BSA) with an average error of 0.77 ppm and a standard deviation of 1 ppm [20]. These impressive data were in most cases acquired on home-built mass spectrometers under conditions, which are not easily reproduced on commercial instruments. However, modern FTICR mass spectrometers offer a resolving power above 100,000 and mass measurement accuracies below 2 ppm on a routine basis. Another point making FTICRMS suitable for analysis of biological samples is that it is a sensitive MS technique. Detection limits for proteins and peptides in low attomole and even zeptomole amounts have been reported [21-24].

The principle underlying FTICR mass spectrometry forms the basis of its unique capabilities. There have been a couple of excellent reviews that give very detailed descriptions about this technique [17;25-27], therefore this article will give only a brief introduction into the basics of FTICRMS. FTMS uses the ICR principle to determine *m/z* values of ions. Ions are trapped in a cell that is located within a static magnetic field (Figure 1). Under the influence of the magnetic field the ions describe a circular motion caused by the Lorentz force and the centrifugal force (inertia) affecting them in opposite directions. Hereby, the angular frequency ω_c of the so-called cyclotron motion depends solely on the *m/z* value of the circulating ions and the magnetic field strength B (Equation 1):

$$\omega_c = B \cdot z/m \quad (1)$$

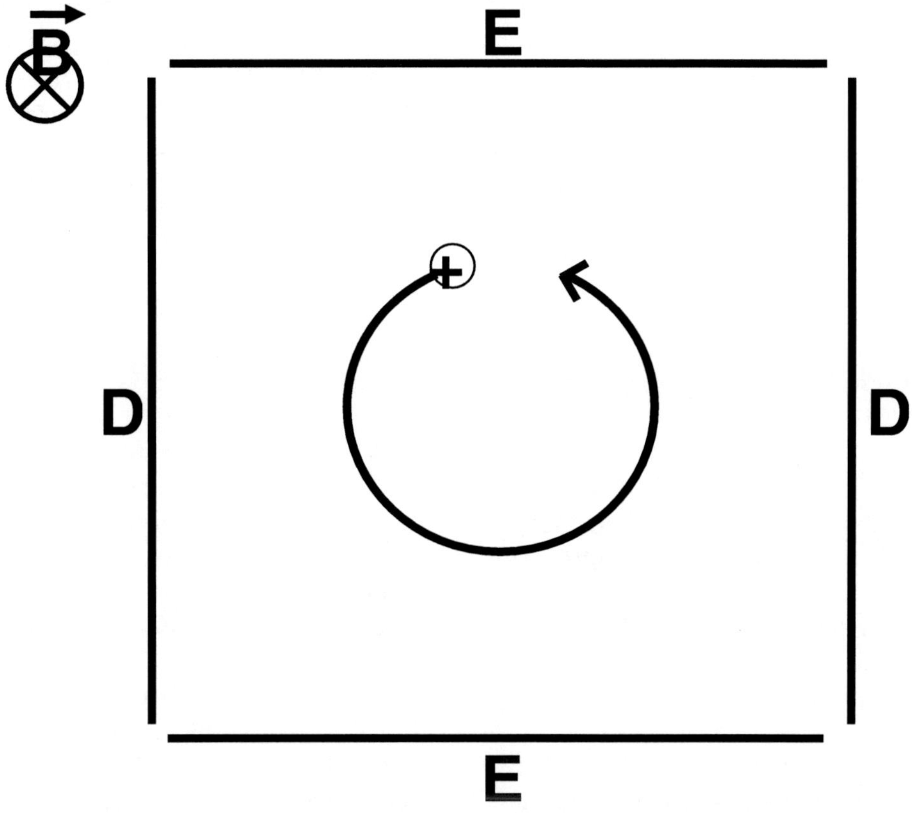

Figure 1. Schematic overview of an ICR cell with excitation (E) and detection electrodes (D).

In order to measure the cyclotron frequency ω_c it is necessary to force the ions of one *m/z* type into coherent motion by applying a frequency pulse through the two opposite excitation electrodes, which is in resonance with the cyclotron frequency of the ions. As an additional result, the kinetic energy of the ions and therefore the radius of the circular motion will increase. After excitation, the circulating ion clouds induce an oscillating image current in the two opposite detection plates whose frequency corresponds to the cyclotron frequency and thus to the *m/z*-values of the ions. A breakthrough in the ICR technique was achieved when Comisarow and Marshall applied inductive detection and Fourier transformation to simultaneously determine the cyclotron frequencies of all ions in the cell after simultaneous excitation by a broadband rf-pulse [8]. The image currents are recorded as a time domain signal, the so-called transient, and after amplification, digitization and Fourier transformation a frequency spectrum is obtained, which is easily converted into a mass spectrum. The high accuracy of this type of mass measurement is primary based on that frequency is not a function of kinetic energy - and hence not influenced by small fluctuations in kinetic energy distribution. Moreover, frequencies can be measured more exactly than any other physical properties.

Another aspect of the ICR technique is that this type of measurement is non-destructive; ions can be stored in the analyzer cell for several minutes or even hours. This offers the opportunity to cool down (either by collisional cooling or by quadrupolar axialization [28;29]) and re-measure the ions in order to increase sensitivity and resolution [30;31]. It is possible to study gas-phase reactions in the cell [32] or to fragment ions by various techniques, e.g., IRMPD [33] (infrared multi-photon dissociation), SORI-CID [34] (sustained off-resonance irradiation collision-induced dissociation), BIRD [35] (blackbody infrared radiative dissociation) or ECD [36] (electron capture dissociation) for MS^n experiments. For this purpose, the ability for selective isolation of ions of one *m/z*-value may also be of great importance. Different isolation techniques have been described, most commonly used are SWIFT [37] (stored-waveform inverse Fourier transformation), CHEF [38] (correlated harmonic excitation waveforms, also as multi-CHEF [39] for selection of multiple ions) and QEA [29;40] (quadrupolar excitation/axialization).

2.1. Tandem Mass Spectrometry

Tandem MS (MS/MS) strategies can be employed to gain structural information about analyte molecules, for example, amino acid sequences can be derived for protein analysis. A comprehensive review about high resolution MS/MS techniques and applications is given by Håkansson *et al.*[41]. Different dissociation techniques are used in connection with FTICRMS in order to fragment isolated ions. Collision-induced dissociation (CID) is performed by pulsing in a neutral gas (Ar, Xe, N_2) before energetic excitation of the ions. Collisions between excited ions and neutral molecules lead to conversion of kinetic energy into internal energy and result in the fragmentation of analyte ions. In the ICR cell, ion activation for CID is mostly achieved by SORI excitation using dipolar radiation with a frequency that is slightly offset from the ions' cyclotron frequency [34]. In comparison to resonant excitation the SORI approach is softer and leads to mass spectra that can be interpreted more easily. Another possibility for energetic fragmentation of ions consists in their irradiation with infrared light. In BIRD this is achieved by thermally heating the environment of the analyzer cell [35]. More frequently used for IR irradiation, however, is the IRMPD approach where activation with a single wavelength infrared laser is used to dissociate the ions [33]. A common feature of all these low-energy dissociation techniques is that they preferably lead to cleavages of the weakest chemical bonds in the molecule. In the case of proteins or peptides, cleavages resulting in the loss of post-translational modifications as well as cleavages of the amide bonds, thus creating b and y type ions [41], occur most frequently. Herein, the relative fragment ion abundances and the fragmentation efficiencies highly depend on the amino acid sequences of the peptides under investigation.

2.2. Electron Capture Dissociation (ECD)

Almost exclusively available for FTICRMS is fragmentation by electron capture dissociation, which was introduced in 1998 by Zubarev *et al.*[36]. This relatively new method offers some unique features for the analysis of peptides and proteins. In contrast to the abovementioned techniques SORI-CID and IRMPD, ECD is believed to proceed by a non-ergodic mechanism, which is still under scientific debate. One theory says that the capture of a low-energy electron (<1 eV, applied through a heated filament electron gun [42] or through an indirectly heated dispenser cathode [43] outside the ICR cell) leads to a reduced molecular ion $[M+nH]^{(n-1)+}$, which can release an energetic $H^{\bullet}$ atom [42]. This $H^{\bullet}$ atom can undergo a radical site reaction with a carbonyl group of the peptide, thus initiating backbone cleavage [44]. A modification of this theory was proposed by Leymarie *et al.*[45] based on ECD studies on cyclic peptides. The authors postulate that nonergodic ECD generates an α-carbon radical species that propagates along the protein backbone. This radical migration initiates multiple radical rearrangements, which cause both multiple backbone cleavages and additional side chain cleavages. The preferred formation of c and z· type ions and the more random distribution of cleavage sites agree well with both proposed mechanisms for ECD [46]. The different cleavage sites observed in ECD compared to other MS/MS techniques and its ability to cleave disulfide bonds in proteins [47;48] make ECD a valuable complementary tool for peptide sequencing. A number of studies show exhaustive sequence coverage for peptides and proteins by ECD [44;49].

The use of higher energy electrons (up to 13eV), termed HECD (hot ECD), even allowed distinguishing between the isomeric amino acids Leu and Ile, based on the presence of different d and w type ions after fragmentation of the respective amino acid side chains [50].

Another important characteristic of ECD is that labile bonds of post-translational modifications are not cleaved, thus allowing to map such modifications in peptides and proteins. Various examples demonstrate the analysis of phosphorylated [51;52], glycosylated [53-55], γ-carboxylated and sulphated [56] as well as fatty acid-modified [57] peptides. Acetylation, methylation, and phosphorylation patterns of intact proteins are also amenable to analysis by ECD [58].

One drawback of the ECD technique consists in its low fragmentation efficiency; usually only 5-20% of the precursor ions are cleaved. Especially larger proteins (>20kD) often show a charge reduction, but no fragmentation [42]. The low yield of fragmentation products from large proteins was proposed to have its origin in that ECD cleaves the protein backbone while not disrupting weak non-covalent interactions of the protein's tertiary gas-phase structure [59]. Energetic activation of the cleaved, but still non-covalently attached proteins enables dissociation and detection of the ECD fragments. Therefore, ECD is used in combination with IRMPD or as activated ion (AI) ECD to improve its fragmentation efficiency for proteins [59;60]. On the other hand, the ability of ECD to fragment the protein backbone without cleaving non-covalent interactions is employed for kinetic and structural studies of proteins in the gas-phase [61-64], as will be discussed later.

3. Proteome Analysis

To date, the standard approach for analyzing proteomes is to separate the proteins by 2-DE [65]. After visualization by staining, the protein spots of interest are cut out from the gel, the proteins are enzymatically digested, and the resulting peptide mixtures are analyzed by mass spectrometry to identify the respective protein. Despite its numerous successful applications during the last decade, this 2-DE/MS approach possesses several limitations, which originate mainly from the gel electrophoresis step. 2-DE is a time-consuming technique, which is difficult to reproduce and to automate, and what is even more important: 2-DE cannot handle "extreme" proteins. It is difficult to separate very small, very large, very basic, very acidic, and last but not least, hydrophobic proteins on 2D-gels. To overcome these limitations, efforts have been made to substitute 2-DE by other separation techniques like liquid chromatography or capillary electrophoresis for protein or peptide separation prior to MS analysis [66-73]. In combination with these liquid-based separation techniques, analysis of increasingly complex mixtures is favored over that of a single protein, which greatly enhances sample throughput. FTICRMS has become an efficient tool for proteomics studies, as its superior mass resolution capabilities allow for a simultaneous measurement of numerous compounds, and its high mass measurement accuracy gives additional confidence for subsequent identification of the respective compounds.

Two different strategies are used for protein analysis by FTICR mass spectrometry:

(1) Analysis of proteolytic digests of proteins or protein mixtures ("Bottom-up" approach)
(2) Analysis of intact proteins ("Top-down" approach)

3.1. "Bottom-Up" Analysis

In the "bottom-up" strategy, proteins are specifically cleaved into peptide fragments (mostly by using enzymatic proteolysis) before the peptide mixtures are analyzed by mass spectrometry. To date, the "bottom-up" strategy is the preferred method for protein identification, mainly because peptides are easier to handle compared to large proteins. After 2-DE, which is still the most commonly used technique for protein separation in proteomics, peptides arising from proteolytic cleavage are readily eluted from the gel – a task which is quite difficult to achieve for proteins. Another reason for the attractiveness of the "bottom-up" strategy is that most liquid chromatographic and mass spectrometric techniques are well established for peptides, whereas they are still difficult to realize or less efficient for intact proteins. Identification of proteins is accomplished by two different "bottom-up" strategies. The first one is entitled peptide mass fingerprinting (PMF), in which the parent protein is specifically cleaved to gain a set of peptides that is unique to this protein and as such can be used as a "fingerprint" to identify the original protein by database searching. PMF is broadly applied with conventional MS instrumentation, but only single proteins or simple mixtures can be confidently analyzed. FTICRMS offers some advantages for this type of analysis: With increasing mass measurement accuracy the portion of unique peptide masses for a protein increases significantly [74;75], thus making database searches more confident and allowing protein identifications based on a smaller number of peptides. Together with its high-resolution capabilities, FTICRMS has been shown to be applicable for the analysis of more complex protein mixtures [76]. The analysis of a crude *E. coli* cell extract by LC/FTICRMS has been described recently [77]. Bergquist *et al.* demonstrated the identification of proteins from human body fluids without prior separation or purification. The tryptically digested body fluids were either directly analyzed by ESI-FTICRMS or in combination with LC or capillary electrophoresis to identify proteins in human plasma [78], cerebrospinal fluid [73], urine and saliva [79].

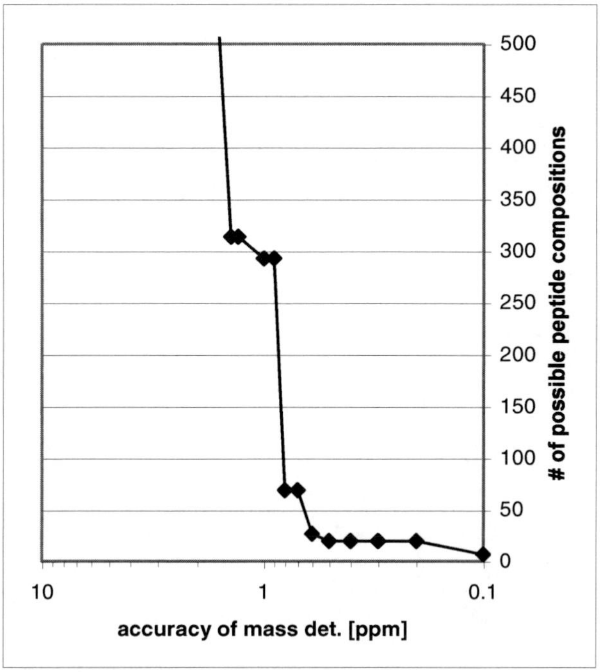

Figure 2*:* Number of possible amino acid compositions of peptides for a given example mass of 1005.4433 u as depending on the accuracy of mass determination. Calculations were performed including all of the 19 amino acids A C D E F G H (I _ L) K M N P Q R S T V W and Y. Mass calculations correspond to the monomolecular isotopomers of the singly positively charged protonated peptides (C terminus -OH, N terminus -H) (reprinted with permission from reference[83]).

In the second type of "bottom-up" approach, additional information on the protein under investigation is obtained by isolation of selected proteolytic peptides and subsequent fragmentation using MS/MS techniques [80-82]. These tandem MS data offer structural information related to the peptide sequence in addition to information on its mass, thus further increasing the specificity of protein identification. Again, high mass measurement accuracy leads to a more confident data interpretation. A recent publication [83] demonstrates the benefit for *de-novo* sequencing of peptides when additional information from accurate mass determination by FTICRMS is employed. Figure 2 shows the dependence of mass measurement accuracy on the number of possible amino acid compositions (AAC) of peptides, exemplified for the mass 1005.4433 u. With increasing accuracy the calculated set of possible AACs can be significantly reduced and the combination with MS/MS data further limits the number of compositions. Starting with in this way reduced datasets, a relative simple algorithm (composition based sequencing) leads to an unambiguous determination of the respective peptide sequence. As novel FTICR instruments allow accurate mass determinations and online tandem MS experiments at the same time in routine analytical work, the resultant level of more and more exact structural information about the analyzed peptides will make FTICRMS even more valuable for protein analysis.

To date, the AMT (accurate mass and time) tag approach, a type of "bottom-up" technique developed by Smith and colleagues [74], has obtained the broadest coverage of one proteome. More than 61% of the predicted proteins of *Deinococcus radiodurans* could be identified with high confidence [84;85] and later improvements even increased the coverage to 83% [85]. For this approach, a library of peptide AMT tags is generated by collecting data from various LC / tandem MS experiments on conventional instruments before peptide masses and LC elution times are validated using LC/FTICRMS. Once a library of AMT tags has been established for an organism, these AMT tags can be used as confident markers for their corresponding proteins in future LC/FTICRMS experiments. This enables high-throughput proteome analysis, given that peptide masses can be determined with sufficient accuracy. As an impressive example, Figure 3 shows the mass versus time plot of ~22,000 detected putative peptides of *D. radiodurans* during a single LC/FTICRMS experiment [75]. About 1,500 AMT tags could be typically matched during one single LC/MS run to about 700 ORFs, corresponding to 15-20% of the predicted proteome of the microorganism.

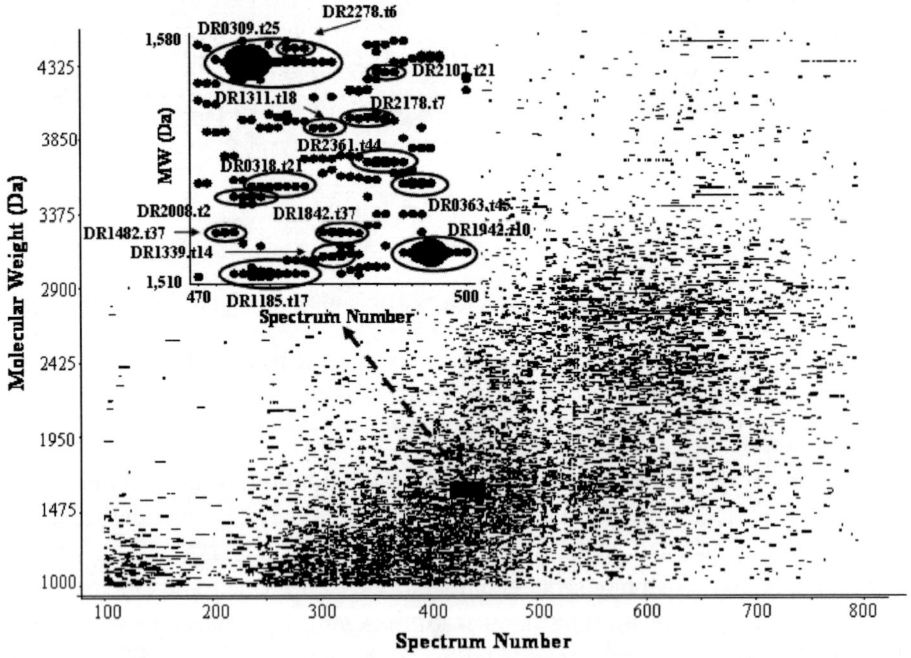

Figure 3*:* Two-dimensional display of identified AMTs from *Deinococcus radiodurans*. In this display, all of the peptides observed in an LC/FTICR mass spectrometric analysis of the *D. radiodurans* proteome are displayed based on their molecular weight and elution order (i.e., FTICR spectrum number). The circled spots labeled in the inset show spots that were identified as AMTs. The spots are labeled based on their annotation within the organism's genome sequence (i.e., t25: the 25th tryptic peptide counting from the amino terminus, based on complete digestion (reprinted with permission from reference[75]).

3.2. "Top-Down" Analysis

The most direct analytical strategy for protein characterization is based on the analysis of intact proteins instead of proteolytic digests. In this approach, the mass of the intact protein is used together with partial sequence information, derived from tandem MS experiments, for protein identification. Proteins or protein mixtures are usually ionized by electrospray ionization, as this method produces multiply charged ions, which can be measured in a convenient *m/z*-range (< 4000) and are easier to dissociate. After determination of the intact protein's mass, the corresponding protein ions are isolated and dissociated by one of the fragmentation methods available for FTICRMS. Determination of the fragment ion masses and subsequent computing and database matching of parent ion and fragment ion masses lead to protein identification. Even *de-novo* sequencing of proteins is possible if full sequence coverage is obtained after fragmentation, as has been shown for the small protein ubiquitin and the peptide melittin [86]. Because of the high resolution that is required for exact mass measurements of larger proteins and because of its unique abilities to store, isolate, and fragment ions [87], FTICRMS is currently the most widely used MS technique for "top-down" proteomics. A number of studies show the successful application of this approach for protein identifications. Kelleher *et al.* compared different "bottom-up" and "top-down" strategies at the example of carbonic anhydrase. The authors were able to achieve 100% sequence coverage for each top-down approach using SORI-CID on the intact protein or on protein fragments after digestion with BrCN or Lys-C, whereas sequence coverages were 95% and 96% in the "bottom-up" strategie for chymotryptic and Lys-C digestion, respectively[88]. Li *et al.* applied an LC/MS approach for "top-down" analysis of a five protein mixture using IRMPD as fragmentation method. Each of the proteins could be identified out of more than 1000 database candidates [89].

An important advantage of the "top-down" strategy is that detected masses of the intact proteins directly point to modifications, which cannot be predicted from DNA sequence data. Especially in higher organisms, proteins often undergo excessive processing after biosynthesis, including proteolysis and a variety of chemical post-translational modifications (PTMs). In addition, DNA splicing or undiscovered errors in DNA databases lead to differences between the amino acid sequences of expressed proteins and predicted ORFs. The knowledge of the intact protein's mass together with sequence information gained from subsequent MS/MS studies has proved to be a valuable tool for identifying and mapping PTMs or DNA sequence errors. Especially ECD has been shown to be very useful for protein dissociation, as even labile PTMs can be localized [90]. A "top-down" approach has been successfully employed in combination with ECD fragmentation to map different PTMs of histone proteins, such as acetylation, methylation and phosphorylation, which encode for the "histone code" (Figure 4) [58].

4. Analysis of Protein Conformation

Another field of protein analysis deals with the determination of the tertiary structures of proteins. Classical high-resolution methods as NMR and X-ray crystallography are well established and have so far solved a large number of three-dimensional protein structures. In recent years, mass spectrometric methods have matured to become a supplement of these high-resolution techniques, as low-resolution structures of proteins can be determined very fast and with very low sample consumption. Again, the unique features of FTICRMS are of great advantage for studying protein conformations.

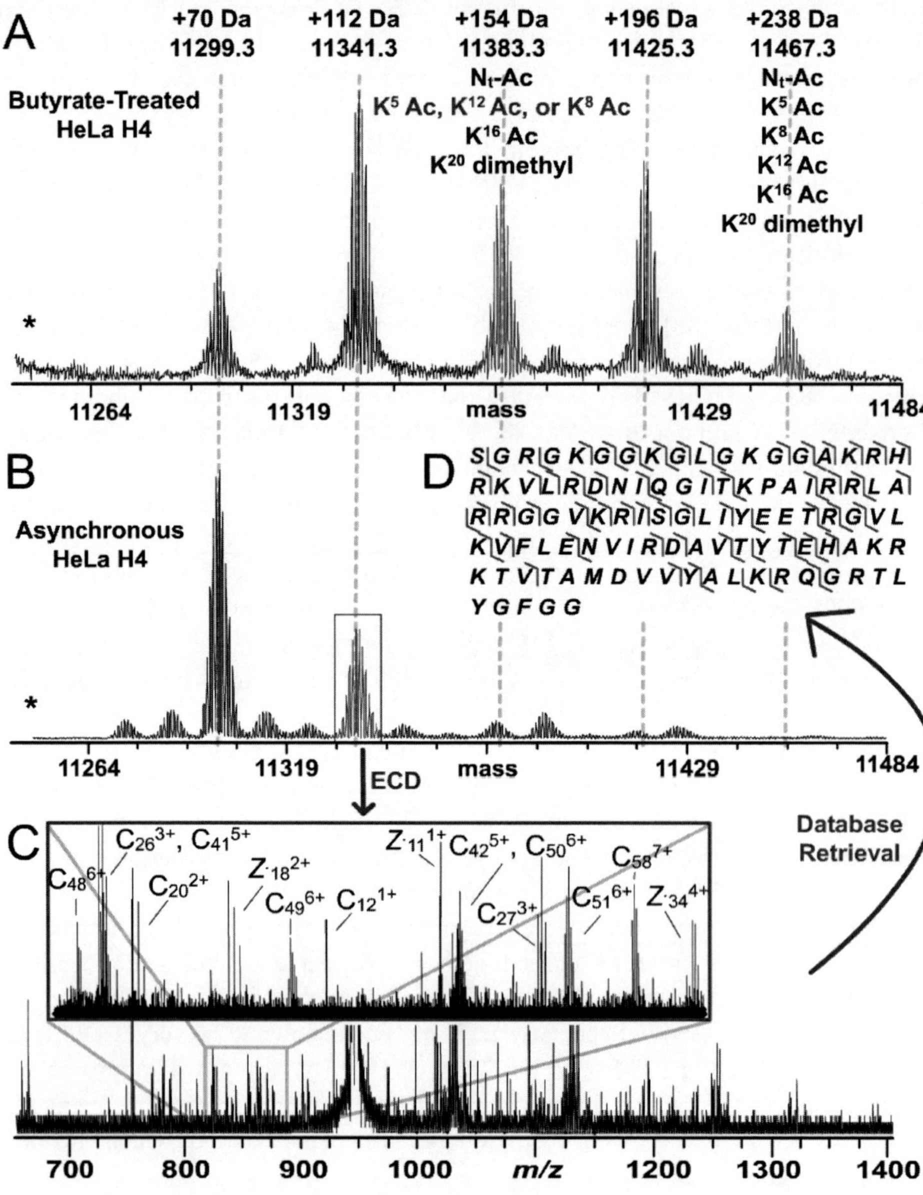

Figure 4*:* (A) ESI-FTICR mass spectrum (15 scans) of intact histone protein H4 isolated from the butyrate-treated HeLa cells (scale set to mass for 12+ ions). An asterisk (*) denotes the position of the unobserved, unmodified histone H4. (B) ESI-FTICR mass spectrum (40 scans) of intact H4 isolated from the nuclei of asynchronous HeLa cells (scale set to mass for 12+ ions). (C) ECD MS/MS spectrum (107 scans) of a +112 Da modified form of histone H4. (D) ECD fragmentation map of histone H4 +112 Da after streamlined protein characterization achieved by retrieval from a highly annotated database. The best match from the database search is N-terminal acetylation, K16 acetylation, and K20 dimethylation (highlighted in red) (reprinted with permission from reference[58]).

4.1. Chemical Cross-Linking

Chemical cross-linking is a process forming covalent bonds between different molecules (intermolecular) or parts of a molecule (intramolecular) which has proven successful in combination with mass spectrometry as a tool for low-resolution structure determination [91;92]. The goal of performing intramolecular chemical cross-linking of a protein is to get hints on its three-dimensional structure [93], whereas the aim of intermolecular cross-linking between different proteins is to answer, whether or not components interact and how and where they physically contact each other [94;95]. A great variety of cross-linking reagents with different spacer lengths and reactivities can be employed to connect functional groups in proteins (e.g. amine or sulfhydryl groups) and thus measure distances between them. For analyzing cross-linking reaction mixtures FTICRMS represents an attractive tool, as its high mass accuracy serves as an additional constraint in the identification of cross-linking reaction products. Additionally advantageous is that FTICRMS allows a ‘gas-phase’ purification and subsequent fragmentation of cross-linking products in the ICR cell.

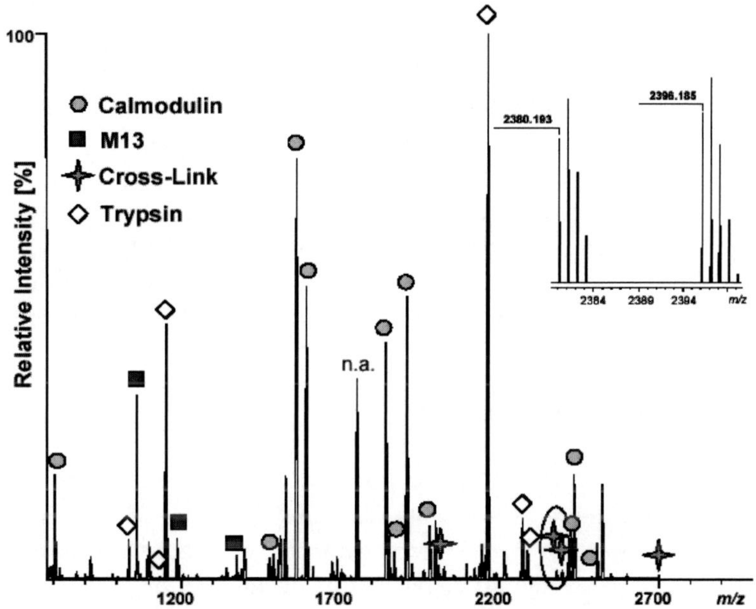

Figure 5: Deconvoluted ESI-FTICR mass spectrum of a tryptic peptide mixture from a complex between calmodulin and the peptide M13, cross-linked with BS^3 (100-fold molar excess over protein/peptide concentration) at an incubation time of 60 min. The insert shows the magnified signal of a cross-linking product at *m/z* 2380.193 between calmodulin residues 75-86 and M13 residues 19-26, which was also detected at *m/z* 2396.185 with one methionine residue oxidized; n.a.: signal not assigned [95].

Both “bottom-up” and “top-down” approaches have been used to analyze cross-linked proteins. The “bottom-up” approach has been applied for mapping protein interfaces, but it has also been successfully employed to determine low-resolution three-dimensional structures of proteins [91]. An ESI-FTICR mass spectrum of a tryptic digestion mixture from a cross-linked complex is presented in Figure 5, which allowed the identification of intermolecular cross-linking products between calmodulin and a peptide derived from skeletal muscle myosin light chain kinase (M13) [95]. The "top-down" approach has so far been demonstrated successful for deducing distance constraints in relatively small proteins, such as ubiquitin [39;96]. The analysis of large protein complexes, however, is likely to be challenging.

4.2. H/D Exchange

H/D exchange employs the relationship between isotopic exchange rates of peptide amide protons in proteins and the protein's secondary and tertiary structure. The theory behind this technique is that the exchange rates of the amide protons are directly related to protein structure characteristics, namely the extent, to which the protons are shielded from the solvent and whether they are involved in intramolecular hydrogen bonding [97-100]. Using ESI and MALDI mass spectrometry, H/D exchange rates are examined by determining the mass increases [101;102]. In order to achieve higher structural specificity, proteins are fragmented after H/D exchange. Again, two strategies are possible; proteins are either proteolytically digested before MS analysis or alternatively, they are physically dissociated inside the MS [103;104]. FTICRMS has been successfully employed for this purpose, e.g., the study of the structural stability of myoglobin has been described by Buijs *et al.*[105]. Another publication deals with the structural investigation of the interface within an enzyme-inhibitor complex [106]. The use of FTICRMS in combination with H/D exchange additionally allows to analyze gas-phase structures, as protein ions can be stored in the analyzer before they are interrogated with gaseous D_2O. Using this method, McLafferty and colleages detected different stable gas-phase conformers of bovine cytochrome *c* [107]. A study of Freitas *et al.* on bradykinins demonstrated the existence of peptide zwitterions in the gas phase [108].

4.3. Fragmentation Studies by ECD

Another possibility to study gas-phase structures of protein ions with FTICRMS consists in performing electron capture experiments. The unique feature of ECD to cleave the amino acid backbone of proteins, whereas non-covalent interactions of tertiary structure are maintained [59] can decrease the efficiency of this technique for protein sequencing, but on the other hand enables the researcher to draw conclusions about gas-phase conformations of proteins. This approach was used for investigation of folding and unfolding characteristics of gaseous ubiquitin depending on charge state and temperature [61;64], as well as for gas-phase folding of cytochrome *c* [63]. A recent publication of Breuker *et al.*[62] described ECD like fragmentation of native cytochrome *c* even without application of external electrons. The fragment ion yield depended on the protein concentration in the ESI solution and could be correlated with cytochrome *c* dimerization. This so-called 'native ECD (NECD)' process was explained by intermolecular electron transfer during protein dimer dissociation. Interestingly, the resulting fragment ions correlated with non-covalent protein-heme interaction in the native structure of cytochrome *c*.

5. Concluding Remarks

In recent years, FTICR mass spectrometry has become an essential tool for analyzing biological macromolecules. In this paper, we described a number of technological and application developments that have strongly contributed for the increasingly important role FTICRMS plays in bioanalytical research. It is likely that the number of applications for solving exciting biological problems using FTICR mass spectrometry will boost in the near future.

Acknowledgments

The junior research group of A.S. is funded by the Saxon State Ministry of Higher Education, Research and Culture. Financial support by the Thermo Electron Corporation (Mattauch-Herzog award of the German Society for Mass Spectrometry to A.S.) is also gratefully acknowledged.

References

[1] Twerenbold, D.; Vuilleumier, J.-L.; Gerber, D.; Tadsen, A.; van den Brandt, B.; Gillevet, P. M. *Appl. Phys. Lett.* 1996, *68*, 3503-05.
[2] Heck, A. J. R.; van den Heuvel, R. H. H. *Mass Spectrom. Rev.* 2004, *23*, 368-89.
[3] Fenn, J. B.; Mann, M.; Meng, C. K.; Wong, S. F.; Whitehouse, C. M. *Science* 1989, *46*, 64-71.
[4] Karas, M.; Hillenkamp, F. *Anal. Chem.* 1988, *60*, 2299-301.
[5] Peng, J.; Gygi, S. P. *J. Mass Spectrom.* 2001, *36*, 1083-91.
[6] Mann, M.; Hendrickson, R. C.; Pandey, A. *Annu. Rev. Biochem.* 2001, *70*, 437-73.
[7] Lottspeich, F. *Angew. Chem. Int. Ed.* 1999, *38*, 2476-92.
[8] Comisarow, M. B.; Marshall, A. G. *Chem. Phys. Lett.* 1974, *25*, 282-83.
[9] Comisarow, M. B.; Marshall, A. G. *Chem. Phys. Lett.* 1974, *26*, 489-90.
[10] Yates, J. R., III *Annu. Rev. Biophys. Biomol. Struct.* 2004, *33*, 297-316.
[11] Aebersold, R.; Mann, M. *Nature* 2003, *422*, 198-207.
[12] Gross, J.H., Mass Spectrometry -- A Textbook 2004, Springer, Heidelberg .
[13] Bogdanov, B.; Smith, R. D. *Mass Spectrom. Rev.* 2004, *Published Online*, Mar. 30.
[14] Page, J. S.; Masselon, C. D.; Smith, R. D. *Curr. Opin. Biotech.* 2004, *15*, 3-11.
[15] Bergquist, J. *Curr. Opin. Molec. Ther.* 2003, *5*, 310-14.
[16] Heeren, R. M. A.; Kleinnijenhuis, A. J.; McDonnell, L. A.; Mize, T. H. *Anal. Bioanal. Chem.* 2004, *378*, 1048-58.
[17] Marshall, A. G. *Int. J. Mass Spectrom.* 2000, *200*, 331-56.
[18] Hughey, C. H.; Rodgers, R. P.; Marshall, A. G. *Anal. Chem.* 2002, *74*, 4145-49.
[19] He, F.; Hendrickson, C. L.; Marshall, A. G. *Anal. Chem.* 2001, *73*, 647-50.
[20] Bruce, J. E.; Anderson, G. A.; Wen, J.; Harkewicz, R.; Smith, R. D. *Anal. Chem.* 1999, *71*, 2595-99.
[21] Baykut, G.; Jertz, R.; Witt, M. *Rapid. Commun. Mass Spectrom.* 2000, *14*, 1238-47.

[22] Belov, M. E.; Gorshkov, M. V.; Udseth, H. R.; Anderson, G. A.; Smith, R. D. *Anal. Chem.* 2000, *72*, 2271-79.

[23] Valaskovic, G. A.; Kelleher, N. L.; McLafferty, F. W. *Science* 1996, *273*, 1199-202.

[24] Moyer, S. C.; Budnik, B.; Pittman, J. L.; Costello, C. E.; O'Connor, P. B. *Anal. Chem.* 2003, *75*, 6449-54.

[25] Amster, I. J. *J. Mass Spectrom.* 1996, *31*, 1325-37.

[26] Marshall, A. G.; Hendrickson, C. L.; Jackson, G. S. *Mass Spectrom. Rev.* 1998, *17*, 1-35.

[27] Marshall, A. G.; Hendrickson, C. L. *Int. J. Mass Spectrom.* 2002, *215*, 59-75.

[28] Williams, E. R.; Henry, K. D.; McLafferty, F. W. *J. Am. Chem. Soc.* 1990, *112*, 6157-62.

[29] Speir, J. P.; Gorman, G. S.; Pitsenberger, C. C.; Turner, C. A.; Wang, P. P.; Amster, I. J. *Anal. Chem.* 1993, *65*, 1746-52.

[30] Solouki, T.; Marto, J. A.; White, F. M.; Guan, S.; Marshall, A. G. *Anal. Chem.* 1995, *67*, 4139-44.

[31] Solouki, T.; Paša-Tolic, L.; Jackson, G. S.; Guan, S.; Marshall, A. G. *Anal. Chem.* 1996, *68*, 3718-25.

[32] Nibbering, N. M. M. *Int. J. Mass Spectrom.* 2000, *200*, 27-42.

[33] Little, D. P.; Speir, J. P.; Senko, M. W.; O'Connor, P. B.; McLafferty, F. W. *Anal. Chem.* 1994, *66*, 2809-15.

[34] Gauthier, J. W.; Trautmann, T. R.; Jacobson, D. B. *Anal. Chim. Acta* 1991, *246*, 211-25.

[35] Price, W. D.; Schnier, P. D.; Williams, E. R. *Anal. Chem.* 1996, *68*, 859-66.

[36] Zubarev, R. A.; Kelleher, N. L.; McLafferty, F. W. *J. Am. Chem. Soc.* 1998, *120*, 3265-66.

[37] Marshall, A. G.; Wang, T.-C. L.; Ricca, T. L. *J. Am. Chem. Soc.* 1985, *107*, 7893-97.

[38] de Koning, L. J.; Nibbering, N. M. M.; van Orden, S. L.; Laukin, F. H. *Int. J. Mass Spectrom. Ion. Proc.* 1997, *165*, 209-19.

[39] Kruppa, G. H.; Schnier, P. D.; Tabei, K.; van Orden, S. L.; Siegel, M. M. *Anal. Chem.* 2002, *74*, 3877-86.

[40] Schweikhard, L.; Guan, S.; Marshall, A. G. *Int. J. Mass Spectrom. Ion. Proc.* 1992, *120*, 71-83.

[41] Håkansson, K.; Cooper, H. J.; Hudgins, R. R.; Nilsson, C. L. *Curr. Org. Chem.* 2003, *7*, 1503-25.

[42] Zubarev, R. A.; Horn, D. M.; Fridriksson, E. K.; Kelleher, N. L.; Kruger, N. A.; Lewis, M. E.; Carpenter, B. K.; McLafferty, F. W. *Anal. Chem.* 2000, *72*, 563-73.

[43] Tsybin, Y. O.; Håkansson, P.; Budnik, B. A. *Rapid. Commun. Mass Spectrom.* 2001, *15*, 1849-54.

[44] McLafferty, F. W.; Horn, D. M.; Breuker, K.; Ge, Y.; Lewis, M. E.; Cerda, B. A.; Zubarev, R. A.; Carpenter, B. K. *J. Am. Soc. Mass Spectrom.* 2001, *12*, 245-49.

[45] Leymarie, N.; Costello, C. E.; O'Connor, P. B. *J. Am. Chem. Soc.* 2003, 8958.

[46] Zubarev, R. A.; Haselmann, K. F.; Budnik, B.; Kjeldsen, F.; Jensen, F. *Eur. J. Mass Spectrom.* 2002, *8*, 337-49.

[47] Zubarev, R. A.; Kruger, N. A.; Fridriksson, E. K.; Lewis, M. E.; Horn, D. M.; Carpenter, B. K.; McLafferty, F. W. *J. Am. Chem. Soc.* 1999, *121*, 2857-62.

[48] Kruger, N. A.; Zubarev, R. A.; Carpenter, B. K.; Kelleher, N. L.; Horn, D. M.; McLafferty, F. W. *Int. J. Mass Spectrom.* 1999, *182/183*, 1-5.

[49] Ge, Y.; Lawhorn, B. G.; ELNaggar, M.; Strauss, E.; Park, J.-H.; Begley, T. P.; McLafferty, F. W. *J. Am. Chem. Soc.* 2002, *124*, 672-78.

[50] Kjeldsen, F.; Haselmann, K. F.; Budnik, B. A.; Jensen, F.; Zubarev, R. A. *Chem. Phys. Lett.* 2002, *356*, 201-06.

[51] Stensballe, A.; Jensen, O. N.; Olsen, J. V.; Haselmann, K. F.; Zubarev, R. A. *Rapid. Commun. Mass Spectrom.* 2000, *14*, 1793-800.

[52] Kjeldsen, F.; Haselmann, K. F.; Budnik, B.; Sørensen, E. S.; Zubarev, R. A. *Anal. Chem.* 2003, *75*, 2355-61.

[53] Mormann, M.; Macek, B.; Gonzalez de Peredo, A.; Hofsteenge, J.; Peter-Katalinic, J. *Int. J. Mass Spectrom.* 2004, *234*, 11-21.

[54] Haselmann, K. F.; Budnik, B. A.; Olsen, J. V.; Nielsen, M. L.; Reis, C. A.; Clausen, H.; Johnsen, A. H.; Zubarev, R. A. *Anal. Chem.* 2001, *73*, 2998-3005.

[55] Håkansson, K.; Cooper, H. J.; Emmett, M. R.; Costello, C. E.; Marshall, A. G.; Nilsson, C. L. *Anal. Chem.* 2001, *73*, 4530-36.

[56] Kelleher, N. L.; Zubarev, R. A.; Bush, K.; Furie, B.; Furie, B. C.; McLafferty, F. W.; Walsh, C. T. *Anal. Chem.* 1999, *71*, 4250-53.

[57] Guan, Z. *J. Am. Soc. Mass Spectrom.* 2002, *13*, 1443-47.

[58] Pesavento, J. J.; Kim, Y.-B.; Taylo, G. K.; Kelleher, N. L. *J. Am. Chem. Soc.* 2004, *126*, 3386-87.

[59] Horn, D. M.; McLafferty, F. W. *Anal. Chem.* 2000, *72*, 4778-84.

[60] Tsybin, Y. O.; Ramström, M.; Witt, M.; Baykut, G.; Håkansson, P. *J. Mass Spectrom.* 2004, *39*, 719-29.

[61] Breuker, K.; Oh, H.; Horn, D. M.; Cerda, B. A.; McLafferty, F. W. *J. Am. Chem. Soc.* 2002, *124*, 6407-20.

[62] Breuker, K.; McLafferty, F. W. *Angew. Chem. Int. Ed.* 2003, *42*, 4900-04.

[63] Horn, D. M.; Breuker, K.; Frank, A. J.; McLafferty, F. W. *J. Am. Chem. Soc.* 2001, *123*, 9792-99.

[64] Oh, H.; Breuker, K.; Sze, S. K.; Ge, Y.; Carpenter, B. K.; McLafferty, F. W. *Proc. Natl. Acad. Sci.* 2002, *99*, 15863-68.

[65] Williams, J. D.; Burinsky, D. J. *Int. J. Mass Spectrom.* 2001, *212*, 111-33.

[66] Schrader, W.; Klein, H.-W. *Anal. Bioanal. Chem.* 2004, *379*, 1013-24.

[67] Manabe, T. *J. Chromatogr. B* 2003, *787*, 29-41.

[68] Shen, Y.; Smith, R. D. *Electrophoresis* 2002, *23*, 3106-24.

[69] Shen, Y.; Jacobs, J. M.; Camp II, D. G.; Fang, R.; Moore, R. J.; Xiao, W.; Davis, R. W.; Tompkins, R. G.; Smith, R. D. *Anal. Chem.* 2004, *76*, 1134-44.

[70] Shen, Y.; Tolic, N.; Masselon, C. D.; Paša-Tolic, L.; Camp II, D. G.; Hixson, K. K.; Zhao, R.; Anderson, G. A.; Smith, R. D. *Anal. Chem.* 2004, *76*, 144-54.

[71] Washburn, M. P.; Wolters, D.; Yates, J. R., III *Nature Biotech.* 2001, *19*, 242-47.

[72] Wetterhall M.; Palmblad M.; Håkansson, P.; Markides K.E.; Bergquist J. *J. Proteome* Conrads, T. P.; Anderson, G. A.; Veenstra, T. D.; Paša-Tolic, L.; Smith, R. D. *Anal. Chem.* 2000, *72*, 3349-54.

[73] Smith, R. D.; Anderson, G. A.; Lipton, M. S.; Masselon, C.; Paša-Tolic, L.; Shen, Y.; Udseth, H. R. *OMICS* 2002, *6*, 61-90.

[74] Ihling, C.; Berger, K.; Höfliger, M. M.; Führer, D.; Beck-Sickinger, A. G.; Sinz, A. *Rapid. Commun. Mass Spectrom.* 2003, *17*, 1240-46.

[75] Ihling, C.; Sinz, A. *Proteomics* 2004, *in press*.

[76] Palmblad M.; Wetterhall M.; Markides K.E.; Håkansson, P.; Bergquist, J. *Rapid. Commun. Mass Spectrom.* 2000, *14*, 1029-34.

[77] Bergquist, J.; Palmblad, M.; Wetterhall, M.; Håkansson, P.; Markides, K. E. *Mass Spectrom. Rev.* 2002, *21*, 2-15.

[78] Huang, E. C.; Henion, J. D. *J. Am. Soc. Mass Spectrom.* 1990, *1*, 158-65.

[79] Eng, J. K.; McCormack, A. L.; Yates, J. R., III *J. Am. Soc. Mass Spectrom.* 1994, *5*, 976-89.

[80] McCormack, A. L.; Schieltz, D. M.; Goode, B.; Yang, S.; Barnes, G.; Drubin, D.; Yates, J. R., III *Anal. Chem.* 1997, *69*, 767-76.

[81] Spengler, B. *J. Am. Soc. Mass Spectrom.* 2004, *15*, 703-14.

[82] Lipton, M. S.; Paša-Tolic, L.; Anderson, G. A.; Anderson, D. J.; Auberry, D. L.; Battista, J. R.; Daly, M. J.; Fredrickson, J.; Hixson, K. K.; Konstandarithes, H.; Masselon, C.; Markillie, L. M.; Moore, R. J.; Romine, M. F.; Shen, Y.; Strittmatter, E.; Tolic, N.; Udseth, H. R.; Venkateswaran, A.; Wong, K.-K.; Zhao, R.; Smith, R. D. *Proc. Natl. Acad. Sci.* 2002, *99*, 11049-54.

[83] Paša-Tolic, L.; Lipton, M. S.; Masselon, C.; Anderson, G. A.; Shen, Y.; Tolic, N.; Smith, R. D. *J. Mass Spectrom.* 2002, *37*, 1185-98.

[84] Horn, D. M.; Zubarev, R. A.; McLafferty, F. W. *Proc. Natl. Acad. Sci.* 2000, *97*, 10313-17.

[85] McLafferty, F. W.; Fridriksson, E. K.; Lewis, M. E.; Zubarev, R. A. *Science* 1999, *284*, 1289-90.

[86] Kelleher, N. L.; Lin, H. Y.; Valaskovic, G. A.; Aaserud, D. J.; Fridriksson, E. K.; McLafferty, F. W. *J. Am. Chem. Soc.* 1999, *121*, 806-12.

[87] Li, W.; Hendrickson, C. L.; Emmett, M. R.; Marshall, A. G. *Anal. Chem.* 1999, *71*, 4397-402.

[88] Meng, F.; Forbes, A. J.; Miller, L. M.; Kelleher, N. L. *Mass Spectrom. Rev.* 2004, *published online*.

[89] Sinz, A. *J. Mass Spectrom.* 2003, *38*, 1225-33.

[90] Back, J. W.; de Jong, L.; Muijsers, A. O.; de Koster, C. G. *J. Mol. Biol.* 2003, *331*, 303-13.

[91] Dihazi, G. H.; Sinz, A. *Rapid. Commun. Mass Spectrom.* 2003, *17*, 2005-14.

[92] Schulz, D. M.; Ihling, C.; Clore, G. M.; Sinz, A. *Biochemistry* 2004, *43*, 4703-15.

[93] Kalkhof, S.; Ihling, C.; Mechtler, K.; Sinz, A. *Anal. Chem.* 2004, *Published online*, Dec. 7.

[94] Novak, P.; Young, M. M.; Schoeninger, J.; Kruppa, G. H. *Eur. J. Mass Spectrom.* 2003, *9*, 623-31.

[95] Zhang, Z.; Smith, D. L. *Prot. Sci.* 1993, *2*, 522-31.

[96] Englander, W. S.; Sosnick, T. R.; Englander, J. J.; Mayne, L. *Curr. Opin. Struct. Biol.* 1996, *6*, 18-23.

[97] Kaltashov, I. A.; Eyles, S. J. *Mass Spectrom. Rev.* 2002, *21*, 37-71.

[98] Eyles, S. J.; Kaltashov, I. A. *Methods* 2004, *34*, 88-99.

[99] Katta, V.; Chait, B. T. *J. Am. Chem. Soc.* 1993, *115*, 6317-21.

[100] Thévenon-Emeric, G.; Kozlowski, J.; Zhang, Z.; Smith, D. L. *Anal. Chem.* 1992, *64*, 2456-58.
[101] Palmblad, M.; Buijs, J.; Håkansson, P. *J. Am. Soc. Mass Spectrom.* 2001, *12*, 1153-62.
[102] Kaltashov, I. A.; Eyles, S. J. *J. Mass Spectrom.* 2002, 565.
[103] Buijs, J.; Ramström, M.; Danfelter, M.; Larsericsdotter, H.; Håkansson, P.; Oscarsson, S. *J. Colloid Interface Sci.* 2003, *263*, 441-48.
[104] Akashi, S.; Takio, K. *Prot. Sci.* 2000, *9*, 2497-505.
[105] McLafferty, F. W.; Guan, Z.; Haupts, U.; Wood, T. D.; Kelleher, N. L. *J. Am. Chem. Soc.* 1998, *120*, 4732-40.
[106] Freitas, M. A.; Marshall, A. G. *Int. J. Mass Spectrom.* 1999, *182/183*, 221-31.

In: Biomacromolecular Mass Spectrometry Yearbook. Volume 1 ISBN: 978-1-60876-167-8
Ed: Simone Koenig

Chapter 3

IDENTIFICATION OF TYROSINE-O-SULFATION AS POST-TRANSLATIONAL MODIFICATION OF MELANOMA INHIBITORY ACTIVITY (MIA) PROTEIN*

Marcel P. de Vries, Nicole C.C.J.M. Rullens-Ligtvoet, Wouter J. Hoeberichts, Alain J. van Gool, Mario J.M. van Zeeland, Milou Kouwijzer, Gijs F. Verheijden, Annemieke M.H. Boots, Peter D.E.M. Verhaert*, and Ebo S. Bos♦

Departments of Target Discovery and Pharmacology, NV Organon, PO Box 20, 5340 BH Oss, the Netherlands

*Flemish Interuniversity Institute for Biotechnology; Laboratoire pour Spectrometrie de Masse, Institut de Chimie B6c, Université de Liège, B-4000 Liège; Biomedical Research Institute, Limburgs Universitair Centrum, B-3590 Diepenbeek; and Laboratory for Neuroplasticity and Neuroproteomics, Naamsestraat 59, B-3000 Leuven, Belgium

ABSTRACT

Matrix assisted laser desorption ionization time-of-flight mass spectrometry (MALDI-TOF MS) and isoelectric focusing (IEF) of pure recombinant human MIA (melanoma inhibitory activity) protein, as expressed in CHO cells, indicated that part of the protein was post-translationally modified.

Whereas standard positive ion mode MALDI analysis of intact MIA revealed an additional peak 80 Da higher than the theoretical mass, IEF showed three major protein bands which could all unequivocally be identified as MIA isoforms by peptide mass fingerprinting.

---

* A version of this chapter was also published in *Biomacromolecular Mass Spectrometry Volume 1, Number 1*, published by Nova Science Publishers, Inc. It was submitted for appropriate modifications in an effort to encourage wider dissemination of research.

♦ Corresponding author: Dr. Ebo Sijbren Bos, NV Organon, Molenstraat 110, PO Box 20, 5340 BH Oss, the Netherlands. Tel: +31 412 662668. Fax: +31 412 662542 E-mail: ebo.bos@organon.com

Negative ionization MALDI mass spectrometry on a quadrupole TOF instrument indicated that the differences between these isoforms were due to one or two 80 Da modifications confined to tryptic fragment 59-76. The identity of the 80 Da post-translational modification (PTM) was determined to be a Tyr-O-sulfation through enzymatic removal of the 80 Da moiety from MIA by arylsulfatase treatment, and through the absence of the 80 Da modification(s) in MIA harvested from cells cultured in the presence of chlorate, which is known to inhibit Tyr-O-sulfation. The PTM being sulfation was further corroborated by its metastability upon MALDI-TOF MS analysis. It is concluded that the CHO produced MIA is partially post-translationally modified by Tyr-O-sulfation at positions 69 and/or 70 (^{59}LFWGGSVQGD**YY**GDLAAR^{76}).

Protein Tyr-O-sulfation, has been reported to serve various functions, including modulation of protein-protein interactions. Combination of this notion with the observation that the eukaryotic MIA protein shows only a weak binding towards laminin and fibronectin leads to the speculation that the putative *in vivo* metastasis promoting activity of MIA may be based on the interaction of MIA with an as yet unknown factor via Tyr-(O)-sulfation enhanced binding kinetics.

INTRODUCTION

Melanoma inhibitory activity (MIA) or cartilage-derived retinoic acid-sensitive protein (CD-RAP) is an 11 kDa secretory protein which is normally expressed by cartilage, but is also produced by malignant melanomas and chondrosarcomas and to a lesser extent by adenocarcinomas including breast and colon cancers [1]. Human MIA/CD-RAP (huMIA) is translated as a 131 amino acid precursor and is processed into a mature, non-glycosylated 107 aa protein by cleavage of a secretion signal peptide. The tertiary structure of *E. coli* produced rec huMIA revealed an SH3-resembling molecule stabilized by two disulfide bridges [2-4], yielding a final protein product with a calculated molecular mass of 12106 Da.

Originally, Blesch *et al.* identified MIA as a factor secreted by melanoma cells that was able to inhibit their growth *in vitro* [5]. However, further studies revealed expression patterns inconsistent with tumor suppression and indicated that MIA may actually promote melanoma metastasis *in vivo* [6]. Expression of the MIA gene was not detected in normal skin, but was associated with progression of melanocytic tumors [6]. Also, increased expression of MIA in melanoma cells was reported to enhance their metastatic potential when injected into hamsters [7].

MIA shares structural homology with the binding pockets of α4β1 and α5β1 integrins as was concluded from the finding that antibodies inhibiting binding of integrin to fibronectin cross-react with MIA [8]. The distal loop comprising residues Asp 68 to Ala 75 of MIA resembles a sequence in laminin that has been implicated in recognition of α2β1 integrin [3]. Furthermore, it is suggested by Bosserhoff *et al.* [9], that the MIA protein specifically inhibits attachment of melanoma cells to fibronectin and laminin by masking the binding sites of integrins to these extracellular matrix proteins, thus promoting invasion and metastasis *in vivo*.

In normal tissues, MIA is expressed primarily by chondrocytes. Its expression begins at chondrogenesis and continues through development and in mature chondrocytes, indicating that MIA is fundamental to the cartilage cell phenotype [10]. The level of MIA/CD-RAP in synovial fluid reflects remodeling of articular cartilage and may be used as a marker for restorative reaction of chondrocytes [11].

Most of the recent studies on MIA structure and metastasis-promoting activity have been performed with recombinant MIA protein produced by *E. coli* [2,3,12]. In order to obtain human MIA that is produced under mammalian expression conditions, thus avoiding the possibly troublesome process of protein denaturation and refolding from inclusion bodies we opted for overexpression of MIA by stable transfection into Chinese hamster ovary (CHO) cells.

Recombinant huMIA purified from conditioned medium of CHO-transfectants showed an extra +80 Da peak upon matrix assisted laser desorption ionization time-of-flight mass spectrometry (MALDI-TOF MS) analysis, as well as multiple bands in isoelectric focusing (IEF). In this report, we characterize these MIA protein forms as specific post-translational modifications (Tyr-O-sulfation). The possible implications of this finding for the putative role of MIA in metastasis formation are discussed.

Material and Methods

Expression and Purification of huMIA

Human MIA (huMIA) cDNA was isolated by PCR and subsequently stably transfected into CHO-K1 cells using the expression vector pNGV-1 to yield transfectants that secreted MIA under serum-free culture conditions. From the pool of transfectants a producer cell line was selected with expression levels of approximately 50 mg/L.

MIA was purified to near homogeneity from culture supernatants by capture on a cation exchange column SP-Sepharose equilibrated in 0.2 mol/L NaCl; 0.02 mol/L sodium phosphate pH 7.4. MIA was eluted with 0.5 mol/L NaCl; 0.02 mol/L sodium phosphate pH 7.4, concentrated and subjected to size exclusion chromatography on a Superdex-75 column equilibrated in 0.5 mol/L NaCl; 0.02 mol/L sodium phosphate pH 7.4 Fractions containing pure MIA as determined by SDS-PAGE were pooled, concentrated to around 2 g/L and stored at -20°C.

SDS- Page

SDS-PAGE was carried out in 4-12% NuPAGE gels (Invitrogen, Breda, the Netherlands) using MOPS as running buffer essentially according to the manufacturer's instructions. After fixation for 30 min in 200 g/L TCA, gels were stained with colloidal Coomassie Brilliant Blue [13].

Determination of Melanoma Inhibitory Activity

Melanoma inhibitory activity of the eukaryotic huMIA was determined on G-361/CRL 1424, HT-144, SK-MEL-1, SK-MEL-3/HTB 69 1, SK-MEL-24 and SK-MEL-31 cells essentially as described [14]. In part of the experiments, MIA had been extensively dialyzed against 0.1 mol/L acetic acid for at least 16h prior to incubation with the melanoma cells [14].

Biomolecular Interaction Analysis (BIA)

All BIA experiments were carried out on a Biacore 3000 analyzer.

For quantitative binding studies of MIA to ECM proteins laminin and fibronectin the following experiment was performed.

Human placental laminin and human plasma fibronectin (both from Sigma, Zwijndrecht, the Netherlands) were immobilized by amine coupling on a CM5 chip. The immobilization levels were 8240 response units (RU) and 9534 RU, respectively. Bovine serum albumin (BSA), immobilized to a level of 11482 RU, was used as a control, which enables proper comparison with previously published results [2]. Subsequently, 50 µL of a dilution series of rec huMIA in HBS-EP buffer (Biacore, Breda, the Netherlands, consisting of 0.01 mol/L Hepes, 0.15 mol/L NaCl, 3 mmol/L EDTA, 50 mg/L P20) with concentrations ranging between 6.25 and 50 mg/L were injected at a flow rate of 10 µL/min, followed by 70 µL of HBS-EP buffer (all at 25°C). Finally, the surface was regenerated with a 30 s injection of 10 mmol/L HCl. Combination plots and K_{aff} calculations were made using BIAevaluation software (Biacore) with the BSA signal subtracted [15]. At the level of MIA binding observed mass transport limitation is not considered an issue.

In order to investigate whether the PTM of MIA was tyrosine-O-phosphorylation, the following experiment was carried out.

Mab 4G10, an anti-phosphotyrosine IgG2b/κ (Upstate, Dundee, UK), and the isotype control were immobilized on a CM5 chip by amine coupling to levels of 5980 and 5664 RU respectively. Twenty-five µL of 25 mg/L rec MIA in HBS-EP buffer was injected at a flow rate of 5 µL/min followed by 70 µL of HBS-EP buffer. A solution of 25 mg/L of a Tyr-O-phosphorylated peptide (viz. biotinylated ITAM1 peptide: KRSGLY(P)NELNKGR REEY(P)DVLD, Mw 2941 Da) was used as a positive control. Regeneration was achieved through a 30 s injection of 10 mmol/L HCl. Combination plots were made using BIAevaluation software (Biacore) with the isotype control signal subtracted [15].

Isoelectric Focusing (IEF)

IEF gels (5%T, 3% C, 125×240×0.5 mm), were cast using 40 g/L Servalyt 3-10 IsoDalt (Serva, Heidelberg, Germany) as ampholytes. The gel contained 4 mol/L urea in order to circumvent solubility problems of MIA at low ionic strength. A 30 min pre-run was carried out at 2000V, 10 mA and 10 W. Subsequently, 25 µg of lyophilized MIA (extensively dialyzed against 0.3 mol/L NH_4Ac) were dissolved in 4 mol/L urea and applied on the gel. Focusing was carried out for 3 h at 2000 V, 10 mA and 10 W and for 30 min at 2000V, 20 mA and 20 W. Gels were stained with colloidal Coomassie Brilliant Blue [13] after fixation in 200 g/L TCA (30 min). The various bands representing MIA isoforms were quantified using a GS-700 imaging densitometer (Bio-Rad Laboratories, Hercules CA, USA).

Identification of IEF Bands with Mass Spectrometry

Coomassie-stained bands, representing MIA isoforms were excised from the IEF gel and processed for Peptide Mass Fingerprinting (PMF). The gel plugs were washed [2x in H_2O], shrunk [2x in acetonitrile] and dried (*in vacuo* centrifugation in a SpeedVac system). In gel protein digestion was allowed to take place overnight (37°C) in a minimal volume (15-30 μL) of 25 mmol/L NH_4CO_3 (pH 8.0) containing 12.5 mg/L porcine trypsin (Promega, Leiden, The Netherlands). The resulting samples were acidified by addition of 1 μL HCOOH prior to concentration and desalting on self-made reversed phase microcolumns [GELoader tip (Eppendorf, Hamburg, Germany) fitted with 0.3-0.5 μL POROS20R2 (Applied Biosystems, Framingham, MA, USA)]. PMF spectra were recorded by MALDI-TOF MS on a Voyager DE Pro (Applied Biosystems, Framingham, MA, USA). The system was used in reflectron mode, with delayed extraction. Alpha-cyano-4-hydroxycinnamic acid (αCHCA) was employed as matrix (2.5 g/L in 300 mL/L acetonitrile; 1mL/L TFA). Mass spectra were externally calibrated using two tryptic BSA fragments (m/z 927.49 and 2045.03). The samples were measured both in positive and negative ionization mode. The spectra were processed using Data Explorer processing software (Applied Biosystems, Framingham, MA, USA).

MALDI-TOF MS Analysis of Intact MIA

MALDI-TOF spectra of the intact (undigested) protein were obtained in sinapinnic acid (SA) with the Voyager DE Pro in linear delayed extraction mode, using a two point external calibration (on myoglobin at m/z 16953 [MH^+] and 8477 [$MH2^+/2$]). Prior to analysis, the samples were desalted and concentrated on μC18-ZipTips (Millipore Corporation, Billerica MA, USA). Elution was directly onto a stainless steel MALDI target in 1 μL of a solution containing 10 g/L SA in 500 mL/L ACN/ 1 mL/L TFA.

MS/MS Analysis of MIA Tryptic Digests

Purified, lyophilized huMIA was taken up in 25 mmol/L NH_4HCO_3 (pH 8.0) containing 12.5 mg/L porcine trypsin (Promega, Leiden, the Netherlands) and digested in solution. MS and MS/MS spectra were recorded on an orthogonal MALDI quadrupole TOF hybrid instrument (QSTAR-Pulsar*i*, Applied Biosystems, Framingham, MA, USA). Of the digest 0.5 μL was directly spotted onto a stainless steel target and 0.5 μL of a highly concentrated matrix solution (160 g/L 2,5 dihydroxybenzoic acid (DHB) in 330 mL/L ACN / 1mL/L TFA) were added. MS spectra were recorded and peptide masses were selected for CID (collision induced dissociation) tandem MS fragmentation analysis. Argon was used as collision gas and collision energies were adjusted for each precursor ion by real time monitoring of the fragment spectrum. The identified peptide sequences were interpreted considering a theoretically calculated digest of MIA (SwissProt accession # Q16674).

Nuclear Magnetic Resonance (NMR)

A ^{31}P-1D experiment to detect ^{31}P-nuclei present in the sample was performed on MIA (7 g/L in 0.5 mol/L NaCl; 20 mmol/L phosphate buffer pH 7.0). The measurement was performed on a 400 MHz Bruker DPX spectrometer and recorded using the os_p31cpd parameter set in a proton-decoupled fashion (8192 spectra).

Sulfatase Treatment of MIA

Fifteen U of aryl-sulfatase type VIII (Sigma, Zwijndrecht, the Netherlands) were added to 32 μg of recMIA in 1 mol/L NH_4Ac buffer (pH 6.0) and the reaction mixture was incubated for 3 h at 37°C. MALDI-TOF spectra of enzyme-treated and control MIA preparations were obtained in SA as described above.

Part of the samples were dialyzed against 0.3 mol/L NH_4Ac, lyophilized overnight and dissolved in 4mol/L urea for IEF analysis as described above.

MIA Expression in the Presence of Chlorate

MIA-expressing CHO transfectants were cultured till confluence in DMEM/F12 containing 50ml/L Fortified, 62.5 mg/L penicillin and 62.5 mg/L streptomycin, corresponding with approximately 10^8 cells per 490 cm^2 roller bottle. The adhering cells were washed with fresh medium and the cell culture was continued in similar medium in the presence and absence of 20 mmol/L $NaClO_3$. After 48h, supernatants were harvested, MIA was purified according to standard procedures and analyzed by IEF and MALDI-TOF MS as described above.

RESULTS

Purification and Functional Analysis of Rec huMIA

Rec huMIA was purified from cell culture medium by cation exchange chromatography on SP-Sepharose FF and size exclusion chromatography on Superdex 75 as described under Materials and Methods. This yielded virtually pure MIA as judged by SDS-PAGE (Fig.1). Based on ELISA measurements, the average total recovery in 5 experiments was around 70%. The molecular mass of the intact protein of 12106 as measured by MALDI-TOF MS, the peptide mass fingerprint and MS/MS analysis of the tryptic peptides identified the isolated protein as human MIA. It was recognized by and could be quantified with the commercially available MIA ELISA (Roche Applied Science, Mannheim, Germany).

As part of the quality control, the final product was assessed for melanoma inhibitory activity on a series of cell lines as outlined in the Materials and Methods section. In contrast to our expectations, none of the MIA batches prepared showed growth-inhibiting activity (data not shown), not even after prolonged acid treatment which is claimed to enhance the *in vitro* activity considerably [14].

As was published by several authors, *E.coli*-produced MIA exhibits high affinity binding to laminin and fibronectin [1-3]. Binding studies of our CHO-produced rec huMIA to these ECM proteins were performed through BIA using laminin and fibronectin as immobilized ligands and MIA as analyte in solution. From the results presented in Fig. 2, it is concluded that both the MIA-laminin and MIA-fibronectin interactions are characterized by fast on/fast off kinetics at binding affinities of 15 μmol/L and 5 μmol/L respectively.

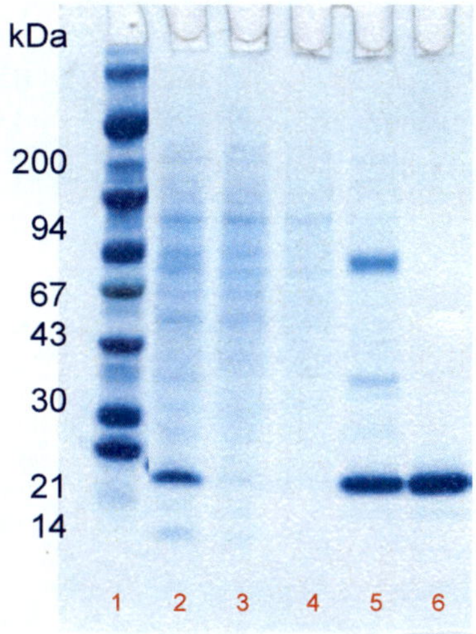

Figure 1. SDS PAGE analysis of huMIA purification
Lane 1, prestained MW markers (Amersham Biosciences); lane 2, conditioned medium of CHO transfectant; lane 3, fall-through of SP column; lane 4, wash SP column; lane 5, eluate SP column; lane 6, huMIA after SEC on Superdex-75.

MALDI-TOF MS Indications for a Post-Translational Modification of huMIA

(1) MALDI-TOF analysis of the purified protein (Fig. 3A) revealed the expected peak at m/z 12105 for mature MIA. However, an additional peak at m/z 12185 was observed corresponding to the theoretical molecular weight plus 80 Da. The size difference between the two peaks suggested a post-translational modification either by (Tyr) sulfation (80.0642 Da) or phosphorylation (79.9799 Da).

(2) Following reduction, alkylation and in-solution trypsin digestion of the MIA preparation, the 80 Da modification was not observed in positive ion mode reflectron MALDI-PMF analysis. However, switching the ionization polarity in the source of the MS instrument (negative ion mode MALDI-TOF PMF), did allow for a +80 Da peak to be detected associated with tryptic peptide ^{59}LFWGGSVQ GDYYGDLAAR^{76}: next to the expected signal for the deprotonated tryptic peptide (AA 59-76) at m/z 1972, an additional signal of m/z 2052 was observed (Fig. 3B). The sequence of this peptide contains two adjacent tyrosine residues (Tyr^{69} and Tyr^{70}), which could be substrates for either phosphorylation or sulfation.

(3) IEF analysis of rec huMIA revealed the presence of three major isoforms (I, IIa and III, as indicated in Fig. 4). The apparent pI of the most basic isoform corresponds with unmodified MIA (predicted pI 9.6). Using densitometry scanning, the relative intensities of these bands were determined to be 78%, 18%, and 5% respectively. Bands were excised from the IEF gel, reduced, alkylated and in-gel digested by trypsin after which the digest was subjected to MS analysis (Fig. 5).

As before, only negative ion mode MALDI-PMF allowed detection of the +80 Da modification in IEF bands IIa and III (Fig. 5). The spectrum of band I, the most basic isoform, showed a major peak at m/z 1972 corresponding with the deprotonated tryptic peptide ($[M-H]^-$) ^{59}LFWGGSVQGDYYGDLAAR^{76}, a non-modified fragment of MIA protein. Bands IIa and III yielded an additional peak at m/z 2052. This corresponds with the mass of fragment 59-76 increased by +80 Da (Fig. 5).

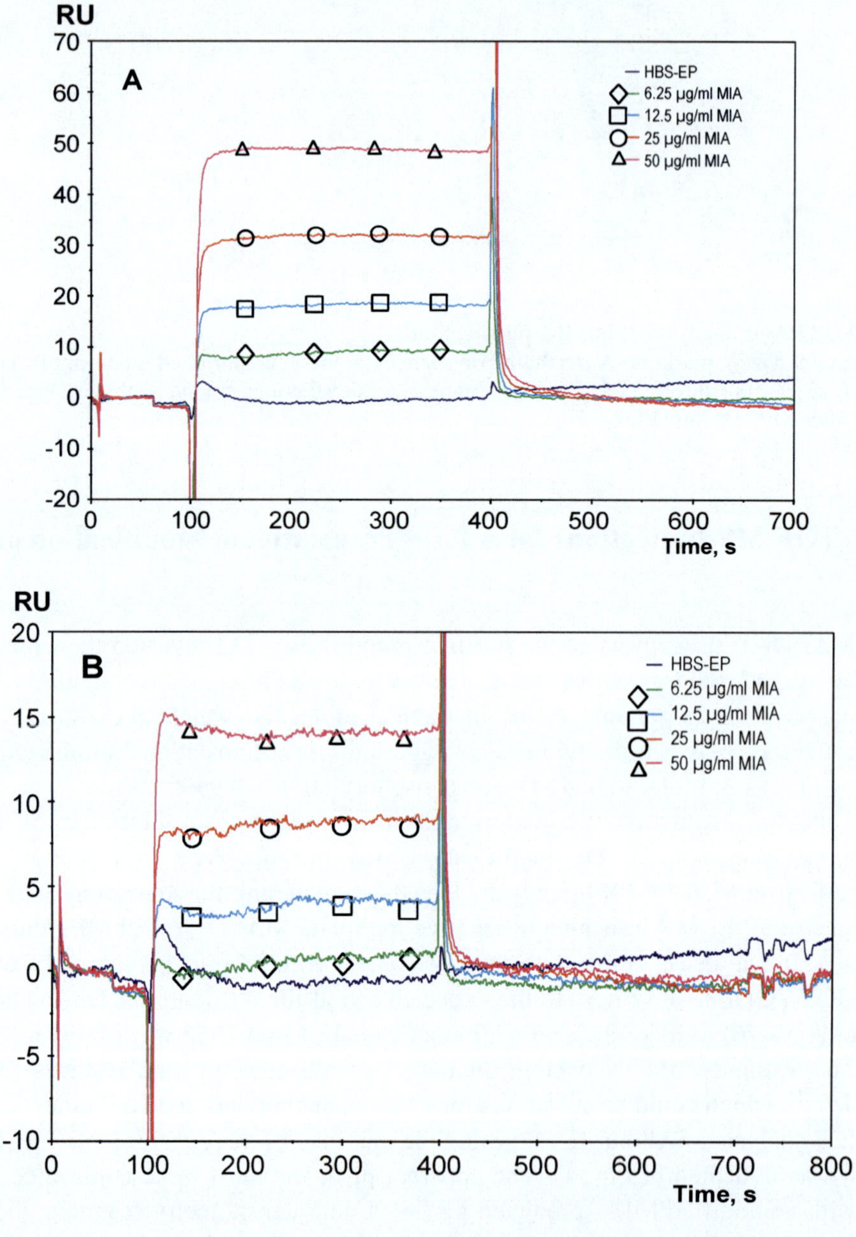

Figure 2. BIA of huMIA binding to ECM proteins laminin (A) and fibronectin (B)
Both proteins were immobilized to the chip, whereas huMIA was employed as soluble analyte. K_{aff} values were calculated according to a 1:1 Langmuir binding model (employing BIAevaluation software) and were 15µmol/L for the MIA-laminin interaction and 5 µmol/L for the MIA-fibronectin interaction, respectively.

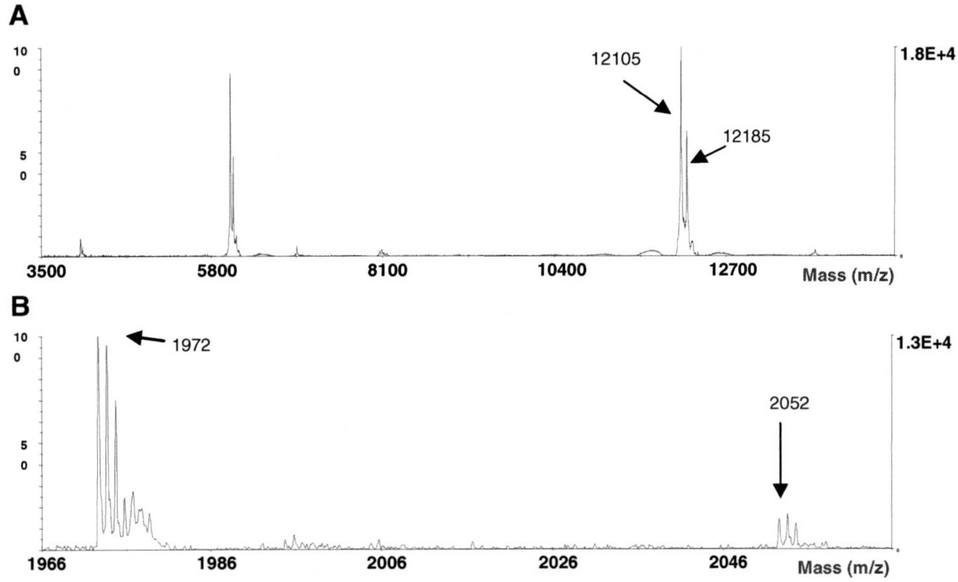

Figure 3. MALDI-TOF analysis of huMIA

A. Intact huMIA analysis. Arrows point to the 12105 Da peak representing unmodified MIA and the 12185 Da peak, representing MIA containing a +80 Da PTM. The peaks at m/z 6052 and 6092 represent the doubly charged ions [M+2H/2]+.

B. Negative-ion MALDI-TOF MS PMF analysis (detail). Both MIA tryptic peptide 59-76 (m/z 1972) and its +80 Da modified variant (m/z 2052) are indicated. (The poorly resolved ion cluster following the well resolved 1972 isotope pattern is indicative of metastability of a fraction of higher mass peptide ions present in the preparation, see text and Fig. 4).

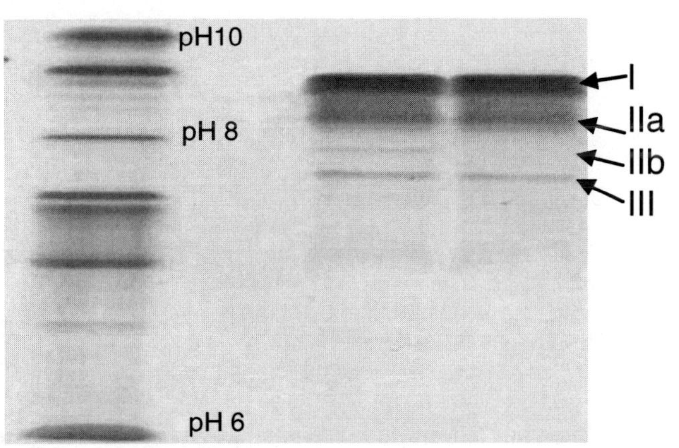

Figure 4. Isoelectric focussing of purified rec huMIA

Lane 1, Serva 3-10 IEF marker; lane 2 and 3, two different MIA preparations. The three major bands representing different isoforms of huMIA are indicated.

(4) A detailed comparative analysis of selected peaks observed in the MALDI-TOF spectra of separately excised bands I to III (Figs. 4-6) is quite revealing for ascribing the +80 Da PTM to either phosphorylation or sulfation.

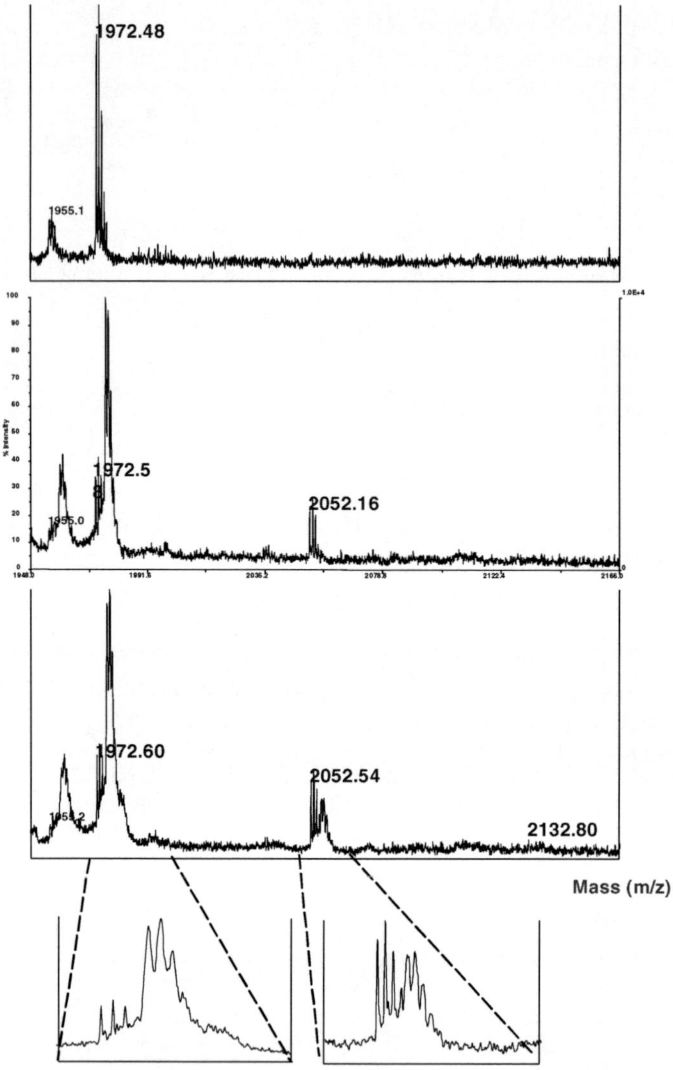

Figure 5. MALDI-TOF MS negative ionisation mode PMF analysis of huMIA IEF bands I, II & III (upper, middle and lower trace respectively; selected mass range)
Details from lower panel show resolved isotopic clusters of MIA tryptic fragment 59-76 with and without a +80 Da modification (right and left panel respectively), with the superimposed metastable ion cluster from the PSD fragmentation of the same tryptic peptide with one and two +80 Da modifications.

(4a) Phosphorylated ions are reported to be relatively stable in MALD-TOF MS. Next to some other characteristic fragments, Ser-, Thr- and Tyr-phosphorylated peptide ions can generally be readily detected as +80 Da peaks in positive ion mode [16]. This, however, does not appear the case for the +80 Da peptide ions observed here.

(4b) Peptide ions containing Tyr-sulfate side chains are described to be considerably more labile as compared to Tyr-phosphorylated ones, which is reflected in their metastable fragmentation behavior [17]. Tyr-sulfated peptide ions readily lose their sulfate moiety (-80 Da) in the mass spectrometer flight tube. As this decay occurs 'post source' somewhere during the ion drift along the flight tube of the MALDI-TOF instrument, the fragment ions will not be focused onto a reflectron detector. This results in an isotopic cluster less resolved than that of in-source generated ions [18]. Therefore, post-source decay (PSD) ions have a higher apparent m/z than the in-source fragmented ions as the former have been traveling undegraded for part of the time-of-flight tube. In a conventional MALDI-TOF instrument, peptide ions losing their sulfate moiety within the MS source are mass spectrometrically indistinguishable from non-sulfated peptide ions, with the typical well resolved isotopic pattern. This observation has formerly been used to discriminate between sulfotyrosine and phosphotyrosine in both electrospray [19] and MALDI [20], and for more general detection of PTMs in positive ion mode reflectron MALDI-TOF MS [21].

Our results show band I with a well resolved isotope pattern at m/z 1972, corresponding with $[M\text{-}H]^-$ (deprotonated peptide ion) of the unmodified tryptic fragment 59-76 (Fig. 5, upper panel). Band IIa exhibits an additional, well resolved peak at 2052, representing the +80 Da modification of the same fragment (Fig. 5 middle panel). In the same spectrum of band IIa, the resolved isotopes at m/z 1972 are overshadowed by a peak cluster of noticeably reduced resolution with an apparent m/z value of 1976 (Fig. 5 lower two panels). These appear as metastable (PSD) ions according to Harvey *et al.* [18]. The data are fully consistent with a Tyr-O-sulfation on the 59-76 fragment, with the resolved signal at m/z 2052 representing a small fraction of the ions reaching the detector intact.

Likewise, MALDI-PMF of band III showed metastable ions at m/z 1976 and 2056, and well resolved signals at m/z 1972 and m/z 2052 respectively (Fig. 5). This would indicate the presence of a second sulfate group on the already sulfated 2052 peptide. As indicated above, two sulfations can indeed occur on two adjacent tyrosine residues in this peptide (^{59}LFWGGSVQGD**YY**GDLAAR^{76}). A similar observation of double sulfation has been reported before for recombinant factor VIII [22] and glycoprotein 1b-α [23]. In the linear MALDI-TOF MS mode (which does not allow the detection of post-source decay fragments), well resolved peaks at m/z 1972, 2052 as well as at m/z 2132 are detectable in band III, supporting the hypothesis of these ions being the result of metastable decay (data not shown).

(4c) In this respect, also the MS analyses in the orthogonal MALDI quadrupole-TOF hybrid instrument are quite enlightening. In this MS instrument, source and analyzer are fully decoupled, with the ions leaving the source efficiently cooled, so that PSD metastability is not likely to be as prominent as in conventional MALDI-TOF systems. Ions which do fragment in the quadrupole section of the instrument prior to entering the TOF analyzer are all focused on the MS detector (see Fig. 6). Negative-ionization MALDI-Q-TOF-MS of band IIa and III (the more acidic isoforms), reveals much less fragmentation than in the MALDI-TOF analyses (Fig. 6). A substantially larger proportion of the ions reach the detector unfragmented. Hence, more pronounced peaks at 2052 in band IIa and III, and even a additional peak at m/z 2132, which corresponds with the mass of AA59-76 MIA peptide with two +80 Da additions are detected (Fig. 7).

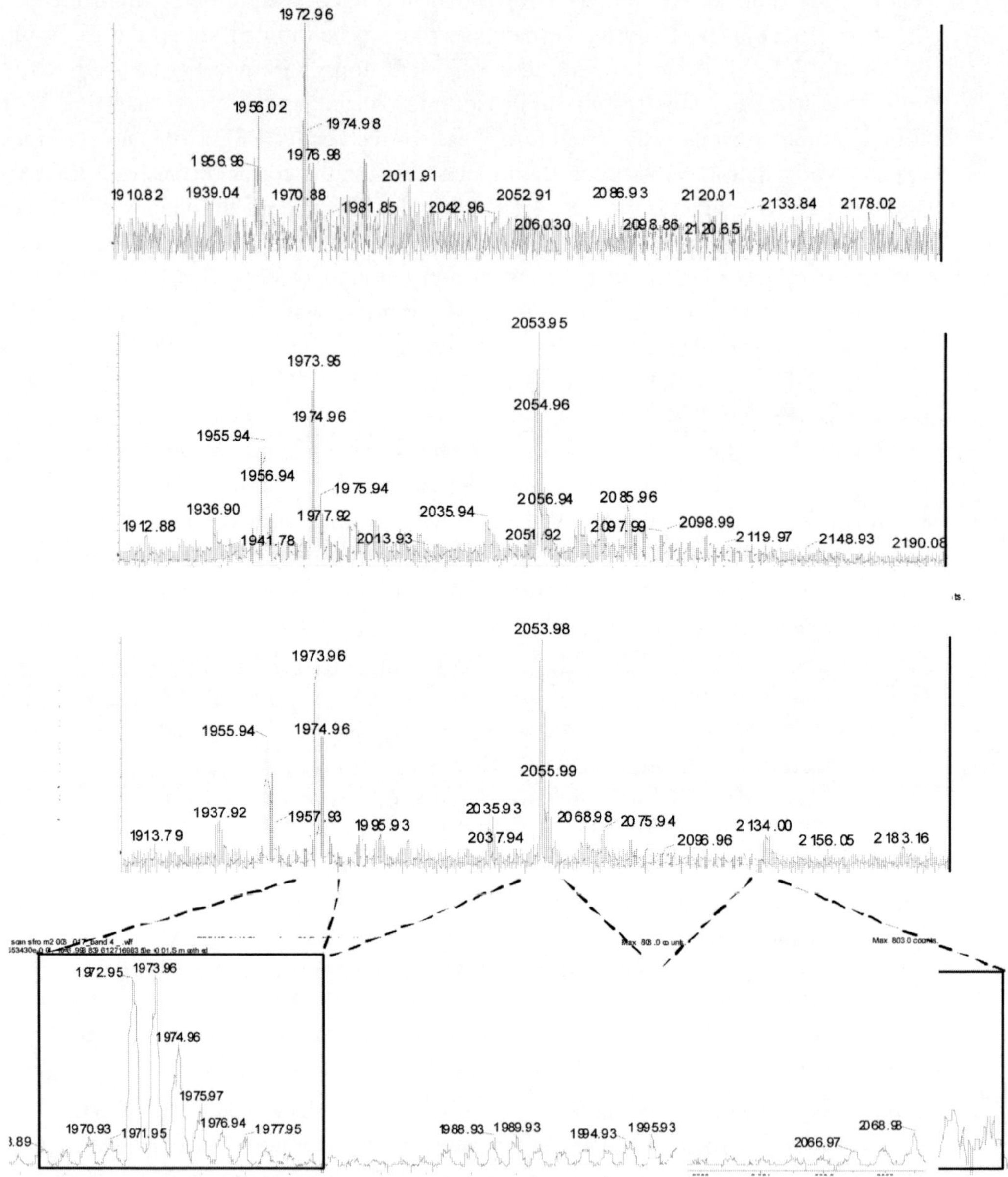

Figure 6. Negative ionisation mode orthogonal MALDI hybrid Q-TOF PMF analysis of MIA IEF bands I, IIa & III Details from lower panel show the resolved isotopic clusters of MIA tryptic fragment 59-76 with no, one and two +80 Da modifications (left, middle and right panel)

(4d) Whereas MS/MS analysis of the +80 Da peptide would seem the most straightforward approach to locate the exact position of the modification involved, this did not yield the results hoped for. As it turns out, the selected modified peptide is so labile that it readily loses its +80 Da moiety in the collision chamber, with no +80 Da containing fragments remaining in the rest of the spectrum.

The poor stability of the +80 Da ions observed in the above MS analyses again argues in favor of a sulfation to be responsible for the +80 Da modification, rather than a phosphorylation [19].

The charge difference between bands IIa and III is likely to be the consequence of one and two sulfations, respectively.

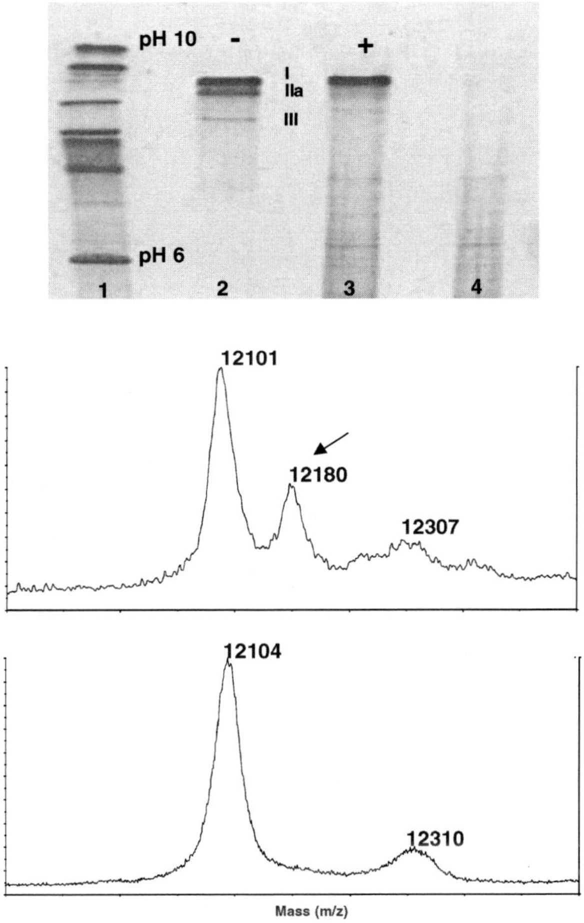

Figure 7. MALDI and IEF analysis of aryl sulfatase-treatment of MIA
A. IEF of aryl sulfatase-treated and untreated MIA. Lane 1: Serva 3-10 IEF marker; lane 2: untreated MIA; lane 3: MIA treated with aryl sulfatase. Dissappearance of bands II and III indicate removal of the +80 Da moiety by the sulfatase; lane 4: aryl sulfatase per se.
B. MALDI spectra of untreated and treated MIA shown in upper and lower trace respectively. Unmodified and +80 Da containing huMIA (arrow) appear as m/z peaks 12101 Da and 12180 Da respectively; additional higher m/z peaks are the result of sinapinnic acid adduct formation. Lower panel shows disappearance of the +80 Da signal after sulfatase treatment.

A minor extra band (designated IIb) discernible between band IIa and III in some preparations (see Fig. 4) is likely the result of an additional modification in a different part of the MIA molecule. A deamidation, for instance, would result in a charge shift similar to the one observed. Since this band represents only a minor fraction of the four isoforms viz. 0.3% by densitometry analysis, it was not further investigated.

Additional Studies on the Modification of MIA

Experiments including NMR and BIA studies aiming to assess whether the observed 80 Da shift could be attributed to phosphorylation, were both negative.

In NMR studies, no ^{31}P-nuclei were detected in the recorded spectrum suggesting that MIA protein was not phosphorylated to a detectable level. Furthermore, rec MIA was not bound to a significant extent (1.7% of the theoretical value) by a monoclonal anti-phosphotyrosine antibody in a Biacore experiment, in contrast with a tyr-phosphorylated control peptide (which was bound to a level of 70% under the same conditions).

Together, these data indicate that rec huMIA is not phosphorylated on tyrosine residues.

The most straightforward way to determine whether the +80 Da PTM of MIA comprises a sulfation, is treatment with arylsulfatase and subsequent product analysis by MALDI-TOF-MS and IEF [24].

The MALDI-TOF spectra of the untreated recMIA showed the expected peaks at m/z 12101 (+/- 10 Da, corresponding to non-modified MIA) and at m/z 12180 (+80 Da modified MIA). In contrast, treatment of MIA with arylsulfatase resulted in disappearance of the +80 Da peak (Fig. 7B). In IEF analysis, untreated rec huMIA yielded the characteristic three major MIA isoforms (Fig. 7A lane 2), whereas enzyme-treated MIA lacked the more acidic bands (Fig. 7A lane 3).

In order to obtain additional evidence for Tyr-O-sulfation, MIA-producing CHO transfectants were grown in the presence of chlorate which is known to block Tyr-O-sulfation by inhibiting tyrosyl protein sulfotransferase [22,25]. The rec huMIA produced under these conditions was purified and analyzed as described above.

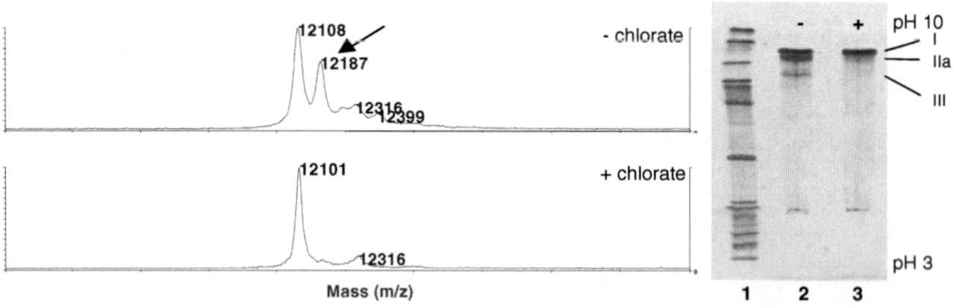

Figure 8. MALDI and IEF analysis of huMIA produced in absence and presence of chlorate
A. MALDI spectra of MIA produced in the absence or presence of chlorate are shown in upper and lower panels respectively. Arrow points to MIA + 80 Da m/z peak.
B. IEF analysis. Lane 1: Serva 3-10 IEF marker, lane 2 MIA produced in absence of chlorate, lane 3: MIA produced in presence of chlorate. In the latter preparation bands IIa and III have disappeared.

MALDI MS analysis of rec huMIA from cells grown in the absence of chlorate yielded peaks at m/z 12108 and 12187, corresponding with non-sulfated and sulfated MIA respectively (Fig. 8A upper panel). In contrast, cells cultured in the presence of chlorate showed a single peak at m/z 12101 (Fig. 8A lower panel).

In addition, IEF analysis of MIA produced in the absence of chlorate revealed the three major characteristic rec huMIA isoforms I-III (Fig. 8B lane 2), whereas MIA produced in the presence of chlorate showed an IEF pattern that lacked the more acidic isoforms (Fig. 8B lane 3), corroborating the previous findings that these isoforms represent Tyr-O-sulfated MIA.

Discussion and Conclusion

The MS and biochemical data presented in this paper support the conclusion that part of rec huMIA produced recombinantly in CHO cells is post-translationally modified by sulfation of tyrosine residues 69 and 70. The charge difference between bands IIa and III is likely to be the consequence of one and two sulfations, respectively.

We would like to emphasize that the substoichiometric sulfation of MIA, viz. approximately 25% as estimated from IEF analyses, does not necessarily reflect the stoichiometry *in vivo*. Incomplete sulfate ester modification of recombinant proteins expressed in CHO cells has been reported by several authors [33,34]. In order to study the extent of Tyr-O-sulfation of MIA *in vivo*, the protein should be purified from natural sources like serum or synovial fluid. However, MIA concentrations range between 10 and 30 μg/L in these biological fluids [11], which is almost prohibitive for the isolation of sufficient amounts of pure, native huMIA.

Sulfation of proteins is less often described than protein phosphorylation. Nevertheless, it is a common PTM, occurring on as much as 1% of all tyrosine residues of total protein in an organism [26]. It is one of the last modifications to occur on proteins transiting the trans-Golgi and thus has been found almost exclusively on secreted and plasma membrane proteins, and this in all metazoan species examined. The modification is present usually at the interface of interacting proteins and is known to modulate extracellular protein/protein and protein/carbohydrate interactions, generally resulting in a substantially increased binding potency. Examples are hirugen/ thrombin [27], heparin cofactorII/heparin [28] and Factor VIII/ von Willebrandt factor interaction [22]. Furthermore, Tyr-(O)-sulfation has been implicated in intracellular protein transport, proteolytic processing (e.g. gastrin processing) and is important in the proteolytical activation of proteins involved in coagulation [25,29,30].

In a recent paper by Bosserhoff *et al.* [8] it is claimed that MIA shares significant structural homology with the fibronectin binding pockets of the α4β1 and α5β1 integrins and that MIA is, therefore, able to mask the respective epitopes on ECM molecules. However, the homology turns out to be poor (i.e. well below 20%) and modeling of the putatively homologous sequences in a PDB structure of α5β3 integrin (67% similarity with α5β1 integrin) [31,32] failed to demonstrate any structural resemblance between MIA and the integrins. Moreover, our studies on the binding kinetics of huMIA and fibronectin reveal an interaction of low affinity with fast on/ fast off characteristics. In our opinion it is highly unlikely, that these binding kinetics could explain the results found by other groups using prokaryotic material in an ELISA-like setting. These arguments make the conclusions of Bosserhoff *et al.* [8] less likely.

Lougheed *et al.* point at the resemblance of the MIA amino acid sequence ^{66}QGDYYGDLAAR^{77} (extruding from the proten moiety in the published 3D structure [2-4, 12]) with the YGYYGDALR sequence in laminin that has been implicated in recognition of α2β1 integrin. However, as these studies were performed with *E. coli* produced MIA (to our knowledge no cases of tyrosine sulfation in prokaryotes have been reported), they did not take into account the consequences of a possible Tyr-(O)-sulfation at this particular site which might change the interaction properties of the distal loop dramatically.

We would like to emphasize here that all essential investigations up to now on structure and function of human MIA have been carried out with protein produced by *E. coli* (denatured and refolded from inclusion bodies). Preliminary NMR experiments on the 3D structure of CHO-produced huMIA suggest a different folding of the protein as compared with the prokaryote product. The present data show that recombinant huMIA protein can be Tyr-O-sulfated but it remains to be established that these data are equivalent in native huMIA. In the latter case it may be speculated that O-tyrosine sulfation of huMIA is a key factor in its biological activity. The postulated metastasis-promoting activity of MIA *in vivo* could be mediated through an as yet unknown factor via Tyr-(O)-sulfation enhanced binding kinetics.

This might imply that, at least, part of the experiments on the function of MIA should be repeated with eukaryotic material and that studies on the effects of tyrosine sulfation on MIA binding and biological activity should be included. Furthermore, resolving the 3D structure of CHO-produced huMIA, either by X-ray crystallography or by NMR, may contribute to a correct interpretation of the *in vitro* and *in vivo* experiments at the molecular level.

Acknowledgments

The authors are much indebted to H. Schellekens for performing the cloning experiments, W. Koot for cell culture experiments, Dr. C. Kloks for performing the NMR analysis, M. Bruisten for protein purification and D. Vu-Pham for carrying out the IEF analysis. Dr. M.-J. van Lierop is acknowledged for carrying out the MIA *in vitro* assays and for valuable discussions. We thank Dr. A. Garritsen for critical reading of the manuscript.

References

[1] Bosserhoff, A.K.; Buettner, R. *Histol. Histopathol.* 2002 17, 289-300.

[2] Stoll, R.; Renner, C.; Zweckstetter, M.; Brueggert, M.; Ambrosius, D.; Palme, S.; Engh, R.A.; Golob, M.; Breibach, I.; Buettner, R.; Voelter, W.; Holak, T.A.; Bosserhoff, A.K. *EMBO J.* 2001 20, 340-349.

[3] Lougheed, J.C.; Domaille, P.; Handel, T.M. *J. Biomol. NMR* 2002 22, 211-223.

[4] Lougheed, J.C.; Holton, J.M.; Alber, T.; Bazan, J.F.; Handel, T.M. *Proc. Natl. Acad. Sci.* USA 2001 98, 5515-5520.

[5] Blesch, A.; Bosserhoff, A.K.; Apfel, R.; Behl, C.; Hessdoerfer, B.; Schmitt, A.; Jachlmczak, P.; Lottscpeich, F.; Buettner, R.; Bogdahn, U. *Cancer Res.* 1994 54, 5695-5701.

[6] Van Groningen, J.J.; Bloemers, H.P.; Swart, G.W. *Cancer Res.* 1995 55, 6237-6243.

[7] Guba, M.; Bosserhoff, A.K.; Steinbauer, M.; Abels, C.; Anthuber, M.; Buettner R.; Jauch, K.W. Brit. *J. Cancer* 2000 83, 1216-1222.

[8] Bosserhoff, A.K.; Stoll, R.; Sleeman, J.P.; Bataille, F.; Buettner, R.; Holak, T.A. *Laboratory Investigation* 2003 83, 1583-1594.

[9] Bosserhoff, A.K.; Golob, M.; Buettner, R.; Landthaler, M.; Hein R. Hautarzt 1998 49, 762-769.

[10] Bosserhoff, A.K.; Buettner, R. *Biomaterials* 2003 24, 3229-3234.

[11] Saito, S.; Kondo, S.; Mishima, N.; Ishiguro, Y.; Hasegawa, L.; Sandell, J.; Iwata H. J. *Bone Joint Surg*. 2002 84-B, 1066-1069.

[12] Stoll, R.; Renner, C.; Buettner, R.; Voelter, W.; Bosserhoff, A.K.; Holak, T.A. Protein *Science* 2003 12, 510-519.

[13] Neuhoff, V.; Stamm, R.; Eibl, H. *Electrophoresis* 1985 6, 427-448.

[14] Bogdahn, U.; Buettner, R; Kaluza, B. European Patent 1999 *EP* 0 710 248 B1.

[15] BIAtechnology Handbook, *Pharmacia Biosensor*, Ed. 1994

[16] Annan, R.S.; Carr, S. *Anal Chem*. 1996 68, 3413-3421.

[17] Nemeth-Cawley, J.F.; Karnik, S.; Rouse, J.C. *J. Mass Spect*. 2001 36, 1301-1311.

[18] Harvey, D.J.; Hunter, A.P.; Bateman, R.H.; Brown, J.; Critchley, G. *Int. J. Mass Spectrom.* 1999 188, 131-146.

[19] Rappsilber, J.; Steen, H.; Mann, M. *J. Mass Spectrom*. 2001 36, 832-833.

[20] Wolfender, J.L.; Chu, F.; Ball, H.; Wolfender, F.; Fainzilber, M.; Baldwin, M.A.; Burlingame, A.L. *J. Mass Spectrom*. 1999 34, 447-454.

[21] Wirth, U.; Mueller, D.; Schindler, P.; Lange, J.; van Oostrum, *J. Proteomics* 2002 2, 1445-1451.

[22] Pittman, D.D.; Wang, J.H.; Kaufman, R.J. *Biochemistry* 1992 31, 3315-3325.

[23] Dong, J.; Li, C.Q.; Lopez, J.A. *Biochemistry* 1994 33, 13946-13953.

[24] Cimino, C.O.; Shelmutt, S.R.; Ronis, M.J.J.; Badger, T.M. *Clin. Chim. Acta* 1999 287, 69-82.

[25] Kehoe, J.W.; Bertozzi, C.R. *Chemistry and Biology* 2000 7, R57-R61.

[26] Huttner, W.B. *Ann. Rev. Physiol*. 1988 50, 363-376.

[27] Skrzypzcak-Jankun, E.; Carperos, V.; Ravichandran, K.G.; Tulinsky, A. *J. Mol. Biol.* 1991 221, 1379-1393.

[28] Boehme, C.; Nimtz, M.; Grabenhorst, E.; Conradt, H.S.; Strathmann, A.; Ragg, H. *Eur. J. Biochem.* 2002 269, 977-988.

[29] Niehrs, C.; Beisswanger, R.; Huttner, W.B. *Chem. Biol. Interact*. 1994 92, 257-271.

[30] Moore, K.L. *J. Biol. Chem*. 2003 278, 27, 24243-24246.

[31] Xiong, J.-P.; Stehle, T.; Diefenbach, B.; Zhang, R.; Dunker, R.; Scott, D.L.; Joachimiak, A.; Goodman, S.L.; Arnaout, M.A. *Science* 2001 294, 339-345.

[32] Xiong, J.P.; Stehle, T.; Zhang, R.; Joachimiak, A.; Frech, M.; Goodman, S.L.; Arnaout, M.A. *Science* 2002 296, 151-155.

[33] Mikkelsen, J.; Thomson, J.; Ezban, M. *Biochemistry* 1991, 1533-1537.

[34] White, G.C.; Beebe, A.; Nielsen, B. Thromb. *Haemost*. 1997 78, 261-265.

In: Biomacromolecular Mass Spectrometry Yearbook. Volume 1 ISBN: 978-1-60876-167-8
Ed: Simone Koenig

Chapter 4

MASS SPECTROMETRY IDENTIFIES HOMOLOGUES OF PHASIN PHAP1 PROTEIN OF *RALSTONIA EUTROPHA* ON POLYHYDROXYBUTYRATE GRANULES*

Martin Zeller*[1], Markus Pötter*[2], Frank Reinecke[2], Alexander Steinbüchel[2], William I. Burkitt[3], Peter J. Derrick[3], and Simone König[1]

[1]Integrated Functional Genomics, Interdisciplinary Center for Clinical Research, Medical Faculty
[2]Institute for Molecular Microbiology and Biotechnology, University of Münster, Germany;
[3]Institute of Mass Spectrometry, University of Warwick, UK
* MZ and MP share first-authorship.

ABSTRACT

Many prokaryotes use polyhydroxyalkanoates (PHA) as storage compounds for carbon and/or energy in the form of water-insoluble inclusion PHA granules in the cytoplasm. The surface of such granules is covered by proteins which also affect their form and structure. Here, phasin PhaP1 protein was identified in high abundance on the grana surface of *R. eutropha* employing gel electrophoretic separation and mass spectrometry. Homologous phasins PhaP3 and Pha4 were visualized using 2D-PAGE in the PhaP1 knock-out mutant, but no evidence for the presence of PhaP2 could be found. All detected phasins exhibit isoforms whose structural differences are still a matter of investigation, although C-terminally truncated isoforms of PhaP1 were detected with FTICR-MS.

* A version of this chapter was also published in *Biomacromolecular Mass Spectrometry Volume 1, Number 1*, published by Nova Science Publishers, Inc. It was submitted for appropriate modifications in an effort to encourage wider dissemination of research.

[1] Address correspondence to: PD Dr. Simone König, Integrated Functional Genomics, Röntgenstr. 21. 48149 Münster, Germany. ph: +49-(0)251-8357164. fax: +49-(0)251-8357255. e-mail: koenigs@uni-muenster.de

Keywords: *Mass spectrometry, protein analysis, PhaP homologues, FTICR*

ABBREVIATIONS

2D-PAGE	**two-dimensional gel electrophoresis**
FTICR	**Fourier-transform ion cyclotron resonance**
FWHM	**full width at half maximum**
MS	**mass spectrometry**
MW	**molecular weight**
m/z	**mass/charge**
NanoESI	**nano electrospray**
PHAs	**polyhydroxyalkanoates**
PhaP	**phasin**
PHB	**poly 3-hydroxybutyrate**
pI	**isoelectric point**

INTRODUCTION

A large variety of prokaryotes use polyhydroxyalkanoates as storage compounds for carbon and/or energy in the form of water-insoluble inclusions PHA granules in the cytoplasm. Due to their biodegradability and origin from renewable resources, PHAs have attracted much interest, and many technical and medical applications have been developed in the last few decades [1-3]. The Gram-negative facultative chemolithoautotrophic hydrogen bacterium *Ralstonia eutropha* is most probably the best studied microorganism with regard to the metabolism of polyhydroxyalkanoic acids. *R. eutropha* belongs to those bacteria which accumulate polyoxoesters of various short carbon-chain-length hydroxyalkanoic acids comprising three to five carbon atoms [4, 5]. PHAs are produced in the form of lipophilic inclusion bodies, the so-called grana which contain approximately 97.5 % PHA, 2 % protein, and 0.5 % phospholipid [6], although some estimates on the lipid contents are considerably higher [7].

Interestingly, a completely different type of PHA with a yet unknown function occurs ubiquitarily, i.e. in prokaryotes and eukaryotes, but contributes only 0.1 % of the cell matter or less. These PHA molecules exhibit a much lower molecular weight of up to ~15 kDa, and they are located in the cytoplasmic membrane or are associated or even covalently bound to other cell constituents [8]. However, enzymes and genes involved in biosynthesis of these PHAs, which are referred to as “complexed PHAs” (cPHA) and which also occur in *R. eutropha*, have not been identified, yet.

The surface of PHA granules is covered by a 2 nm membrane. It was recognized that PHA granules contain proteins at their surface which are responsible for their structure, size and function [7]. Four different types of granule-associated proteins were found in PHA producing bacteria. These are (i) the PHA synthase (PhaC), (ii) the intracellular PHA depolymerases (PhaZi), (iii) the phasins (PhaP), and (iv) a regulator protein of the phasin expresssion (PhaR). In *R. eutropha*, it was shown that PhaC is a soluble protein in the cytoplasm, which becomes only insoluble and PHA granule bound when the cells are cultivated under conditions permissive for PHA accumulation and after the onset of PHA accumulation. This is due to the covalent linkage of the growing PHA chain to the enzyme [9-11]. In *R. eutropha*, the first two enzymes of the pathway, β-ketothiolase PhaA and acetoacetyl-CoA reductase PhaB, are found in the cytoplasm during active PHA synthesis and do not bind to the granules [12, 13]. Since the PHA operon, comprising the genes *phaA*, *phaB*, and *phaC* was cloned during the eighties, many other genes involved in PHA metabolism have been subsequently identified, and the functions of the encoded proteins were at least partially analysed. Whereas *phaC* is essential for PHA biosynthesis in *R. eutropha*, *phaA* and *phaB* can be replaced by other enzymes. Phasins are considered as by far the major protein constituents and structural proteins of the layer at the surface of the PHA granules. The structural protein, PhaP, is localized at the surface of the PHB-granules, and it is synthesized in very large quantities under storage conditions, representing as much as 5 % of the total protein [14] . PhaP adheres very tightly to native as well as to artificial poly 3-hydroxybutyrate (PHB) granules. In *Rhodococcus ruber*, the anchoring region was located at the carboxy terminus; and it was demonstrated that phasin molecules truncated at the carboxy terminal region lost their capability to bind to PHA granules [15].

PHA granules can be generated *in vitro* simply by incubating purified PHA synthase protein and a suitable substrate. The addition of phasins accelerates the PHA synthesis rate. Furthermore, the size of these inclusions can be regulated by the addition of PhaP, although electron microscopic studies revealed that the granules do not have the same surface structure as native inclusions [16].

R. eutropha H16 produces PHB in the cytoplasma under certain conditions (lack of nitrogen-sources and at the same time excess of a carbon-source such as fructose). PHB is biodegradable, thermoplasctically deformable and suited for usage as packing material on an industrial scale [17, 18]. The currently available genome sequence of *R. eutropha* H16 revealed three homologues (PhaP2, PhaP3, Pha4) to the phasin protein PhaP described previously [14]. Phasins are between 14 and 28 kDa in size and form a layer at the surface of the hydrophobic core of the PHB granules with PhaP being predominant [14, 19]. PhaP1. PhaP2, PhaP3, and PhaP4 exhibit 42, 49, and 45 % identity or 61, 62, and 63 % similarity to PhaP1, respectively [20]. Phasins influence the number and size of PHB granules [14, 15, 21]. In this work, proteomic analysis of the isolated and solubilised grana of the *R. eutropha* wild-type and a PhaP1 knock-out was performed to determine, if the homologous phasins are also located at the grana surface.

MATERIALS AND METHODS

PhaP1 from *R. eutropha* was isolated from native granules of fructose-grown cells of *R. eutropha* as described by [14]. The preparation of the native granules from *R. eutropha* strain H16 and the *phaP1* negative mutant *R. eutropha* Re1052 [22], solubilization of the granule-associated proteins, and their separation using 2D-PAGE is described in ref. [20].

For testing *in vivo* phasin phosphorylation cells were cultivated in the presence of ^{32}P-phosphoric acid (500 μCi) in a phosphorous-poor Tris-based mineral salts medium [23]. Granule proteins were prepared as above and the corresponding 2D-gel was dried on filter paper using a vacuum gel slab dryer at 60°C for four hours. The dried gel was covered with a thin cling film and photographic film was placed on top for several days.

For protein identification, spots of interest were cut out of the gel, washed and tryptically digested according to a procedure slightly modified from ref. [24]. Peptides were extracted and purified using ZipTips C18 (Millipore, Bedford, MA, USA). Peptide maps were generated using TofSpec-2E instrument (Micromass, Manchester, UK) in reflectron mode. Data analysis was performed with Mascot (MatrixScience, London, UK) in-house searching a proprietary *R. eutropha* database. Isoforms were scrutinized manually.

For FTICR, lyophilised sample of PhaP1 purified from *R. eutropha* wild-type (10 μmol in 100 mM Tris according to Bradford test) was reconstituted with 100 μl destilled water. 10 μl was acidified with 2 μl formic acid and desalted using ZipTips C4 into 10 μl 70 % acetonitrile containing 0.1 % acetic acid. After dilution with 10 μl 0.1 % acetic acid FTICR experiments were carried out on a BioAPEX-94e mass spectrometer with a passively shielded 9.4 Tesla magnet and a standard hexapole-based interface (Analytica of Branford, MA, USA). For nanoelectrospray [25] metal-coated glass capillaries from Protana (Protana, Odense, Denmark) were used. The *m/z* scale was externally calibrated with an Agilent calibrant.

RESULTS AND DISCUSSION

2D-PAGE was employed to separate the phasin proteins in crude cell extracts or poly(3HB) granules (Fig. 1). Difficulties due to the similarities in the MWs of the proteins (Table 1) had to be observed and, therefore, pH gradient strips 5-8 were used for better resolution. PhaP1 occurs in large amounts (appproximately 5 % of the protein content under storage conditions) in the wild-type preventing the detection of phasins present in lower amounts. Three spots were resolved in the 2D-PAGE experiment of the wild-type strain, which could be all assigned to PhaP1 (Fig. 1a). Fig. 2 shows the peptide maps of the left and the right spot the spectrum of the middle spot being fairly identical with that of the left spot. As is highlighted in Fig. 3, signals for 91 % of the expected sequence could be found for the intense spots. For the fainter right spot still 62 % of the sequence were assigned. In both left spots most peptide peaks were accompanied by signals 57 and 75 Da higher in mass. It remains to be clarified what modification or contamination was responsible for that effect and if it caused any shift in protein pI. No evidence for the addition of hydroxy butyric acid derivatives could be found but phosphorylation was considered. Due to the introduction of additional negative charges by phosphorylation proteins shift towards a more acidic pI on the gel. This effect has been successfully employed for comparative phosphoproteomics based on enzymatic dephosphorylation [26].

Table1: GenBank entries for phasin homologues

accession number	phasin	MW	pI
AAC78327	PhaP1	19966,54	5,96
AAP85954	PhaP2	20223,04	5,95
AY489113	PhaP3	19595,94	6,12
AY489114	PhaP4	20183,85	5,91

To check for phosphorylation, two approaches were followed in parallel. *In vivo* phosphorylation of phasins was questioned by cultivation of cells in the presence of ^{32}P-phosphoric acid in a phosphorous-poor Tris-based mineral salt medium [23]. However, no signals were observed in the phasing region.

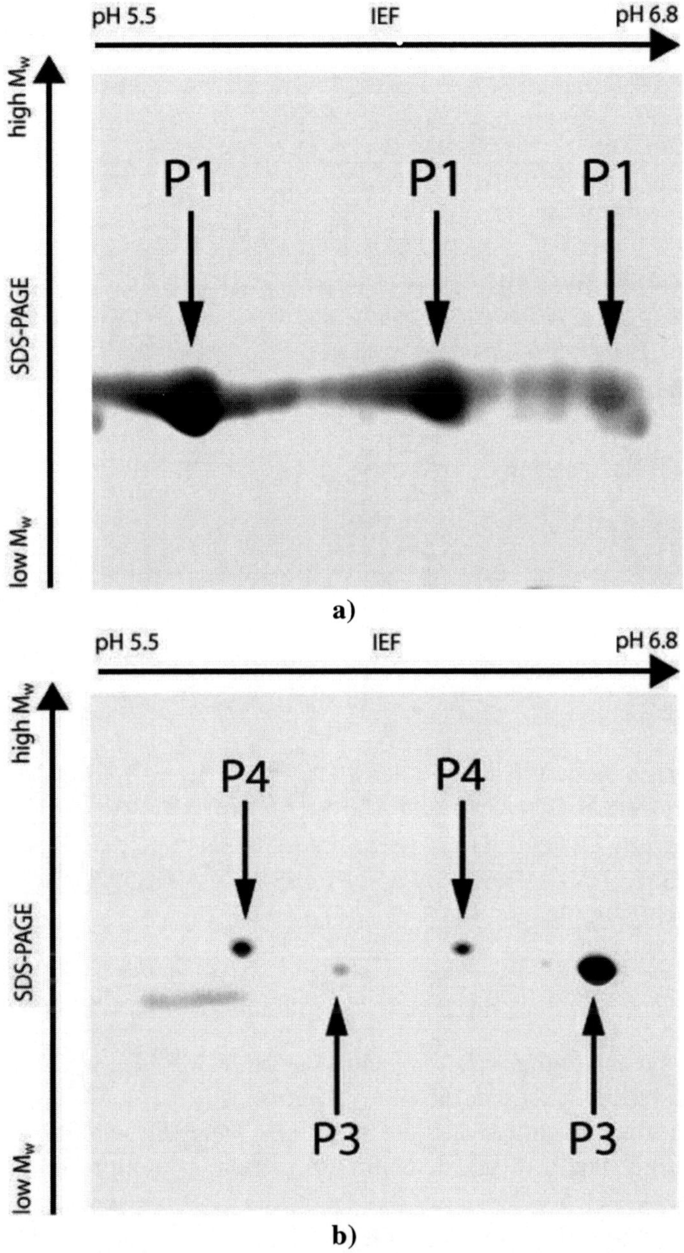

b)

Figure 1: Gel images of the separated phasing homologues of a) *R. eutropha* wild-type; b) knock-out mutant *ΔphaP1*. Sections of the gel photographs presented in ref. [20]

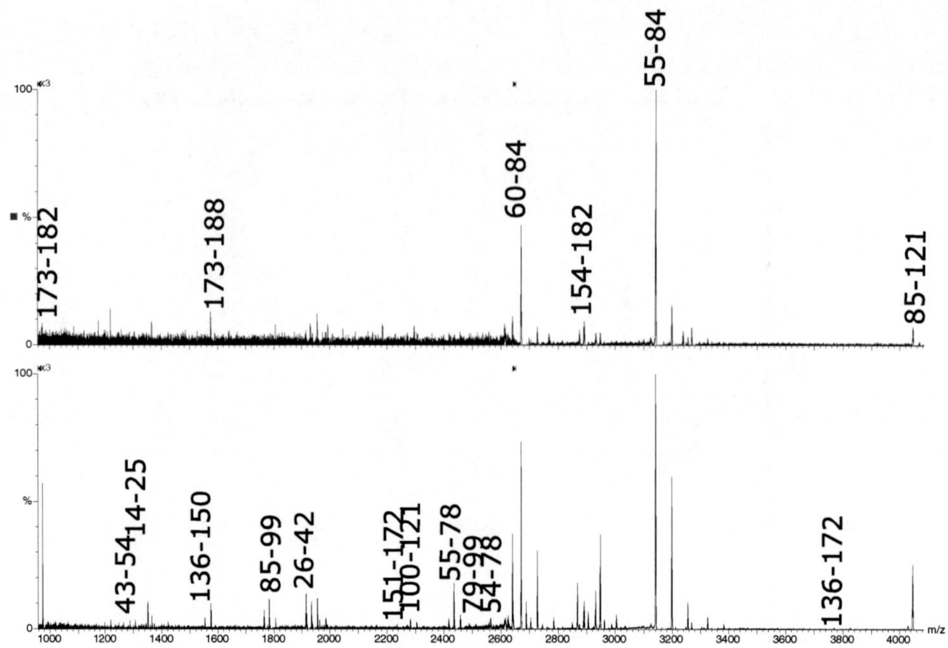

Figure 2: Peptide maps (relative intensity *versus m/z*) of PhaP1-spots shown in Fig. 1a. Top trace: Right-most spot showing lowest abundance. Bottom trace: abundant left spot. Signals labeled with the corresponding amino acid residues. See Fig. 3.

1 MILTPEQVAA AQKANLETLF GLTTKAFEGV EKLVELNLQV
41 VKTSFAEGVD NAKKALSAKD AQELLAIQAA AVQPVAEKTL
81 AYTRHLYEIA SETQSEFTKV AEAQLAEGSK NVQALVENLA
121 KNAPAGSEST VAIVKSAISA ANNAYESVQK ATKQAVEIAE
161 TNFQAAATAA TKAAQQASAT ARTATAKKTT AA

Figure 3: PhaP1 sequence showing assignments of peptide maps in Fig. 2. Underlined: Fig. 1a right spot, spectrum Fig. 2 top trace. Italicized: Fig. 1a left spot, spectrum Fig. 2 bottom trace.

In a second approach high resolution nanoESI-FTICR-MS [27] was used to determine the molecular weights of the intact phasin wild-type proteins [28-30]. With this technique it was expected to both find modifications of PhaP1 *R. eutropha* wild-type and any of the homologous isoforms PhaP2-4 which could be present as minor compounds due to the purification protocol [14]. Fig 4a shows the nanoESI-FTICR mass spectrum with a zoom-in on the 14+ peak of PhaP1 in Fig. 4b. Approximately 250 scans were accumulated. The resolution ($m/\Delta m_{FWHM}$) of the peaks is > 100.000 and the isotopic peaks are baseline-resolved which is a prerequisite for the exact mass determination. The peaks for the different charge states of the proteins all have satellites indicating methionine oxidation which is likely caused by sample handling. Since there is only one Met in the PhaP1 sequence at the N-terminus, the other protein species were expected to be either other phasins or C-terminally modified PhaP1 [1-192] (Table 2). Predominant in the spectrum are the peaks for PhaP1 [1-192] in the charge states 16+ to 12+. Minor peaks could be assigned to PhaP1 [1-190] and [1-182]. There is also evidence for the presence of PhaP1 [1-184] and [1-115]. However, no signals for PhaP2-4 or phosphorylation of any of the species were detected (Table 3).

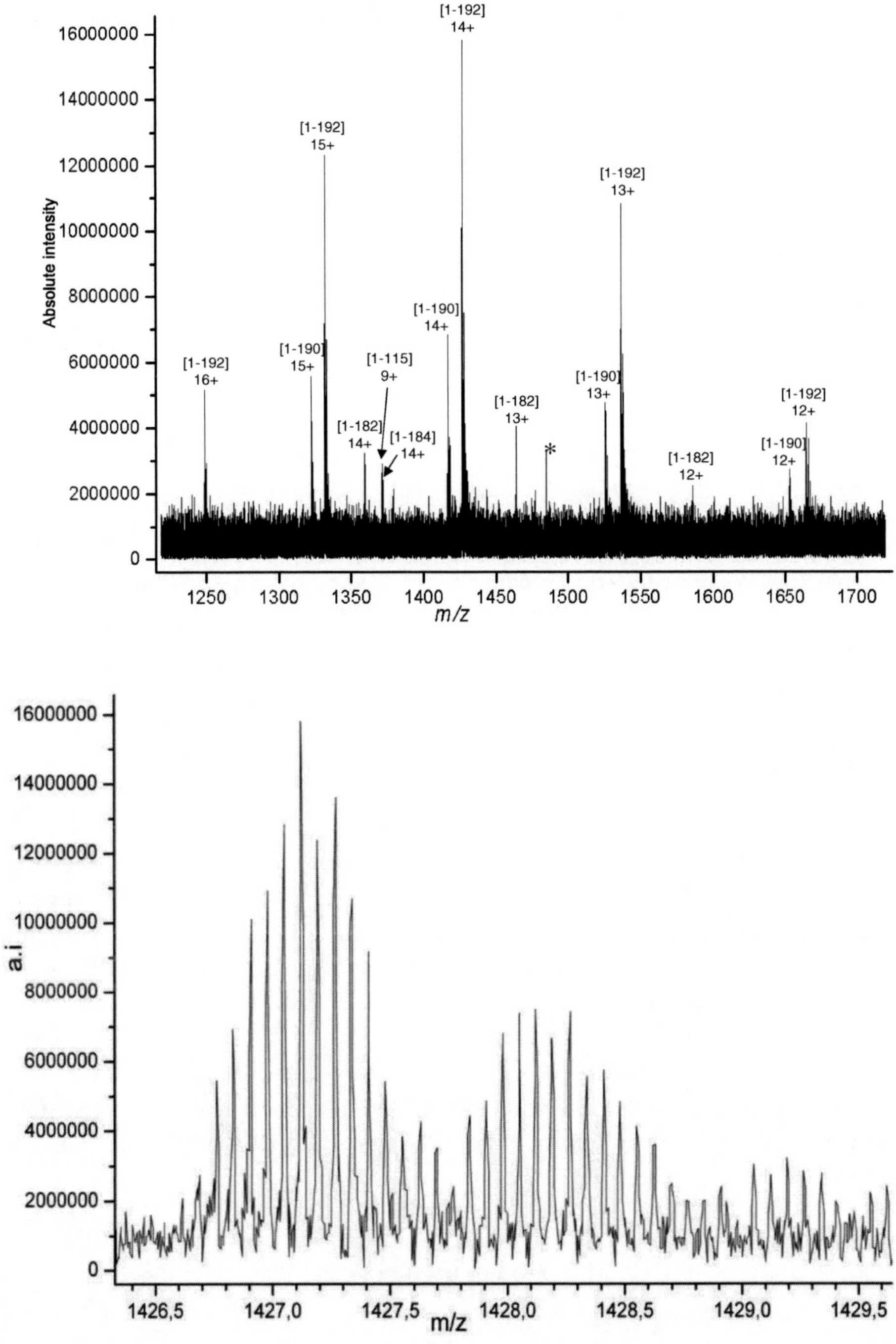

Figure 4: a) NanoESI-FTICR mass spectrum of PhaP1. b) Zoom in on the +14-charge state. (* denotes a spike)

Table 2: Assignment of most abundant isotopomere FTICR peaks observed the PhaP1 preparation

Charge state	Observed mass (m/z)	Calculated mass (m/z)	Δ ppm
$[^{13}C_{11}{}^{32}S_1]$ PhaP1 [1-192]			
+ 12	1664,81220	1664,80273	5,68
+ 13	1536,82294	1536,81852	2,88
+ 14	1427,12073	1427,11776	2,08
+ 15	1332,04555	1332,04376	1,34
+ 16	1248,85536	1248,85401	1,08
$[^{13}C_{11}{}^{32}S_1]$ PhaP1 [1-190]			
Charge state	Observed mass (m/z)	Calculated mass (m/z)	Δ ppm
+ 12	1652,96790	1652,96323	2,82
+ 13	1525,89299	1525,88973	2,14
+ 14	1416,97200	1416,96960	1,69
+ 15	1322,58153	1322,57215	7,09
$[^{13}C_{11}{}^{32}S_1]$ PhaP1 [1-182]			
Charge state	Observed mass (m/z)	Calculated mass (m/z)	Δ ppm
+ 12	1586,09909	1586,09198	4,48
+ 13	1464,08756	1464,08529	1,00
+ 14	1359,58219	1359,57976	1,79

Table 3: MWs (calculated according to ref. [32]).and pIs for the PhaP1 isoforms detected with FTICR

Amino acid residues	MW (average)	pI
[1-192]	19967	6,0
[1-190]	19825	6,0
[1-184]	19194	5,3
[1-182]	19022	5,3
[1-115]	12333	5,0

The FTICR results do not offer an explanation for the pI shift, since peptides were found in all three maps up to amino acid residue 188. While the suggested loss of the two C-terminal alanines might be possible according to the above experimental results, it should not cause a pI shift because their side chains do not carry charges. It is evident that differences in sample preparation need to be investigated.

Was the abundant PhaP1 protein knocked-out in mutant Δ*phaP1*, 2D-PAGE demonstrated the presence of PhaP3 and PhaP4 beside BktB and GroEL in the granule proteome (Fig. 1b) [20]. OPhaP2 was not detected here, although others showed its synthesis in *R. eutropha* [31].

Conclusion

PhaP1 protein was identified in high abundance on the grana surface of *R. eutropha.* Homologous phasins PhaP3 and Pha4 could be visualized on 2D-PAGE in the PhaP1 knock-out mutant. No evidence for the presence of PhaP2 was found in these experiments. All detected phasins exhibit isoforms whose structural differences are still a matter of investigation, although C-terminally truncated isoforms of PhaP1 were detected with FTICR-MS.

ACKNOWLEDGMENTS

M. Zeller was funded by a Marie Curie Training Fellowship of the European Union. Studies on PhaP1 and other phasins were founded by the Deutsche Forschungsgemeinschaft (Ste386/6-2) and by the German Federal Ministry of Education and Research (BMBF) within the framework of the Competence Network Göttingen "Genome research on bacteria (GenoMik). We thank A. J. Sinskey (MIT, Cambridge, USA) for providing the *R. eutropha* strain Re1052.

This work is part of the PhD-theses of M. Zeller, M. Pötter, and F. Reinecke.

REFERENCES

[1] Anderson, A. J., Dawes, E. *A. Microbiol. Rev*. 1990, 54, 450-472.

[2] Hocking, P. J., Marchessault, R. H. Biopolyesters. *In Chemistry and Technology of Biodegradable Polymers;* Chapman and Hall: London, 1994.

[3] Williams, S. F., Martin, D. P. *In Biopolymers*; Doi, Y., Steinbüchel A., Ed.; WILEY-VCH: Weinheim, 2002; Vol. 4 (Polyesters III - Applications and commercial products), pp 91-127.

[4] Schlegel, H. G., Gottschalk, G. , Bartha, V. *Nature* 1961, 29, 463-465.

[5] Steinbüchel, A., Valentin, H. E. *FEMS Microbiol. Lett*. 1995, 128, 219-228.

[6] Griebel, R., Smith, Z., Merrick, J.M. *Biochemistry* 1968, 7, 3676-3681.

[7] Steinbüchel, A., Aerts, K., Babel, W., Föllner, C., Liebergesell, M., Madkour, M. H., Mayer, F., Pieper-Fürst, U., Pries, A., Valentin H. E., Wieczorek, R. Can. *J. Microbiol.* 1995, 41, Suppl. 1, 94-105.

[8] Reusch, R. N. In Polyesters I - *Biological systems and biotechnological production*; WILEY-VCH: Weinheim, 2002; Vol. 3A, pp 123-172.

[9] Gerngross, T. U., Reilly, P., Stubbe, J., Sinskey, A.J. and Peoples, O.P. J. *Bacteriol.* 1993, 175, 5289-5293.

[10] Liebergesell, M., Schmidt, B., Steinbüchel, A. *FEMS Microbiol. Lett*. 1992, 78, 227-232.

[11] Liebergesell, M., Sonomoto, K., Madkour, M., Mayer, F., Steinbüchel, A. *Eur. J. Biochem*. 1994, 226, 71-80.

[12] Haywood, H. W., Anderson, A. J., Chu, L. , Dawes, A. E. *FEMS Microbiol. Lett*. 1988, 52, 91-96.

[13] Haywood, H. W., Anderson, A. J., Chu, L. , Dawes, A. E. *FEMS Microbiol. Lett*. 1988, 52, 259-264.

[14] Wieczorek, R.; Pries, A.; Steinbüchel, A.; Mayer, F. J. *Bacteriol*. 1995, 177, 2425-2435.

[15] Pieper-Fürst, U.; Madkour, M. H.; Mayer, F.; Steinbüchel, A. J. *Bacteriol*. 1995, 177, 2513-2523.

[16] Jossek, R., Steinbüchel, A. *FEMS Microbiol. Lett*. 1998, 168, 319-324.

[17] Steinbüchel, A. *In Biomaterials*; Byrom, D., Ed.; Macmillan: Basingstoke, UK, 1991, pp 123-213.

[18] Müller, H. M.; Seebach, D. Angew. *Chem.* 1993, 105, 483-509.

[19] Pötter, M.; Madkour, M. H.; Mayer, F.; Steinbüchel, A. *Microbiology* 2002, 148, 2413-2426.

[20] Pötter, M.; Müller, H.; Reineke, F.; Wieczorek, R.; Fricke, F.; Bowien, B.; Friedrich, B.; Steinbüchel, A. *Microbiology* 2004, 150, 2301-2311.

[21] Steinbüchel, A.; Aerts, K.; Babel, W.; Follner, C.; Liebergesell, M.; Madkour, M. H.; Mayer, F.; Pieper-Fürst, U.; Pries, A.; Valentin, H. E. Can. J. *Microbiol.* 1995, 41, Suppl 1: 94-105.

[22] York, G. M., Junker, B. H. H., Stubbe, J. A. , Sinskey, A. J. J *Bacteriol* 2001, 183, 4217-4226.

[23] Mergeay, M., Nies, D., Schlegel, H. G., Gerits, J., Charles, P. ,Van Gijsegem, F. J *Bacteriol.* 1985, 162, 328-334.

[24] Shevchenko, A.; Wilm, M.; *Vorm, O.; Mann, M. Anal. Chem. 1996, 68, 850-858.*

[25] *Wilm, M.; Mann, M. Anal.* Chem. 1996, 68, 1-8.

[26] Yamagata, A.; Kristensen, D. B.; Takeda, Y.; Miyamoto, Y.; Okada, K.; Inamatsu, M.; Yoshizato, K. *Proteomics* 2002, 2, 1267-1276.

[27] Palmblad, M.; Hakansson, K.; Hakansson, P.; Feng, X.; Cooper, H. J.; Giannakopulos, A. E.; Green, P. S.; Derrick, P. J. *Eur. J. Mass Spectrom.* 2000, 6, 267-275.

[28] Green, M. K.; Vestling, M. M.; Johnston, M. V.; Larsen, B. S. *Anal. Biochem.* 1998, 260, 204-211.

[29] Lafitte, D.; Heck, A. J.; Hill, T. J.; Jumel, K.; Harding, S. E.; Derrick, P. J. *Eur. J. Biochem.* 1999, 261, 337-344.

[30] Johnson, K. A.; Amster, I. J. J. *Am. Soc. Mass Spectrom.* 2001, 12, 819-825.

[31] Srinivasan, S. B., G.C.; Gerngross, T.U. Appl. Environ. Microbiol. 2002, 68, 5925-5932.

[32] Yergey, J. A. *Int. J. Mass Spectrom. Ion Phys.* 1983, 52, 337-349.

In: Biomacromolecular Mass Spectrometry Yearbook. Volume 1 ISBN: 978-1-60876-167-8
Ed: Simone Koenig

Chapter 5

MÜNSTER CONFERENCE ON SINGLE CELL ANALYSIS

PROGRESS IN RESEARCH AND TECHNOLOGY MÜNSTER, NOVEMBER 16, 2004

Organizers: Interdisciplinary Center of Clinical Research, Core Facility Integrated Functional Genomics, Münster, Germany Clemens Sorg, Peter Josten, Kurt Sieberns, Kristina Riehemann, Simone König
Chairs: Clemens Sorg, Simone König, Hans Oberleithner, Solon Thanos

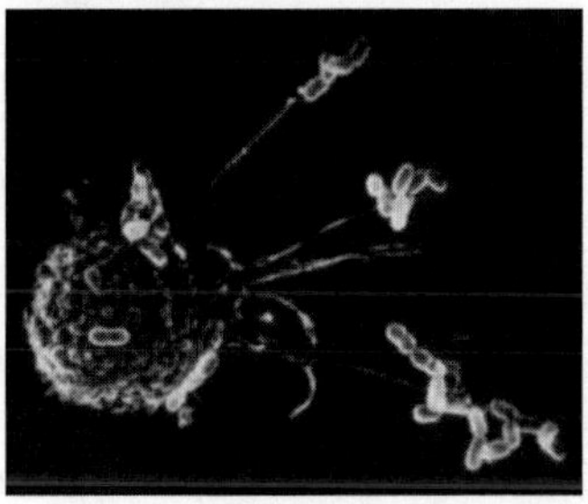

SPEAKER ABSTRACTS

MASS SPECTROMETRY IN SINGLE CELL ANALYSIS

Simone König

As technologies are refined, individual cells become increasingly the focus of functional investigations and science-political interest. Mass spectrometry has only a very limited role to play in that area yet, essentially being restricted to small molecule detection and tissue imaging. However, the technique advances fast and will, therefore, be monitored in this journal.

The 1st Conference on Single Cell Analysis drew much interest from scientists of very different fields. The organizing committee tried to address the broad spectrum of methods applicable to single cell or single molecule analysis so that mass spectrometry was represented only as a small part. Nevertheless, the invited contributions by Bernhard Spengler and Markus Stöckli on tissue imaging techniques created excitement among the present non-mass spectrometrists and science managers. The interplay of mass detection with peripheric techniques such as capillary electrophoresis and laser microdissection was underscored as in the presentation by Norman Dovichi. To be watched is also the "toponomics" developed by the group of Walter Schubert, a technique based on antibodies which allows protein localization with the cell environment.

While the 2004 conference was not at all comprehensive in terms of mass spectrometry applications to single cell analysis, and the pioneers in the field (see Garden et al. *J. Mass Spectrom.* 1996, 31, 10: 1126-1130; Li et al. *Anal. Chem* 1997, 69: 563-565; Chiu & Zare *PNAS* 1998, 95, 7: 3972-3977; Jiminez *et al. Biochemistry* 1998, 37, 7: 2070-2076) could not even be present, there is potential to improve on that this year. A show of hands among the participants confirmed the general feeling that monitoring progress in single cell analysis is timely and meets scientific interest. The participants called for a repetition of the conference which will take place in Münster in a larger setting in September 2005.

COMPUTATIONAL SYSTEMS BIOLOGY: TOWARDS 'IN SILICO' CELL MODELS?

Hergo Holzhütter
Humboldt-University Berlin

After sequencing of complete genomes it has become apparent that single genes or their products do not cause most of the biological phenomena observed. These findings have drawn researchers to the conclusion that the most interesting phenomena in biology result from the interrelated actions of many components within the system as a whole. For elucidating and quantifying these molecular interrelations efficient high-throuput techniques have been developed to measure in parallel concentration profiles of many different mRNA's, proteins and metabolites (=molecular variables). The central problem is currently the causal interpretation of such 'molecular snapshots'. Computational systems biology is a rapidly growthing new field in biosciences that aims at formulating both molecular interactions and molecular states into mathematical models of various complexity. The final goal of this approach to set up computer models of whole cells that allow to the predict the temporal and spatial behavior of a large set of molecular variables at various external conditions (e.g. changes in temperature, pH or supply of substrates) or alterations on the molecular level (e.g. inherited enzyme defects, inhibition of processes with drugs). Currently there exist basically two alternative approaches to reach this ambitious goal. The bottom-up approach continues the traditional way of settling models on a detailed kinetic description of the underlying elementary processes. The top-down approach tries to identify and quantify the relevant interrelations among molecular variables directly from large amounts of high-troughput data.

In my talk I will outline the specific strengths and weaknesses of these approaches and the role that single cell analysis may have in contributing to both of them.

THE NECESSITY OF CELL- AND CELL TYPE-SPECIFIC EXPRESSION ANALYSES IN CARDIOPULMONARY RESEARCH

Ludger Fink

Justus-Liebig-University, Giessen

The morphological study and the identification of cell differentiation within tissues are the basis for assessment of physiological and pathological processes. Applying molecular techniques, the analysis of commonly used tissue homogenates bears the risk of masking genetic deviations or expression changes of an individual cell type by the bulk of surrounding cells. To overcome the problem of tissue heterogeneity, cells have to be harvested selectively for further analysis. Therefore, microdissection techniques were developed making retrieval of target cells simple, rapid and precise. Accordingly, the use of these techniques has increased exponentially within the last few years.

Presenting three examples from cardiopulmonary research, it is demonstrated that single cell isolation is often preconditional for the investigation of splicing isoform expression (1). Combining microdissection of few cells and real-time RT-PCR allows to determine relative mRNA expression in a cell-type specific manner to reveal gene regulation (2). Combination with RNA preamplification techniques and micro arrays results in cell type- or compartment-specific expression profiles that especially differ from those of tissue homogenates when minor represented cell types are investigated (3). In consequence, the isolation of cells or cell types allows a more accurate investigation of complex tissues and gives deeper insight to regulation processes and crosstalk of the respective cells.

IMMUNOGENETICS AT SINGLE CELL LEVEL

Jan Buer

Institute of Medical Microbiology, Hannover Medical School, Hannover & Mucosal Immunity Group, German Research Centre for Biotechnology, Braunschweig

The immune response is governed by a complex array of cellular populations and sub-population. Gene expression profiling is proving an important means of understanding and classifying physiologic and pathophysiologic processes by identifying genes, gene pathways and pathway networks not previously known to be associated with inflammation and autoimmunity. However, disease-associated gene expression can be obscured by surrounding 'normal' cells / tissue. The ability to study and compare gene usage at the cellular level will therefore provide valuable new insights into the biology of a disturbed immune response. A summary of developments in the area of expression profiling in single cells of the immune system will be described, and the rationale for these types of studies will be presented based on our own experience in studying the maintenance and break down of tolerance in both mice and humans.

SINGLE CELL PROTEOMICS

Norman Dovichi
University of Washington, Seattle, USA

It is now possible to generate one- and two-dimensional electrophoresis data from single mammalian cells. These separations are capable of resolving hundreds of components from the cell, and the data can be correlated with cell cycle and other properties of the cells. Examples have been generated from breast, prostate, colon, andesophageal cancer cell lines, from astrocytes and neurons, and fromosteoprecursors and myoblasts. Instrumentation is under development that can characterize tens of thousands of cells per day in one-dimensional electrophoresis and thousands of cells per day in two-dimensional electrophoresis. Related technology is under development to monitor post-translation modifications of specific proteins in single cells at the single copy level.

HIGHER MASS ACCURACY, HIGHER LATERAL RESOLUTION: A CLOSER VIEW OF SINGLE CELLS

Bernhard Spengler, Werner Bouschen, Dieter Kirsch and Martin Köstler
Institute of Inorganic and Analytical Chemistry, Justus Liebig University, Giessen

With the combination of soft ion formation techniques and ultraprecise mass spectrometers, the accurate mass of biomolecules is becoming a physical quantity of high interest in bioanalytical methodology. FT-ICR mass spectrometry with ESI or MALDI sources now allows to easily determine the amino acid composition and sequence of medium size unknown peptides. This new method, called "Composition-based Sequencing (CBS)" [1] appears to open new perspectives in proteomics. On the single cell level, another feature of FT-ICR mass spectrometry will be of importance in the future, which is the high mass resolving power for being able to analyse highly complex, non-fractionated samples.

Investigating single cells directly by MALDI requires not only the high mass resolving power and high mass accuracy, but also a sufficient lateral focusing of the laser beam. MALDI imaging [2] is a new technique of visualizing biomolecule distributions in tissue by mass spectrometry. Most instruments developed for this approach, however, lack the necessary lateral resolution for reaching the single cell level. All of them so far lack the necessary mass resolving power and accuracy to identify and characterize proteins from single cells by this approach.

A goal for the future is to combine micrometer-resolution MALDI imaging with high accuracy FT-ICR mass spectrometry. The prospects and limitations of this approach will be discussed and first results will be presented.

[1] B. Spengler; *J. Am. Soc. Mass Spectrom.* 15 (2004) 703 - 714.
[2] B. Spengler, M. Hubert; *J. Am. Soc. Mass Spectrom.* 13 (2002) 735-748.

MALDI Mass Spectrometric Imaging Applied in the Race to Find New Cures for Diseases

Markus Stöckli
Novartis Institutes for BioMedical Research Basel

Molecular imaging techniques play an increasingly important role in biomedical research. With the ability to localize changes in organisms on the molecular level, we have a tremendous amount of information in hands to support the effort to find new cures for diseases. Widely applied molecular imaging techniques in research such as MRI and PET utilize molecular probes to report the presence of the analytes of interest in living organisms. With matrix-assisted laser desorption/ionization mass spectrometric imaging (MALDI MSI) a technique is presented which allows de-novo spatial analysis of biological samples. Sections of the tissues are introduced in mass spectrometer, where a laser is used to raster over the tissue while acquiring mass spectra of the ablated ions at every image point. From one array of spectra, hundreds of analyte-specific images can be generated, based on the selected masses. This allows a detailed tissue analysis with a resolution in the micrometer range for protein, compounds and metabolites. The techniques has been applied to target finding, biomarker discovery and drug delivery studies. Applications shown in this presentation cover different stages in biomedical research, from localization of high-mass proteins in tissue sections to drug imaging in dosed animals.

Laser Micromanipulation and Capture of Single Cells – New Approaches in Cell and Molecular Biology

Yilman Niyaz, M. Stich, B. Sägmüller, R. Gangnus, K. Schütze
P.A.L.M Microlaser Technologies AG, Bernried

The understanding of the molecular mechanisms of cellular interactions depends on identification, isolation and finally characterization of a single cell or a population of cells and thereafter of their specific subsets of biomolecules. Thus, sample generation is a very crucial step for high-resolution downstream applications. Laser Microdissection and Pressure Catapulting (LMPC) is a well-known method to isolate and collect selected cells from complex animal or plant tissues, resulting in an eminent increase in the specificity of downstream molecular analyses like PCR and RT-PCR amplification, DNA, RNA and protein microarray analysis, and MALDI/SELDI spectrometry. With LMPC, only the force of focused laser light is utilized to excise selected single cells or large tissue areas of a specimen down to a resolution of subcellular components like organelles or chromosomes, respectively. After microdissection, the separated sample is directly lifted up into an appropriate collection device by using the same laser (Laser Pressure Catapulting, LPC). As both processes are working entirely without any mechanical contact, LMPC enables pure and homogeneous sample retrieval from e.g., paraffin and cryosections, cell smears, cytospins and chromosome preparations. An important innovation is the laser driven isolation of cultured cells. The work with selected living cells is extremely facilitated with this new approach and opens a wide field of new applications and research possibilities. Individual or small groups of cultured cells can be separated by morphology or fluorescent label for direct molecular analysis, clonal expansion or re-cultivation (e.g., stem cells), creation of mixed cell populations (e.g., tissue engineering), and maintaining specific cell type ratios in mixed cultures. The efficacy of this method was recently shown by LMPC-mediated isolation of transgenic murine embryonic stem cells, which were successfully used to generate chimeric mice. Furthermore, the same laser system can be used to open the cell and nuclear membrane of living cells, or drill holes into the solid wall of plant cells. This enables new methods for microinjection of drugs or genetic material into living cells without any involvement of mechanical, chemical or viral components. In addition, the controlled fusion of different cell types becomes possible with Laser MicroBeams, as could be demonstrated for B-cell hybridoma. The applied laser wavelength of 337 nm does not affect the viability of the specimens: There is no impairment of the biological information, even after several times of catapulting and recultivation, as is shown by, e.g., Comparative Genomic Hybridization (CGH) analysis. With the specificity of LMPC, the sensitivity of downstream analyzing methods becomes a critical issue. Especially in gene expression arrays but also for protein array analysis enormous efforts have been carried out to reduce the required molecule amounts. On the one hand, biased amplification should be minimized but also biomolecule detection should become feasible below femptomolar amounts. For a proof-of-principle we extracted RNA of heat shock treated versus untreated cells, both isolated by LMPC. After one single round of T7 RNA amplification, the amplified message was hybridized on a microarray chip from Zeptosens AG (Switzerland). We could demonstrate a significant and reliable gene expression result from as few as 500 cells using the extremely sensitive planar wave-guide SensiChip™ technology.

IN SITU PROTEOMICS IMAGING: ANALYSING "REAL WORLD" PROTEIN NETWORKS WITHIN INTACT CELLS AND TISSUES

Walter Schubert

Meltec Magdeburg

The hierarchy of cell function comprises at least four distinct functional levels: genome, transcriptome, proteome, and toponome. The toponome (rule of protein arrangement) is the entirety of all proteins, protein-complexes, and protein networks traced out directly on the single cell level in the natural environment of the cells in situ (e.g. tissues). Whole cell protein-fingerprinting (WCPF/MELK) enables the quantitative mathematical description and exploration of the toponome by using large tag libraries localizing proteins (hundreds simultaneously) sub- and supracellularly.

Thereby the technology addresses the fact that each protein must be at the right time at the right place at the right concentration in a cell to interact with other proteins assembled in a spatially organized network. To encode the myriads of cellular functionalities, cells appear to having at their disposal a large, albeit finite and highly non-random repertoire of toponome units, i.e. a system of rules to construct topological hierarchies in a cell's proteome. The entirety of these toponome units represents the total functional code of a cell. The complete toponome is therefore as fundamental a data set as the genome or the proteome.

NANOBIOLOGY: FROM SINGLE MOLECULE DETECTION AND MANIPULATION TOWARDS SINGLE CELL ANALYSIS IN MICROFLUIDIC NETWORKS

Dario Anselmetti

Experimental Biophysics & Applied Nanoscience, Physics Faculty, Bielefeld University, Bielefeld Institute for BioPhysics & NanoScience (BINAS)

Our concepts to detect and analyze minute amount of substance (single molecules) via mechanical and optical schemes in order to 1) fundamentally investigate the nature and related physical mechanisms of specific biomolecular interaction and 2) to ambitiously analyze the proteome of single cells in nanofluidic microdevices (lab-on-a-chip) will be presented.

a) Specific binding of single receptor and ligand molecules can be investigated in dynamic force spectroscopy experiments by atomic force microscopy (AFM) or optical laser tweezers (OT) and give access to intermolecular forces, interaction mechanism, kinetics of the reaction and the binding energy landscape. The ability to measure inter- and intramolecular forces and elasticities under physiological conditions in biological transcriptional regulators (DNA-protein) [1-3], in bioorganic synthetized peptide-DNA systems [4] and in supramolecular guest-host systems (cation-calixarene) [5] with piconewton-, nanometer- and millisecond resolution allows investigation of single ligand-receptor interaction in a broad affinity range from $K_d=10^{-5}$-10^{-15} M at a discrimination level of single point mutations.

b) The understanding and control of cellular adhesion plays an enormous role in many applications like implantation, transplantation, tissue engineering or inflammation. Our most recent results where molecular interaction and cellular adhesion was investigated on individual cells downto the single molecule level with OT [6,7] will be presented.

c) We have recently demonstrated that periodical geometric structuring of microchannels allow separation of long DNA-fragments (with sizes > 40kbp) in less than a minute [8], a fragment size range which is rarely accessible by standard separation mechanisms in microchannels. We extended our investigations on single-molecule DNA-fragment migration by means of computer simulations [9] and exploited how such structures can be used to generate DNA separation devices based on ratchet effects. In microfluidic devices governed by conditions far from thermodynamic equilibrium unexpected migration of particles and other mechanisms such as negative mobility [10,11] have been predicted theoretically. We experimentally prove this paradox migration mechanism for the first time in a structured microfluidic device with colloidal particles in solution.

d) Our concept of single-cell analytics and the first results of our microchip-integrated separation and ultra-sensitive label-free detection of minute amounts of analyte molecules is presented and discussed [12].

[1] F.W. Bartels et al., *J. of Struc. Biol.*, **143**, 145-152, (2003).
[2] R. Eckel et al., *Biophysical Journal*, **85**, 1968-1973, (2003).
[3] A. Sischka et al., *Biophys. J*, in press (2004).
[4] R. Ros, et al., *J. of Biotechnology*, **112**, 5-12 (2004).
[5] R. Eckel et al., *Angew. Chem. Intl. Ed.*, in press (2004).
[6] A. Sischka et al. *Rev. Sci. Instrum.*, **74**, 4827-4831 (2003).
[7] A. Sischka et al., *Appl. Phys. Lett.*, submitted (2004).
[8] A. Ros et al., *J.Biotechn.*, *122*, 65-67 (2004)
[9] M. Streek et al., *J.Biotechn.*, *122*, 79-89. (2004).
[10] Eichhorn, et al., *Phys.Rev.Lett.*, *88*, 190601 (2002).
[11] Eichhorn et al., *Phys.Rev.E*, *66*, 066132 (2002).
[12] J. Roegener et al., *Anal. Chem.*, **75**, 157-159, (2003).

Fluorescence Spectroscopic Techniques to Study Protein/Protein Interactions and Conformational Changes on the Single Molecule Level

Jörg Fitter, I. Gregor and J. Enderlein
Institute of Biological Information, Processing (IBI) IBI-2:
Structural Biology, Research Center Juelich

Fluorescence spectroscopy is an experimental technique with a great potential of elucidating inter- and intramolecular interactions of biomolecules. Especially the application of fluorescence techniques to the study of single (individual) molecules has several striking advantages, if complex systems (like biological systems) with static and dynamic heterogeneity have to be investigated. This technique makes it possible to directly probe distributions of static properties as well as dynamical transitions between different states (e.g. structural conformations), without the necessity to synchronize the transitions in an ensemble. This ability becomes crucial when studying biological systems where synchronizing molecules in an ensemble is problematic due to the complex or even branched reaction pathways. Although single-molecule investigations of biological systems are still in their infancy, several promising studies in this direction have already been reported.

The fluorescent dyes used in those studies change their emission properties (e.g., emission wavelength, emission intensity, life-time, anisotropy) in accordance with changes of the direct environment of the dyes (local pH, hydrophobicity, energy transfer, etc.) during the different states of the working cycle. A more sophisticated and powerful approach for studying conformational changes on a single-molecule level as well as the interaction between different molecules provides Förster resonance energy transfer (FRET), electron transfer, and fluorescence quenching mechanisms. These techniques can be employed to study conformational changes in a time-resolved manner. An overview of these techniques and some applications will be presented.

TOWARDS SCANNING (NEAR FIELD) INFRARED LASER MICROSCOPY OF CELLS

Erik Bründermann, Andreas Bergner, Götz Wollny, Jean-Sebastien Samson and Martina Havenith-Newen
Ruhr-Universität Bochum, Physikalisch Chemie

We have developed a chemical nanoscope in the chemically important fingerprint region to monitor label free functional groups. Using our unique cw Infrared Opto Parametrc Oscillator (with output powers of 2 W) we have been able to obtain 2-dim images of cells in the spectral fingerprint bands. As a first example we present pictures of hepatocytes scanned in the range of the lipid band at 2920 cm^{-1}. The method allows a label free detection method and provides a very fast scanning method (each pixel is scanned in less than a msec). IR spectroscopy offers moreover the advantage of yielding absolute concentrations. This was demonstrated by a near infrared microscope which offers a cheap solution for the quantitative measurement of water in living cells. It is shown that IR imaging is able to monitor the influence of minimum dosages of pharmaceuticals on the intracellular water concentration on a single cell level. We were able to increase the resolution considerably when measuring in the near field limit. Using the tip of an AFM as nano-antenna we sampled the infrared spectra of the surface by the evanescent field. In a first test we achieved resolutions of 30 nm (corresponding to $\lambda/100$) in our scanning near field infrared microscope (SNIM) setup.

VOLUME, SURFACE AND STIFFNESS OF INDIVIDUAL CELLS MEASURED WITH ATOMIC FORCE MICROSCOPY

Hans Oberleithner
Institute of Physiology II, University of Muenster

Atomic force microscopy (AFM) is a tactile instrument that can be used as a nano-scope as well as a manipulative nano-tool. In principle AFM consists of a fine tip that physically scans surfaces. At low interaction forces (low pN range) the AFM tip images surfaces at sometimes molecular resolution. At high interaction forces (nN range) the AFM tip more or less strongly interacts with the surface and thus distorts it. Since forces can be controlled and measured some unique parameters as e.g. the stiffness of living cells can be quantified. We applied AFM to human endothelial cells and measured single cell volume, single cell surface and single cell stiffness in response to steroid hormones. Aldosterone, a steroid hormone known to regulate salt and water transport in kidney and to control blood pressure in humans was found to exert dramatic effects on cell volume, surface and stiffness. AFM allows the quantification of these fundamental cell characteristics. The instrument is easy to handle, commercially available and well equipped with software that supports many different applications, from molecular imaging to force-volume measurements. In conclusion, AFM is a nanotechnique with an exciting perspective in biomedical research.

Oberleithner H, Ludwig T, Riethmuller C, Hillebrand U, Albermann L, Schafer C, Shahin V and Schillers H. Human endothelium: target for aldosterone. *Hypertension* 43: 952-956, 2004.

Schneider SW, Matzke R, Radmacher M and Oberleithner H. Shape and volume of living aldosterone-sensitive cells imaged with the atomic force microscope. *Methods Mol Biol* 242: 255-279, 2004.

In: Biomacromolecular Mass Spectrometry Yearbook. Volume 1 ISBN: 978-1-60876-167-8
Ed: Simone Koenig

Chapter 6

REFLECTIONS ON MASS SPECTROMETRY AND OTHER ISSUES*#

Stories from Henry M. Fales collected by Simone König

ABBREVIATIONS

CI - chemical ionization
GC - gas chromatography
IR - infrared
MS - mass spectrometry
NIH - National Institutes of Health, Bethesda, MD, USA
NMR - nuclear magnetic resonance
UV - ultraviolet

FOREWORD

Some of the best times of my five year postdoc at the National Institutes of Health in Bethesda, Maryland, I had during lunch with my boss, Henry M. Fales (Hank). About seventy at the time, he had fantastic stories to share about his life with organic chemistry and mass spectrometry. They are too good to be forgotten and Hank kindly agreed to put them on paper. I collected some information shortly after I left his lab in summer of 1999 and worked

* Address correspondence to: PD Dr. Simone König, Integrated Functional Genomics, Röntgenstr. 21, 48149 Münster, Germany, ph: +49-(0)251-8357164, fax: +49-(0)251-8357255 , e-mail: koenigs@uni-muenster.de

A version of this chapter was also published in *Biomacromolecular Mass Spectrometry Volume 1, Number 2*, published by Nova Science Publishers, Inc. It was submitted for appropriate modifications in an effort to encourage wider dissemination of research.

it up for this publication in summer of 2005. It might interest readers among the mass spectrometrists who find the history of their technique as fascinating as I do. A bibliography is attached as a timeline.

Hank is one of the pioneers in biological mass spectrometry and he worked at the National Heart, Lung, and Blood Institute since 1954 after graduating from Rutgers University. He helped to shape a line of enthusiastic scientists in the process and he is one of the most kind and bright human beings I have ever met. On September 16, 1998 his friends and colleagues celebrated his work with a symposium in his honor at The Cloister/Mary Woodard Lasker Center, NIH Howard Hughes Medical Institute. The photographs shown were taken on that day. An impressive number of famous scientists (Figs. 1-8) participated including his major professor Roderick Barnes (Fig. 1), who came from Brazil for the event. Sadly, Prof. Barnes died 2000 from a massive stomach cancer.

Figure 1. Roderick Barnes (center) with his wife (left) and Henry M. Fales.

Hank showed me how "to do science", both scientifically and socially. I am more than grateful to him in many ways and I also appreciate that he agreed to supervise this journal project.

KILLING TIME?

Hank's Beginnings at NIH

At the time NIH hired us as civil service right away. GS-11 I think. There was no real post-doc or fellows program yet. I was given a lab and an assistant (Laura Guiffrida, I remember) and I worked on the synthesis of lycorine with Bill Wildman. It turned out that everyone had the structure wrong, so that was no good. But it did interest me in structure rather than synthesis so that was useful. This was all pretty much pure research. There was the idea that these alkaloids had some blood pressure lowering activity and that was thought sufficient to justify the work. No more though! Still, my idea was just to kill a year at NIH while looking around for a good job at Dupont, etc.. After several of the most interesting years of my life working out these structures by reactions, IR, UV and then GC, Evan Horning (lab chief) left for Baylor because of his fame in GC and Bill Wildman left for the University of Iowa and others in the lab left for other places. I was left without a boss, along with Bob Highet, and we were assigned to Dan Steinberg (Fig. 2). I became a section chief in his lab, and we began work on Refsum's disease. This then led to chemical ionization.

Figure 2. Daniel Steinberg (right) and Hee-Yong Kim.

INFECTION

Mass spectrometry is an analytical technique with the properties of a virus: It sneaks up on you and it is contagious. For me as someone who became involved with mass spectrometry simply due to chance it was interesting to know how Hank became infected. Here is his view:

I got into MS *via* Klaus Biemann who was brought to the attention of us at NIH while we were working on the structures of Amaryllidaceae alkaloids with Ed Warnhoff, now Professor Emeritus. Klaus was invited down to give several days of lectures which he later used as the basis for his book as he mentions in the front of it. I remember that he came to our house for dinner with the group and we discussed GC which he was putting onto his MS instrument at the time. We were very good in this area, having been the first, under Evan Horning, Chuck Sweeley and Bill Vandenheuvel, to run steroids and alkaloids on GC and we showed him how good our silicone phases were when the support was properly deactivated with dichlorodimethylsiloxane. I guess in retrospect, the fact that we were able to get GC peaks on "high" molecular weight compounds did a lot to prepare us for the fact that they should work also on MS and GC-MS.

Figure 3. Fred McLafferty.

The only tools for structure that we had at the time were IR and UV spectrophotometers and a spectropolarimeter. We did everything by chemical reactions, crystallization and recrystallization and combustion analysis to get the formulas. Bob Highet and I were having trouble with an alkaloid called amaryllisine which simply refused to give good combustion analysis so that its formula was in doubt. We sent this to Klaus Biemann who gave it to Al Burlingame, then a student of Klaus, who showed us what the correct mass was and we all published this as a message that MS could help in such cases. No big deal really, but it convinced us that we needed a MS of our own.

I knew nothing about MS and called up several people who were suggested to me by the salesmen from Nuclide, Consolidated, Perkin Elmer, and AEI (England). Unfortunately, I have forgotten who they were, but they all encouraged me as an organic chemist to go into the field. At the time I think there were only a few papers on organic compounds by McLafferty, Biemann, Stenhagen and Ryhage in Sweden and Myerson (Djerassi et al. had just started). Then the shopping began. We nearly bought a Nuclide, but the visit to their factory in State College in an old movie house worried me a little. The Consolidated like Biemann had, was fine but the photoplate seemed to be a chore to use. Bob Highet in our lab went out to learn about the Consolidated that Al Burlingame had just acquired and although he got good data on a series of alkaloids that we all published I got the impression that it was not easy to run. Then AEI showed us a film of their MS-9 which wowed us; it seemed so easy to use. And they had a resolution of 10000! Foolishly, they said they could also run a complete high resolution spectrum on 1 µg of cholesterol measuring all peaks accurately as Biemann showed he could do with the photoplate. And with their high resolution they ran a spectrum showing deuterium incorporation into cholesterol of a sample supplied by my boss at the time Dan Steinberg (a Lab Chief, I was a Section Chief) by resolving the peaks for $C_2{}^{13}CH_7$ and C_3H_6D at m/z 44. I remember that Fred McLafferty (Fig. 3) did not quite believe this so we never published it, but AEI put it out in an advertising brochure.

Right here I should add that there was never any doubt about buying the MS-9 because Dan Steinberg was 100% behind it. The truth is that I owe so much to this guy in advancing my career that I can never repay it (we still collaborate). So we bought it and I remember that they were not quite sure the elevator could stand the weight of the magnet! But when we began to use it, it became apparent that it was not able to work as advertised and we discussed returning it. This panicked AEI and their salesmen Howard Danaceau and Jerry Hearns, who suggested one last try by their top engineer. He showed up and changed the resistors in the filament supply so that the filament was almost melting, calculated all of the approximate masses and their ratios to the PFK standard, stripped to the waist on a hot day and actually did successfully measure all of the cholesterol peaks. I signed off on it.

Figure 4. Hank and Marvin Vestal (right).

As soon as the MS-9 arrived it was planned to be used in all sorts of ways by Dan Steinberg and myself. I hired Bill Milne from Arthritis (National Institute of Arthritis and Musculoskeletal and Skin Diseases - SK). At the time we were working on Refsum's disease (lack of the enzyme that alpha oxidizes phytanic acid from chlorophyll) and we were using the MS-9 to this end. The vacuum requirements were high and of course, like most of those new to the field, we often used too much sample, badly contaminating the source. Natural product chemists really needed mass spec more than anyone at the time because we had so little sample compared, e.g. , to synthetic chemists. This all would have been ca. 1963?

There was not and still is not much known about electron impact, but actually, it was quasiequilibrium theory that got me into CI. We had invited Marvin Vestal (Fig. 4) down to talk on the subject, since he was an authority in the area. But at the end he admitted it was not very good at predicting structures and giving parent ions. So afterwards I asked him about the CI method that Frank Field had just discussed at the ASMS (American Society for Mass Spectrometry – SK) meeting. He agreed that it was interesting and mentioned that his company had been licensed by Field (ESSO) to make the device and were we interested in having him build a source for the MS? I said yes and that was the beginning.

In the middle of the CI work I was invited to Rockefeller to give a talk on it. Dr. Lyman Craig, of countercurrent chromatography fame, was about to retire and they were looking for someone with MS experience since no one there had any. I strongly considered it, and went back a few times to talk with the senior investigators there, who were most encouraging. But having been from there, I knew New York and was not exactly thrilled with working in the city. And the costs would have been high with not much help from Rockefeller. Besides NIH was offering me a lab chief job, so I decided to stay at NIH. I have not regretted it. But at that time I had quite a bit of contact with Frank Field and I was aware that he wanted to move to academia so I suggested to Rockefeller that they talk to him. They did, and the rest is history.

Parenthetically, there were in fact two mass specs already at NIH. One was a Consolidated 103, I believe, which was used for measurement of water and heavy water, run by Bill Comstock working for Scotty Pratt later head of DCRT. Bill was later to work with me on the LKB-9000 for many years. The other was a Bendix TOF that had the perhaps unfair reputation of being a dog. It was used by the group under Bob Bowman who were actually trying to get it to do GC-MS, but it never did work out.

The Vanishing Filament

In the early days of biomolecular mass spectrometry the instruments were far from user-friendly, a property which guides many purchase decisions these days. Here is an example which Hank wrote up at 4 am one day:

Ron Macfarland had offered to build a Californium plasma desorption machine for us after a visit I made to his lab. They were not commercially available and I think no one else had one then. He would use the money for a year off from teaching, to catch up on things in the lab. The big day came and Ron arrived with the machine in pieces in the back of a small truck. He was on his way to vacation in Canada. He assembled the instrument and pumped it down. He then peered into the window that gave visual access to the sample holder and the Cf film. The Cf was deposited as Cf sulfate (I think) on a very thin piece of stainless steel which was then covered with a gold foil that was welded shut trapping the Cf sulfate permanently. If you do not do this, it is well known to creep all over the outside of containers. Ron had had trouble with this in his earlier work and had had to deal with the radiation safety people in Texas A and M. Cf is like plutonium or other fissionable materials in that ingestion of very small samples (< 1µg) can be fatal since it lodges in the lungs and slowly drills holes in the tissue.

I suddenly saw Ron turn white when he peered into the window. He grabbed a wrench and opened up the source, breathed a sigh of relief, and closed it up again. He then said "OK Hank, it's all yours. Let it pump down tonight and start running it tomorrow." Then he took off for his mother's house. The next morning I went in and started to work with it. When I looked in the window I saw NO Cf FOIL! I got on the phone and called Ron at his mother's house and he laughed and said that he had seen the same thing the previous day, but when he opened it up the foil was still there and he was sure it was an optical illusion. OK, so I asked

Figure 5. David Davis (left), Hank, Catherine Fenselau and Sanford Markey.

everyone in the lab to look in. It was no optical illusion ... the foil was missing! I panicked slightly and called radiation safety who came over and sealed the doors to the room shut and sniffed the air with a detector. The idea of evacuating the NIH Clinical Center was in the back of our minds bringing all the patients in their beds up the hill .... But there was no radioactivity in the air. Good. So I was elected to enter the room and get a bit of the pump oil to count in case the foil had crumbled and been pumped out. Still no radioactivity. Good. So now I let air into the machine and pumped it out again catching the air in a detector. Still no radioactivity. So I took it apart and poked a detector in and found no radioactivity except in the region of the detector holder which I expected from the foil having been there. Still no radioactivity. BUT NO FOIL EITHER! I put the foil holder in a vacuum desiccator and went home still worried about where the foil was. The Cf source and its holder were always stored in a vacuum desiccator since the high energy radiation caused N_2 and water vapor to combine to form nitric acid which ate away the gold coating. The next morning I came in looked at the desiccator and to my utter amazement the foil was there! On closer (not too close!) examination, it was clear that the foil had been pushed/punched out of the holder, perhaps by a rush of air, etc. and was simply hanging by a top edge. Then the charging action or something had caused it to roll up like a window shade forming a tight little roll at the top which we had not seen earlier. On standing all night it somehow discharged and rolled down again and was swinging now in the breeze! We had to get a new source, but all was well after all and I could sleep again.

OVERDOSES

A common complaint about method-oriented people is that they do not work on one important scientific problem, but seem to drift across fields applying their skill. They are often looked down at from the "real" scientists and funding organisations fail to see the need to support technological projects. This shortsighted view does not take in account that it takes many, many years of learning and practical experience to become good in one of those advanced techniques such as MS, NMR or X-ray crystallography. Then running samples of varying origin is less of a problem. Looking at Hank´s publication list, one can see that he is a prime example. From fire ants to human diseases and water clusters – there are many areas of research he has made a contribution to. With 335 publications and dozens of entries in the Archives of Mass Spectral data there can hardly be any doubt about the efficiency of a "method man". As Hank said: "I am probably guilty of everything, including an article in Skinned Knuckles about antique cars.". *When we were asked what we were working on he used to reply:* "Everything"*. The story below is a fascinating example of applying new and innovative scientific approaches to real-life problems. MS on call:*

My next door neighbor was Norman Law who was the clinical chemist at Suburban hospital. We got talking and it seemed logical to try to identify the overdose substances by MS. Norman had told me of the cases where the doctors had treated people overdosing with, if I remember correctly, glutethamide and having the patient wake up after having their stomachs pumped, only to have them die later from a second influx of the drug from their intestines. Apparently this is a common problem with certain lipid-soluble drugs. If it had been known that they actually ingested this compound, other steps could be taken such as a lavage of the intestines with olive oil, which is messy but supposed to work. Norm and I, and sometimes Bill Milne, would come in on call at all hours to run messy fluid (blood, urine, vomit) extracts through the LKB GC-MS. This always seemed to be at 2 a.m. Norm got a well-earned award for that work from the local hospital association and his own private grant from the National Institute of Mental Health to buy a mass spectrometer as a demonstration of the method to other hospitals. He bought an early CI quadrupole version from Scientific Research Instruments, the company where Marvin Vestal and Al Yergey (Fig. 6) were working on CI spectrometers after building mine for me. I gave a talk on the drug work at an ACS (American Chemical Society – SK) meeting in Montreal and surprisingly, Klaus Biemann had also run an overdose case so our lectures overlapped.

But it has a sad ending: Norm depended very heavily on a knowledgeable pathologist at Suburban Hospital with whom he worked to use his information wisely. Unfortunately, the pathologist left Suburban and Norm was on his own. Identifying drugs at the time was done (and still is today, for the most part) by crude qualitative procedures that simply listed a drug as "barbiturate", etc. but did not say which one it was. This is critical, since the treatment for a long acting barb is different from a short acting one. But they were able to "quantitate" it by UV absorption, etc.. To make a long story longer, the hospital required Norm to evaluate the MS method against the established technique at the time that gave numbers, not identity. However, identification, not quantitation, is the mass spec's strength. Naturally, Norm lost out on the test since the pathologist was gone and Norm did not have enough status (no PhD) to refuse to be involved in this crazy test. As we learned later, it seems that the emergency room physicians are not all that keen about having the drug identified in the first place. Partly this is because they are afraid of being sued, but it is also logical to treat the patient for his symptoms rather than for a drug that happens to appear in his vomit that he, therefore, may have gotten rid of. So that was the end of the drug stuff, but during the time we did have some

Figure 6. Catherine Fenselau, Richard Caprioli (center) and Alfred Yergey.

interesting experiences such as the time when a fellow decided to commit suicide by downing a "cocktail" of every drug in the house and just to top it off added a small can of engine oil. However, this absorbed all of the worst drugs in his stomach and saved his life!

THE CHINESE SOLUTION

Any analytically working scientist is forced to use computers, databases and software tools as a basic requirement. This was not so in the early days where one was already lucky to have access to a "teletypewriter". Hank about a development for the Merck Index, a popular encyclopedia of chemicals, drugs, and biologicals published by Merck & Co., Inc.:

In a way, the overdose cases were the start of the data base setup. We needed a list of drugs and their major ions for quick reference, so this was done by hand. But a computer seemed the way to go and Bill Milne and Steve Heller set this up and began to look into larger lists. At the same time CI appeared to offer a lot in simplifying the spectra of these messy fluids and we began to apply this as well. Now it became apparent that the Merck Index which contained all the drugs would be useful if it had a molecular weight index. I mentioned this to my Chinese postdoc Tung Sun and somewhat to my surprise, a few days later he had done it by hand! This proved to be so popular when we made it available that Merck produced their own version the next year.

THE NIH SHIFT

This was a problem worked on by Gordon Guroff (who sadly just died from a car crash) and Gene Renson, a post doc from Belgium. He had been assigned the task of developing a test for the enzyme, phenylalanine hydroxylase that converts Phe to Tyr. He was supplied with 4-tritiophenylalanine which, reasonably, would give him tritiated water when the hydroxyl from OH^+ attacked the para position and displaced it. Only it did not happen ... the tritium stayed mostly with the tyrosine formed. They did not understand that and Gene was finally sent back to Belgium with some regret. Then Gordon tried it himself and got the same result. He worked right around the corner from us so one day he asked me how it might be explained. At that time I was very much aware, from CI, of how positively charged things rearrange and I suggested that the OH^+ attacked the para position as expected, but then the tritium at that position migrated to the adjacent position as the oxygen electrons came in to stabilize the ion by putting the charge on oxygen. Then the molecule could lose either tritium or hydrogen to become tyrosine. The isotopic effect of tritium is quite high, up to 40, which I knew having worked on isotope effects earlier in the Amaryllidaceae, and its loss was discriminated against. At that time Gordon and the group around the corner under Sid Udenfriend was collaborating with the organic chemists under Bernhardt Witkop and John Daly instead of us (a result of some difficulty with our boss, Evan Horning, who had had a run in with Witkop when Witkop worked for him) so Guroff brought the problem to Witkop, Daly and also Don Jerina there. They asked me if I wanted to be involved, and stupidly I said, no thanks. But in the end they all did an excellent job proving my suggested mechanism and working on its further ramifications in the arene oxide (cancer-causing aromatic hydrocarbons) area. Gordon did insist on giving me an acknowledgement in his paper on the work, which I very much appreciate.

A TRIP TO THE AMAZON

I felt honoured and lucky that I was treated to a slide show in Hank´s house of his trip to the Amazon where he had participated as natural products chemist and mass spectrometrist. It contained a collection of great stories, only one of which is repeated here:

The Amazon trip was organized by Jim Wheeler of Howard University, Bo Holmsted, a pharmacologist in the Karolinska, and Dick Schulties, a very famous ethnic biologist at Harvard. Also, I had collaborated with several of the others including Murray Blum from the University of Georgia with whom I have done many insect studies, and John Daly of NIH with his poisonous frogs, and Don Jerina who did all of the arene oxide work at NIH, and even Holmsted who got me involved with "Dr X" in another murder case. It was an NSF (National Science Foundation – SK) funded trip and I was invited because they needed a mass spectrometrist to run their spectrometer supplied by LKB (who had some connection with Holmsted) that was on board the "Alpha Helix" boat at the time. The idea was to do limited examination of many of the natural products down there before they were lost due to the encroachment of modern life. And we did do that. The problem was that we rediscovered the wheel several times because the literature was not available to us. The only communication with the outside was by Morse code over a radio, (no voice).

Unfortunately when the MS was installed, the ground point on the transformer was not connected correctly so that its frame was up at 220V! One day the local kids (in Pebas, the little village we were at) were invited over to see the ship. I was aware of the fact that there was voltage on the machine, but had been told by LKB that this was just a little "leakage" voltage that was not significant, but I thought it would frighten the kids if they touched the instrument, so I decided to ground it with a piece of stout wire. When I connected it, the wire blew up and all of the power on the ship went down. After resetting the boat´s circuit breakers, I found the spectrometer, which was the key to the whole voyage (as I was strongly admonished), would not scan. Consultations by Morse code with LKB were not much good because the radioman had been bitten on his "tapping" finger by a bee and it was swollen, so the other mass spectrometrist-pharmacologist and I had to repair it by the book. Fortunately we had the required power transistors and we were able to, thus restoring me in good graces with Schulties.

THE LEMONADE KILLER

MS is an established technique in toxicology and, therefore, can assists in the solution of crimes. Hank in court:

This was a result of contact with an excellent pathologist in the Center for Disease Control, Renata Kimbrough. She had received a routine liver sample from two strange deaths of children in Omaha, Nebraska, and in examining them felt there was evidence for ingestion of a cancer-causing chemical. The police investigated and found that the boyfriend of an older sister of the two girls worked at the Eppley Cancer Center in Omaha. Furthermore, he had taken several of his dogs to the vet with strange illnesses, and yes, he did have access to chemicals, specifically N-nitrosodimethylamine. Somehow they learned that the guy and his girl friend had had a spat and then he offered to make up and invited her family to a picnic where it seems he had dosed the lemonade with the nitrosodimethylamine (nice guy!). Several of the other members of the family were also sick. The guy had gone to Texas. The police followed him with instructions to "twist his arm" to find out the nature of the chemical which at that time was not known. They did this and brought him back for trial, but now they could not use this data against him at the trial because of the manner in which it was obtained. So could I find any evidence for the compound in the glass jars from his house? The answer was no, because it is a volatile liquid and had all evaporated. But there was some white powder on the threads of the jar which when heated very high in the mass spectrometer gave a strange series of compounds with unusual isotope ratios. These turned out to be arsenic oxides! So I had to testify to that fact at his trial and I suppose the prosecution tied it to the poisoning of his dogs etc.. He was put away for life for the death of the children.

Figure 7. Tapani Luukkainen (left), Suzanne Miyamoto and Hank in the front room to his office where he would love to boil and stir his organic concoctions.

Glass-Excreting Feet, Solitary Bees, and Other Anecdotes

Having failed to write the stories down at the time when I still remembered them I have only two more to share:

I remember another time we got a sample of "ground glass" that an older woman had found covering her feet. She felt she was excreting it and the doctor did indeed recover it from her feet. Jim Silverton looked at it by powder x-ray and found it was magnesium sulfate that she had used to bath her tired feet in!

With Suzanne Batra of the USDA we figured out the structures of the little sacs that these bees line their nests with. They turned out to be polyesters that they made by first making omega hydroxyoctadecanoic (and higher) acid and then somehow polymerizing it. We still do not know how they did this and it would be an interesting biological study, but maybe not at NIH.

Epilogue

I much appreciate Hank's willingness to share some his stories with us. At the time, I learned more about mass spectrometry during lunch than during the rest of the day. There were many more of them like these, some of which I have forgotten, some of which did not get written down, some of which were not supposed to be written down (there was something about people drinking D_2O for an experiment....). As Hank said: Have I covered all now? Probably not.

Figure 8. Lewis Pannell, Richard Caprioli, Hank, and Fred Mc Lafferty (from left).

Bibliography of Henry M. Fales

as obtained using SciFinder and sorted by year, excluding Mass Spectral Archives 1-335

(1) Barnes, R. A.; Fales, H. M. *Journal of the American Chemical Society* 1953, 75, 3830-3831.

(2) Barnes, R. A.; Fales, H. M. *Journal of the American Chemical Society* 1953, 75, 975-977.

(3) Fales, H. M.; Barnes, R. A. *Journal of the American Chemical Society* 1954, 76, 1947-1948.

(4) Fales, H. M.; Warnhoff, E. W.; Wildman, W. C. *Journal of the American Chemical Society* 1955, 77, 5885-5890.

(5) Fales, H. M. *Journal of the American Chemical Society* 1955, 77, 5118-5121.

(6) Fales, H. M.; Wildman, W. C. *Journal of the American Chemical Society* 1956, 78, 4151-4153.

(7) Fales, H. M.; Giuffrida, L. D.; Wildman, W. C. *Journal of the American Chemical Society* 1956, 78, 4145-4150.

(8) Briggs, L. H.; Colebrook, L. D.; Fales, H. M.; Wildman, W. C. *Anal. Chem.* 1957, 29, 904-911.

(9) Fales, H. M.; Wildman, W. C. *Journal of the American Chemical Society* 1958, 80, 4395-4404.

(10) Fales, H. M.; Wildman, W. C. *Chemistry & Industry* (London, United Kingdom) 1958, 561-562.

(11) Uyeo, S.; Fales, H. M.; Highet, R. J.; Wildman, W. C. *Journal of the American Chemical Society* 1958, 80, 2590-2591.

(12) Wildman, W. C.; Fales, H. M. *Journal of the American Chemical Society* 1958, 80, 6465-6466.
(13) Fales, H. M.; Horn, D. H. S.; Wildman, W. C. *Chemistry & Industry* (London, United Kingdom) 1959, 1415.
(14) Fales, H. M.; Wildman, W. C. *Journal of the American Chemical Society* 1960, 82, 3368-3373.
(15) Fales, H. M.; Wildman, W. C. *Journal of the American Chemical Society* 1960, 82, 197-205.
(16) Inubushi, Y.; Fales, H. M.; Warnhoff, E. W.; Wildman, W. C. *Journal of Organic Chemistry* 1960, 25, 2153-2164.
(17) Lloyd, H. A.; Fales, H. M.; Highet, P. F.; VandenHeuvel, W. J. A.; Wildman, W. C. *Journal of the American Chemical Society* 1960, 82, 3791.
(18) Banerjee, S. K.; Chakravarti, D.; Chakravarti, R. N.; Fales, H. M.; Klayman, D. L. *Bulletin of the Calcutta School of Tropical Medicine* 1961, 9, 116.
(19) Banerjee, S. K.; Chakravarti, D.; Chakravarti, R. N.; Fales, H. M.; Klayman, D. L. *Tetrahedron* 1961, 16, 251-254.
(20) Battersby, A. R.; Fales, H. M.; Wildman, W. C. *Journal of the American Chemical Society* 1961, 83, 4098-4099.
(21) Fales, H. M.; Wildman, W. C. *Journal of Organic Chemistry* 1961, 26, 881-886.
(22) Fales, H. M.; Wildman, W. C. *Journal of Organic Chemistry* 1961, 26, 1617-1621.
(23) Fales, H. M.; Wildman, W. C. *Journal of Organic Chemistry* 1961, 26, 181-187.
(24) Haahti, E.; Fales, H. M. *Chemistry & Industry* (London, United Kingdom) 1961, 507-508.
(25) Lloyd, H. A.; Kielar, E. A.; Highet, R. J.; Uyeo, S.; Fales, H. M.; Wildman, W. C. *Tetrahedron Letters* 1961, 105-109.
(26) Fales, H. M.; Haahti, E. O. A.; Luukkainen, T.; VandenHeuvel, W. J. A.; Horning, E. C. *Analytical Biochemistry* 1962, 4, 296-305.
(27) Fales, H. M.; Pisano, J. J. *Analytical Biochemistry* 1962, 3, 337-342.
(28) Fales, H. M.; Robertson, A. V. *Tetrahedron Letters* 1962, 111-115.
(29) Kirschner, M. A.; Fales, H. M. *Anal. Chem.* 1962, 34, 1548-1551.
(30) Lloyd, H. A.; Kielar, E. A.; Highet, R. J.; Uyeo, S.; Fales, H. M.; Wildman, W. C. *Journal of Organic Chemistry* 1962, 27, 373-377.
(31) Miles, H. T.; Fales, H. M. *Anal. Chem.* 1962, 34, 860-861.
(32) Wildman, W. C.; Fales, H. M.; Battersby, A. R. *Journal of the American Chemical Society* 1962, 84, 681-682.
(33) Wildmann, W. C.; Fales, H. M.; Highet, R. J.; Breuer, S. W.; Battersby, A. R. *Proc. Chem. Soc.* 1962, 180-181.
(34) Fales, H. M.; Mann, J.; Mudd, S. H. *Journal of the American Chemical Society* 1963, 85, 2025-2026.
(35) Fales, H. M.; Wildman, W. C. *Journal of the American Chemical Society* 1963, 85, 784-791.
(36) Jones, G.; Fales, H. M.; Wildman, W. C. *Tetrahedron Letters* 1963, 397-400.
(37) Mann, J. D.; Fales, H. M.; Mudd, S. H. *Journal of Biological Chemistry* 1963, 238, 3820-3823.
(38) Naegelli, P.; Warnhoff, E. W.; Fales, H. M.; Lyle, R. E.; Wildman, W. C. *Journal of Organic Chemistry* 1963, 28, 206-210.

(39) Banerjee, S. K.; Chakravarti, R. N.; Fales, H. M. *Bulletin of the Calcutta School of Tropical Medicine* 1964, 12, 23-24.
(40) Barchas, J.; Katz, E.; Beaven, V.; Weissbach, H.; Fales, H. *Biochemistry* (Moscow, Russian Federation) 1964, 3, 1684-1687.
(41) Battersby, A. R.; Binks, R.; Breuer, S. W.; Fales, H. M.; Wildman, W. C.; Highet, R. J. *Journal of the Chemical Society*, Abstracts 1964, 1595-1609.
(42) Burlingame, A. L.; Fales, H. M.; Highet, R. J. *Journal of the American Chemical Society* 1964, 86, 4976-4979.
(43) Fales, H. M.; Pisano, J. J. Biomed. *Appl. Gas Chromatog., Lectures Ann. Gas Chromatog.* Inst., 5th, Buffalo (Herman A. Szymanski, editor, Plenum Press, New York, N.Y.) 1964, 39-87.
(44) Fales, H. M.; Wildman, W. C. *Journal of the American Chemical Society* 1964, 86, 294-295.
(45) Haahti, E.; Lagerspetz, K.; Nikkari, T.; Fales, H. M. *Comparative Biochemistry and Physiology* 1964, 12, 435-437.
(46) Laiho, S. M.; Fales, H. M. *Journal of the American Chemical Society* 1964, 86, 4434-4438.
(47) Shelton, E.; Otani, T. T.; Fales, H. M. *Nature* (London, United Kingdom) 1964, 202, 1229-1230.
(48) Arnold, J. E.; Fales, H. M. *Journal of Gas Chromatography* 1965, 3, 131-133.
(49) Fales, H. M. *Gas Chromatog. Steroids Biol. Fluids*, Proc. Workshop, Warrenton, Va. 1965, 297-300.
(50) Fales, H. M.; Luukkainen, T. *Anal. Chem.* 1965, 37, 955-957.
(51) Joshi, B. C.; May, E. L.; Fales, H. M.; Daly, J. W.; Jacobson, A. E. *Journal of Medicinal Chemistry* 1965, 8, 559-563.
(52) Penttila, A.; Kapadia, G. J.; Fales, H. M. *Journal of the American Chemical Society* 1965, 87, 4402-4403.
(53) Bowers, W. S.; Fales, H. M.; Thompson, M. J.; Uebel, E. C. *Science* 1966, 154, 1020-1021.
(54) Brot, N.; Goodwin, J. S.; Fales, H. M. *Biochemical and Biophysical Research Communications* 1966, 25, 454-461.
(55) Fales, H. M. *Anal. Chem.* 1966, 38, 1058-1059.
(56) Lloyd, H. A.; Warren, K. S.; Fales, H. M. *Journal of the American Chemical Society* 1966, 88, 5544-5549.
(57) Mize, C. E.; Steinberg, D.; Avigan, J.; Fales, H. M. *Biochemical and Biophysical Research Communications* 1966, 25, 359-365.
(58) Mize, C. E.; Avigan, J.; Baxter, J. H.; Fales, H. M.; Steinberg, D. *Journal of Lipid Research* 1966, 7, 692-697.
(59) Penttila, A.; Fales, H. M. *Chemical Communications* (London) 1966, 656-657.
(60) Penttila, A.; Fales, H. M. *Journal of the American Chemical Society* 1966, 88, 2327-2328.
(61) Steinberg, D.; Avigan, J.; Mize, C. E.; Baxter, J. H.; Gammermeyer, J.; Fales, H. M.; Highet, P. F. *Journal of Lipid Research* 1966, 7, 684-691.
(62) Fales, H. M.; Warren, K. S. *Journal of Organic Chemistry* 1967, 32, 501-503.
(63) Haahti, E.; Fales, H. M. *Journal of Lipid Research* 1967, 8, 131-137.

(64) Holtzman, J. L.; Gillette, J. R.; Fales, H. M. *Journal of the American Chemical Society* 1967, 89, 708.
(65) Lloyd, H. A.; Fales, H. M. *Tetrahedron Letters* 1967, 4891-4895.
(66) Steinberg, D.; Mize, C. E.; Avigan, J.; Fales, H. M.; Eldjarn, L.; Try, K.; Stokke, O.; Refsum, S. *Journal of Clinical Investigation* 1967, 46, 313-322.
(67) Tsai, S.-C.; Herndon, J. H., Jr.; Uhlendorf, B. W.; Fales, H. M.; Mize, C. E. *Biochemical and Biophysical Research Communications* 1967, 28, 571-577.
(68) Gordon, P. G.; Penttila, A.; Fales, H. M. *Journal of the American Chemical Society* 1968, 90, 1376-1377.
(69) Kapadia, G. J.; Fales, H. M. *Chemical Communications* (London) 1968, 1688-1689.
(70) Kapadia, G. J.; Fales, H. M. *Journal of Pharmaceutical Sciences* 1968, 57, 2017-2018.
(71) Steinberg, D.; Avigan, J.; Mize, C. E.; Herndon, J. H., Jr.; Fales, H. M.; Milne, G. W. A. *Pathologia Europaea* 1968, 3, 450-458.
(72) Wolff, J.; Horisaka, K.; Fales, H. M. *Biochemistry* 1968, 7, 2455-2457.
(73) Daasch, L. W.; Fales, H. M. *Organic Mass Spectrometry* 1969, 2, 1043-1048.
(74) Fales, H. M.; Milne, G. W. A. *Journal of the American Chemical Society* 1969, 91, 3682-3685.
(75) Faux, A.; Horn, D. H. S.; Middleton, E. J.; Fales, H. M.; Lowe, M. E. *Chemical Communications* (London) 1969, 175-176.
(76) Guarino, A. M.; Conway, W. D.; Fales, H. M. *European Journal of Pharmacology* 1969, 8, 244-252.
(77) Herndon, J. H., Jr.; Steinberg, D.; Uhlendorf, B. W.; Fales, H. M. *Journal of Clinical Investigation* 1969, 48, 1017-1032.
(78) Lloyd, H. A.; Sokoloski, E. A.; Strauch, B. S.; Fales, H. M. *Chemical Communications* (London) 1969, 299-301.
(79) Mize, C. E.; Avigan, J.; Steinberg, D.; Pittman, R. C.; Fales, H. M.; Milne, G. W. A. Biochimica et Biophysica Acta, *Lipids and Lipid Metabolism* 1969, 176, 720-739.
(80) Strauch, B. S.; Fales, H. M.; Pittman, R. C.; Avigan, J. *Journal of Nutrition* 1969, 97, 194-202.
(81) Fales, H. M.; Milne, G. W. A.; Axenrod, T. *Analytical Chemistry* 1970, 42, 1432-1435.
(82) Fales, H. M.; Lloyd, H. A.; Milne, G. W. A. *Journal of the American Chemical Society* 1970, 92, 1590-1597.
(83) Fales, H. M.; Milne, G. W.; Axenrod, T. Analytical chemistry FIELD Publication Date: Oct 1970, 42, 1432-1435. *FIELD Reference Number: FIELD Journal Code: 0370536 FIELD Call Number.*
(84) Greten, H.; Levy, R. I.; Fales, H.; Fredrickson, D. S. *Biochimica et Biophysica Acta, Lipids and Lipid Metabolism* 1970, 210, 39-45.
(85) Kapadia, G. J.; Rao, G. S.; Leete, E.; Fayez, M. B. E.; Vaishnav, Y. N.; Fales, H. M. *Journal of the American Chemical Society* 1970, 92, 6943-6951.
(86) MacConnell, J. G.; Blum, M. S.; Fales, H. M. *Science* (Washington, DC, United States) 1970, 168, 840-841.
(87) Milne, G. W. A.; Axenrod, T.; Fales, H. M. *Journal of the American Chemical Society* 1970, 92, 5170-5175.
(88) Perel, J. M.; Cunningham, R. F.; Fales, H. M.; Dayton, P. G. *Life Sciences* 1970, 9, 1337-1343.
(89) Sarfaty, G. S.; Fales, H. M. *Analytical Chemistry* 1970, 42, 288-291.

(90) Steinberg, D.; Mize, C. E.; Herndon, J. H., Jr.; Fales, H. M.; Engel, W. K.; Vroom, F. Q. *Archives of internal medicine* 1970, 125, 75-87.
(91) Ziffer, H.; Fales, H. M.; Milne, G. W. A.; Field, F. H. *Journal of the American Chemical Society* 1970, 92, 1597-1600.
(92) Beggs, D.; Vestal, M. L.; Fales, H. M.; Milne, G. W. A. *Review of Scientific Instruments* 1971, 42, 1578-1584.
(93) Fales, H. M. Fundam. Drug Metab. *Drug Disposition* 1971, 437-457.
(94) Fales, H. M.; Milne, G. W. A.; Law, N. C. *Archives of Mass Spectral Data* 1971, 2, 620-621.
(95) Fales, H. M.; Milne, G. W. A.; Law, N. C. *Archives of Mass Spectral Data* 1971, 2, 630-631.
(96) Fales, H. M.; Milne, G. W. A.; Law, N. C. *Archives of Mass Spectral Data* 1971, 2, 624-625.
(97) Fales, H. M. Mass Spectrom.: *Tech. Appl.* 1971, 179-215.
(98) Fales, H. M.; Milne, G. W. A.; Nicholson, R. S. *Analytical Chemistry* 1971, 43, 1785-1789.
(99) Fales, H. M.; Beisler, J. A.; Silverton, J. V.; Penttila, A.; Horn, D. H. S. *Journal of the American Chemical Society* 1971, 93, 4850-4855.
(100) Fales, H. M.; Nagai, Y.; Milne, G. W. A.; Brewer, H. B., Jr.; Bronzert, T. J.; Pisano, J. J. *Analytical Biochemistry* 1971, 43, 288-299.
(101) Kiryushkin, A. A.; Fales, H. M.; Axenrod, T.; Gilbert, E. J.; Milne, G. W. A. *Organic Mass Spectrometry* 1971, 5, 19-31.
(102) Law, N. C.; Aandahl, V.; Fales, H. M.; Milne, G. W. A. *Clinica Chimica Acta* 1971, 32, 221-228.
(103) Longevialle, P.; Devissaguet, P.; Khuong Huu, Q.; Fales, H. M. Comptes Rendus des Seances de l'Academie des Sciences, Serie C: *Sciences Chimiques* 1971, 273, 1533-1535.
(104) MacConnell, J. G.; Blum, M. S.; Fales, H. M. *Tetrahedron* 1971, 27, 1129-1139.
(105) Milne, G. W. A.; Fales, H. M.; Axenrod, T. *Analytical Chemistry* 1971, 43, 1815-1820.
(106) Blum, M. S.; Brand, J. M.; Wallace, J. B.; Fales, H. M. *Life Sciences* 1972, 11, 525-531.
(107) Brand, J. M.; Blum, M. S.; Fales, H. M.; MacConnell, J. G. *Toxicon* 1972, 10, 259-271.
(108) Cohen, A. I.; Dreyfuss, J.; Fales, H. M. *Journal of Medicinal Chemistry* 1972, 15, 542-545.
(109) Fales, H. M.; Milne, G. W. A.; Pisano, J. J.; Brewer, H. R., Jr.; Blum, M. S.; MacConnell, J. G.; Brand, J.; Law, N. *Recent Progress in Hormone Research* 1972, 28, 591-626.
(110) Fales, H. M. *Pharmacology* 1972, 8, 142-147.
(111) Fales, H. M.; Blum, M. S.; Crewe, R. M.; Brand, J. M. *Journal of Insect Physiology* 1972, 18, 1077-1088.
(112) Hedler, L.; Kaunitzi, H.; Marquardt, P.; Fales, H.; Johnson, R. E. *N-Nitroso Compounds Anal. Form., Proc. Work. Conf.* 1972, 71-73.
(113) Heller, S. R.; Fales, H. M.; Milne, G. W. A. *Journal of Chemical Education* 1972, 49, 725.
(114) Law, N. C.; Fales, H. M.; Milne, G. W. *Clinical toxicology* 1972, 5, 17-21.
(115) Miles, E. W.; Fales, H. M.; Gin, J. B. *Biochemistry* 1972, 11, 4945-4953.

(116) Blum, M. S.; Wallace, J. B.; Fales, H. M. *Insect Biochemistry* 1973, 3, 353-357.
(117) Blum, M. S.; Mac Connell, J. G.; Brand, J. M.; Duffield, R. M.; Fales, H. M. *Annals of the Entomological Society of America* 1973, 66, 235.
(118) Brand, J. M.; Fales, H. M.; Sokoloski, E. A.; MacConnell, J. G.; Blum, M. S.; Duffield, R. M. *Life Sciences* 1973, 13, 201-211.
(119) Brand, J. M.; Blum, M. S.; Fales, H. M.; Pasteels, J. M. *Journal of Insect Physiology* 1973, 19, 369-382.
(120) Brand, J. M.; Duffield, R. M.; MacConnell, J. G.; Blum, M. S.; Fales, H. M. *Science* (Washington, DC, United States) 1973, 179, 388-389.
(121) Dunlop, P.; Marini, M. A.; Fales, H. M.; Sokoloski, E.; Martin, C. J. *Bioorganic Chemistry* 1973, 2, 235-247.
(122) Fales, H. M. *New Methods Environ. Chem. Toxicol., Collect. Pap. Res. Conf. New Methodol. Ecol. Chem.* 1973, 11-30.
(123) Fales, H. M.; Jaouni, T. M.; Babashak, J. F. Analytical Chemistry 1973, 45, 2302-2303.
(124) Heller, S. R.; Feldmann, R. J.; Fales, H. M.; Milne, G. W. A. *Journal of Chemical Documentation* 1973, 13, 130-133.
(125) Heller, S. R.; Fales, H. M.; Milne, G. W. A. *Organic Mass Spectrometry* 1973, 7, 107-115.
(126) Kingston, D. G. I.; Fales, H. M. *Tetrahedron* 1973, 29, 4083-4086.
(127) Korn, E. D.; Dearborn, D. G.; Fales, H. M.; Sokoloski, E. A. *Journal of Biological Chemistry* 1973, 248, 2257-2259.
(128) Lonevialle, P.; Fales, H. M.; Highet, R. J.; Burlingame, A. L. *Organic Mass Spectrometry* 1973, 7, 417-430.
(129) Longevialle, P.; Milne, G. W. A.; Fales, H. M. *Journal of the American Chemical Society* 1973, 95, 6666-6669.
(130) Longevialle, P.; Smith, D. H.; Burlingame, A. L.; Fales, H. M.; Highet, R. J. *Organic Mass Spectrometry* 1973, 7, 401-415.
(131) Maxwell, T.; Milne, G. W. A.; Fales, H. M.; Law, N. *Environmental Health Perspectives* 1973, 139-140.
(132) Milne, G. W. A.; Fales, H. M.; Colburn, R. W. *Analytical Chemistry* 1973, 45, 1952-1954.
(133) Sun, T.; Pedder, D. J.; Fales, H. M. *Analytical Chemistry* 1973, 45, 2297.
(134) Tsai, S.-C.; Fales, H. M.; Vaughan, M. *Journal of Biological Chemistry* 1973, 248, 5278-5281.
(135) Tsai, S.-C.; Steinberg, D.; Avigan, J.; Fales, H. M. *Journal of Biological Chemistry* 1973, 248, 1091-1097.
(136) Das, K. G.; Fales, H. M.; Beckey, H. D. *International Journal of Mass Spectrometry and Ion Physics* 1974, 15, 110-112.
(137) Fales, H. M.; Heller, S. R.; Milne, G. W. A.; Sun, T. *Biomedical Mass Spectrometry* 1974, 1, 295.
(138) Grantham, P. H.; Mohan, L. C.; Weisburger, E. K.; Fales, H. M.; Sokoloski, E. A.; Weisburger, J. W. *Xenobiotica* 1974, 4, 69-76.
(139) Heller, S. R.; Fales, H. M.; Milne, G. W. A.; Heller, R. S.; McCormick, A.; Maxwell, D. C. *Biomedical Mass Spectrometry* 1974, 1, 207-208.
(140) Heller, S. R.; Fales, H. M.; Milne, G. W. A.; Feldmann, R. J.; Daly, N. R.; Maxwell, D. C.; McCormick, A. *Advances in Mass Spectrometry* 1974, 6, 1037-1042.

(141) Heller, S. R.; Koniver, D. A.; Fales, H. M.; Milne, G. W. A. *Analytical Chemistry* 1974, 46, 947-950.
(142) Nutting, W. L.; Blum, M. S.; Fales, H. M. Psyche, *Revue Internationale des Sciences de l'Homme et de Psychanalyse* 1974, 81, 166-177.
(143) Pant, R.; Fales, H. M. *Phytochemistry* (Elsevier) 1974, 13, 1626-1628.
(144) Assmann, G.; Fredrickson, D. S.; Sloan, H. R.; Fales, H. M.; Highet, R. J. *Journal of lipid research* 1975, 16, 28-38.
(145) Fales, H. M.; Milne, G. W. A.; Winkler, H. U.; Beckey, H. D.; Damico, J. N.; Barron, R. *Analytical Chemistry* 1975, 47, 207-219.
(146) Gross, M.; Baer, H.; Fales, H. M. *Phytochemistry* (Elsevier) 1975, 14, 2263-2266.
(147) Hiranaka, P. K.; Kleinman, L. M.; Sokoloski, E. A.; Fales, H. M. *American Journal of Hospital Pharmacy* 1975, 32, 928-930.
(148) Irreverre, F.; Wilson, E.; Fales, H. M.; Sun, T.; Sokoloski, E. *Lloydia* 1975, 38, 178-180.
(149) Kapadia, G. J.; Paul, B. D.; Silverton, J. V.; Fales, H. M.; Sokoloski, E. A. *Journal of the American Chemical Society* 1975, 97, 6814-6819.
(150) Rowe, V. D.; Fales, H. M.; Pisano, J. J.; Andersen, A. E.; Guroff, G. *Biochemical Medicine* 1975, 12, 123-136.
(151) Aldrich, J. R.; Blum, M. S.; Duffey, S. S.; Fales, H. M. *Journal of Insect Physiology* 1976, 22, 1201-1206.
(152) Cagen, L. M.; Fales, H. M.; Pisano, J. J. *Journal of Biological Chemistry* 1976, 251, 6550-6554.
(153) Fales, H. M.; Fenselau, C.; Duncan, J. H. *Organic Mass Spectrometry* 1976, 11, 669-674.
(154) Goren, M. B.; Brokl, O.; Roller, P.; Fales, H. M.; Das, B. C. *Biochemistry* 1976, 15, 2728-2735.
(155) MacConnell, J. G.; Blum, M. S.; Buren, W. F.; Williams, R. N.; Fales, H. M. *Toxicon* 1976, 14, 69-78.
(156) Pedder, D. J.; Fales, H. M.; Jaouni, T.; Blum, M.; MacConnell, J.; Crewe, R. M. *Tetrahedron* 1976, 32, 2275-2281.
(157) Shukla, Y. N.; Sokoloski, E. A.; Fales, H. M.; Kapadia, G. J. *Phytochemistry* (Elsevier) 1976, 15, 1788.
(158) Cagen, L. M.; Fales, H. M.; Bowden, R. E.; Pisano, J. J. *Monographs of the Physiological Society of Philadelphia* 1977, 3, 149-157.
(159) Chen, P. S.; Fales, H. M.; Levenbook, L.; Sokoloski, E. A.; Yeh, H. J. C. *Biochemistry* 1977, 16, 4080-4085.
(160) Duffey, S. S.; Blum, M. S.; Fales, H. M.; Evans, S. L.; Roncadori, R. W.; Tiemann, D. L.; Nakagawa, Y. *Journal of Chemical Ecology* 1977, 3, 101-113.
(161) Fales, H. M.; Wright, G. J. *Journal of the American Chemical Society* 1977, 99, 2339-2340.
(162) Kapadia, G. J.; Shukla, Y. N.; Chowdhury, B. K.; Basak, S. P.; Fales, H. M.; Sokoloski, E. A. *Journal of the Chemical Society*, Chemical Communications 1977, 535-536.
(163) Aldrich, J. R.; Blum, M. S.; Hefetz, A.; Fales, H. M.; Lloyd, H. A.; Roller, P. *Science* (Washington, DC, United States) 1978, 201, 452-454.

(164) Aldrich, J. R.; Blum, M. S.; Lloyd, H. A.; Fales, H. M. *Journal of Chemical Ecology* 1978, 4, 161-172.
(165) Blum, M. S.; Fales, H. M.; Tucker, K. W.; Collins, A. M. *Journal of Apicultural Research* 1978, 17, 218-221.
(166) Blum, M. S.; Wallace, J. B.; Duffield, R. M.; Brand, J. M.; Fales, H. M.; Sokoloski, E. A. *Journal of Chemical Ecology* 1978, 4, 47-53.
(167) Kapadia, G. J.; Shukla, Y. N.; Basak, S. P.; Fales, H. M.; Sokoloski, E. A. *Phytochemistry* (Elsevier) 1978, 17, 1444-1445.
(168) Lloyd, H. A.; Evans, S. L.; Fales, H. M. *Lloydia* 1978, 41, 494-496.
(169) Aldrich, J. R.; Blum, M. S.; Fales, H. M. *Journal of Chemical Ecology* 1979, 5, 53-62.
(170) Hefetz, A.; Fales, H. M.; Batra, S. W. T. *Science* (Washington, DC, United States) 1979, 204, 415-417.
(171) Jones, T. H.; Blum, M. S.; Fales, H. M. *Tetrahedron Letters* 1979, 1031-1034.
(172) Milne, G. W. A.; Fales, H. M.; Law, N. C. *Instrum. Appl. Forensic Drug Chem., Proc. Int. Symp.* 1979, 91-101.
(173) Weiss, U.; Fales, H. M.; Weisgraber, K. H. *Liebigs Annalen der Chemie* 1979, 914-919.
(174) Baer, H.; Hooton, M.; Fales, H.; Wu, A.; Schaub, F. *Phytochemistry* (Elsevier) 1980, 19, 799-802.
(175) Blum, M. S.; Jones, T. H.; Hoelldobler, B.; Fales, H. M.; Jaouni, T. *Naturwissenschaften* 1980, 67, 144-145.
(176) Fales, H. M.; Jaouni, T. M.; Schmidt, J. O.; Blum, M. S. *Journal of Chemical Ecology* 1980, 6, 895-903.
(177) Fales, H. M.; Highet, R. J. *Journal of Organic Chemistry* 1980, 45, 1699-1700.
(178) Fales, H. M.; Comstock, W.; Jones, T. H. *Analytical Chemistry* 1980, 52, 980-982.
(179) Hunninghake, G. W.; Gadek, J. E.; Fales, H. M.; Crystal, R. G. *Journal of Clinical Investigation* 1980, 66, 473-483.
(180) Jaouni, T. M.; Leon, M. B.; Rosing, D. R.; Fales, H. M. *Journal of Chromatography* 1980, 182, 473-477.
(181) Jones, T. H.; Blum, M. S.; Fales, H. M. *Tetrahedron Letters* 1980, 21, 1701-1704.
(182) Jones, T. H.; Blum, M. S.; Fales, H. M.; Thompson, C. R. *Journal of Organic Chemistry* 1980, 45, 4778-4780.
(183) Jones, T. H.; Franko, J. B.; Blum, M. S.; Fales, H. M. *Tetrahedron Letters* 1980, 21, 789-792.
(184) Kapadia, G. J.; Shukla, Y. N.; Basak, S. P.; Sokoloski, E. A.; Fales, H. M. *Tetrahedron* 1980, 36, 2441-2447.
(185) Levenbook, L.; Boctor, F. N.; Fales, H. M. *Journal of Insect Physiology* 1980, 26, 381-383.
(186) Lundgren, D. W.; Fales, H. M. *Journal of Biological Chemistry* 1980, 255, 4481-4486.
(187) Norden, B.; Batra, S. W. T.; Fales, H. M.; Hefetz, A.; Shaw, G. J. *Science* (Washington, DC, United States) 1980, 207, 1095-1097.
(188) Jones, T. H.; Blum, M. S.; Fales, H. M. *Synthetic Communications* 1981, 11, 889-894.
(189) Blum, M. S.; Jones, T. H.; Snelling, R. R.; Overal, W. L.; Fales, H. M.; Highet, R. J. *Biochemical Systematics and Ecology* 1982, 10, 91-94.
(190) Fales, H. M. *Chem. Nat. Prod., Proc. Sino-Am. Symp.* 1982, 212-229.
(191) Jones, T. H.; Blum, M. S.; Fales, H. M. *Tetrahedron* 1982, 38, 1949-1958.

(192) Jones, T. H.; Blum, M. S.; Howard, R. W.; McDaniel, C. A.; Fales, H. M.; DuBois, M. B.; Torres, J. *Journal of Chemical Ecology* 1982, 8, 285-300.
(193) Kelner, L.; Fales, H. M.; Markey, S. P.; Crawford, C. K. *International Journal of Mass Spectrometry and Ion Physics* 1982, 43, 249-259.
(194) Waddell, T. G.; Thomasson, M. H.; Moore, M. W.; White, H. W.; Swanson-Bean, D.; Green, M. E.; Van Horn, G. S.; Fales, H. M. *Phytochemistry* (Elsevier) 1982, 21, 1631-1633.
(195) Blum, M. S.; Fales, H. M.; Jones, T. H.; Rinderer, T. E.; Tucker, K. W. *Comparative Biochemistry and Physiology, Part B: Biochemistry & Molecular Biology* 1983, 75B, 237-238.
(196) Blum, M. S.; Jones, T. H.; Overal, W. L.; Fales, H. M.; Schmidt, J. O.; Blum, N. A. *Comparative Biochemistry and Physiology, Part B: Biochemistry & Molecular Biology* 1983, 75B, 15-16.
(197) Fales, H. M. *International Journal of Mass Spectrometry and Ion Physics* 1983, 53, 59-67.
(198) Fales, H. M.; Weiss, U.; Jaouni, T. *Liebigs Annalen der Chemie* 1983, 367-371.
(199) Hill, R.; Fales, H.; McNeil, C.; Macfarlane, R. *Biomedical Mass Spectrometry* 1983, 10, 505-506.
(200) Howard, D. F.; Blum, M. S.; Fales, H. M. *Science* (Washington, DC, United States) 1983, 220, 335-336.
(201) Jones, T. H.; Fales, H. M. *Tetrahedron Letters* 1983, 24, 5439-5440.
(202) Kelner, L.; Markey, S. P.; Fales, H. M.; Crawford, C. K.; Cole, P. A. *International Journal of Mass Spectrometry and Ion Physics* 1983, 46, 3-6.
(203) Waddell, T. G.; Osborne, C. B.; Collison, R.; Levine, M. J.; Cross, M. C.; Silverton, J. V.; Fales, H. M.; Sokoloski, E. A. *Journal of Organic Chemistry* 1983, 48, 4450-4453.
(204) Bian, Z.; Fales, H. M.; Blum, M. S.; Jones, T. H.; Rinderer, T. E.; Howard, D. F. *Journal of Chemical Ecology* 1984, 10, 451-461.
(205) Fales, H. M.; Blum, M. S.; Bian, Z.; Jones, T. H.; Don, A. W. *Journal of Chemical Ecology* 1984, 10, 651-665.
(206) Fales, H. M. *Controlled clinical trials* 1984, 5, 576-578.
(207) Jones, T. H.; Highet, R. J.; Blum, M. S.; Fales, H. M. *Journal of Chemical Ecology* 1984, 10, 1233-1249.
(208) Kelner, L.; Markey, S. P.; Fales, H. M.; Lundquist, T. R. *International Journal of Mass Spectrometry and Ion Processes* 1984, 62, 237-251.
(209) Schmuff, N. R.; Phillips, J. K.; Burkholder, W. E.; Fales, H. M.; Chen, C. W.; Roller, P. P.; Ma, M. *Tetrahedron Letters* 1984, 25, 1533-1534.
(210) Blum, M. S.; Jones, T. H.; Lloyd, H. A.; Fales, H. M.; Snelling, R. R.; Lubin, Y.; Torres, J. *Journal of Entomological Science* 1985, 20, 254-257.
(211) Fales, H. M.; McNeal, C. J.; Macfarlane, R. D.; Shimohigashi, Y. *Analytical Chemistry* 1985, 57, 1616-1621.
(212) Fales, H. M.; Pannell, L. K.; Sokoloski, E. A.; Carmeci, P. *Analytical Chemistry* 1985, 57, 376-378.
(213) Lloyd, H. A.; Fales, H. M.; Goldman, M. E.; Jerina, D. M.; Plowman, T.; Schultes, R. E. *Tetrahedron Letters* 1985, 26, 2623-2624.
(214) Panell, L. K.; Fales, H. M.; Scovill, J. P.; Klayman, D. L.; West, D. X.; Tate, R. L. *Transition Metal Chemistry* (Dordrecht, Netherlands) 1985, 10, 141-147.

(215) Pannell, L. K.; Sokoloski, E. A.; Fales, H. M.; Tate, R. L. *Analytical Chemistry* 1985, 57, 1060-1067.
(216) Phillips, J. K.; Walgenbach, C. A.; Klein, J. A.; Burkholder, W. E.; Schmuff, N. R.; Fales, H. M. *Journal of Chemical Ecology* 1985, 11, 1263-1274.
(217) Schmidt, J. O.; O'Neill, K. M.; Fales, H. M.; McDaniel, C. A.; Howard, R. W. *Journal of Chemical Ecology* 1985, 11, 895-901.
(218) Yang, Y. M.; Fales, H. M.; Pannell, L. *Analytical Chemistry* 1985, 57, 1771-1772.
(219) Bagnasco, S.; Balaban, R.; Fales, H. M.; Yang, Y. M.; Burg, M. *Journal of Biological Chemistry* 1986, 261, 5872-5877.
(220) Jacobson, K. A.; Pannell, L. K.; Kirk, K. L.; Fales, H. M.; Sokoloski, E. A. *Journal of the Chemical Society, Perkin Transactions 1: Organic and Bio-Organic Chemistry* (1972-1999) 1986, 2143-2149.
(221) Liav, A.; Goren, M. B.; Yang, Y.; Fales, H. M. *Carbohydrate Research* 1986, 155, 223-228.
(222) Pitha, J.; Milecki, J.; Fales, H.; Pannell, L.; Uekama, K. *International Journal of Pharmaceutics* 1986, 29, 73-82.
(223) Sangameswaran, L.; Fales, H. M.; Friedrich, P.; De Blas, A. L. *Proceedings of the National Academy of Sciences of the United States of America* 1986, 83, 9236-9240.
(224) Yang, Y. M.; Sokoloski, E. A.; Fales, H. M.; Pannell, L. K. *Biomedical & Environmental Mass Spectrometry* 1986, 13, 489-494.
(225) Yang, Y. M.; Lloyd, H. A.; Pannell, L. K.; Fales, H. M.; Macfarlane, R. D.; McNeal, C. J.; Ito, Y. *Biomedical & Environmental Mass Spectrometry* 1986, 13, 439-445.
(226) Atlas, D.; Diamant, S.; Fales, H. M.; Pannell, L. *Journal of Cardiovascular Pharmacology* 1987, 10, S122-S127.
(227) Blum, M. S.; Morel, L.; Fales, H. M. *Comparative Biochemistry and Physiology, Part B: Biochemistry & Molecular Biology* 1987, 86B, 251-252.
(228) Dhariwal, K. R.; Yang, Y.; Fales, H. M.; Goren, M. B. *Journal of General Microbiology* 1987, 133, 201-209.
(229) Phillips, J. K.; Miller, S. P. F.; Andersen, J. F.; Fales, H. M.; Burkholder, W. E. *Tetrahedron Letters* 1987, 28, 6145-6146.
(230) Pitha, J.; Szabo, L.; Fales, H. M. *Carbohydrate Research* 1987, 168, 191-198.
(231) Tomalski, M. D.; Blum, M. S.; Jones, T. H.; Fales, H. M.; Howard, D. F.; Passera, L. *Journal of Chemical Ecology* 1987, 13, 253-263.
(232) Blum, M. S.; Fales, H. M. *Journal of Chemical Ecology* 1988, 14, 2099-2107.
(233) Fales, H. M.; Blum, M. S.; Southwick, E. W.; Williams, D. L.; Roller, P. P.; Don, A. W. *Tetrahedron* 1988, 44, 5045-5050.
(234) Fukunaga, K.; Szabo, L.; Fales, H. M.; Pitha, J. *Journal of Pharmaceutical Sciences* 1988, 77, 640-642.
(235) Irie, T.; Fukunaga, K.; Yoshida, A.; Uekama, K.; Fales, H. M.; Pitha, *J. Pharmaceutical Research* 1988, 5, 713-717.
(236) Jones, T. H.; Blum, M. S.; Andersen, A. N.; Fales, H. M.; Escoubas, P. *Journal of Chemical Ecology* 1988, 14, 35-45.
(237) Kaufman, E. E.; Nelson, T.; Fales, H. M.; Levin, D. M. *Journal of Biological Chemistry* 1988, 263, 16872-16879.

(238) Massio, G.; Lavaud, C.; Le Men-Olivier, L.; Van Binst, G.; Miller, S. P. F.; Fales, H. M. *Journal of the Chemical Society, Perkin Transactions 1: Organic and Bio-Organic Chemistry* (1972-1999) 1988, 3071-3079.

(239) Page, R. E.; Blum, M. S.; Fales, H. M. *Experientia* 1988, 44, 270-271.

(240) Rinderer, T. E.; Blum, M. S.; Fales, H. M.; Bian, Z.; Jones, T. H.; Buco, S. M.; Lancaster, V. A.; Danka, R. G.; Howard, D. F. *Journal of Chemical Ecology* 1988, 14, 495-501.

(241) Russell, W. L.; Pandey, R. C.; Schaffner, C. P.; Fales, H. M. *Journal of Antibiotics* 1988, 41, 149-156.

(242) Taber, D. F.; Deker, P. B.; Fales, H. M.; Jones, T. H.; Lloyd, H. A. *Journal of Organic Chemistry* 1988, 53, 2968-2971.

(243) Fales, H. M.; Yang, Y. M.; Pannell, L. K.; Lamoureux, C. J. H.; Feil, V. J. *Biomedical & Environmental Mass Spectrometry* 1989, 18, 71-76.

(244) Irie, T.; Fukunaga, K.; Pitha, J.; Uekama, K.; Fales, H. M.; Sokolowski, E. A. *Carbohydrate Research* 1989, 192, 167-172.

(245) Mason, R. T.; Fales, H. M.; Jones, T. H.; O'Brien, L. B.; Taylor, T. W.; Hogue, C. L.; Blum, M. S. *Insect Biochemistry* 1989, 19, 737-740.

(246) Mason, R. T.; Fales, H. M.; Jones, T. H.; Pannell, L. K.; Chinn, J. W.; Crews, D. *Science* (Washington, DC, United States) 1989, 245, 290-293.

(247) Pannell, L. K.; Pu, Q. L.; Fales, H. M.; Mason, R. T.; Stephenson, J. L. *Analytical Chemistry* 1989, 61, 2500-2503.

(248) Raina, A. K.; Jaffe, H.; Kempe, T. G.; Keim, P.; Blacher, R. W.; Fales, H. M.; Riley, C. T.; Klun, J. A.; Ridgway, R. L.; Hayes, D. K. *Science* (Washington, DC, United States) 1989, 244, 796-798.

(249) Showell, J. S.; Fales, H. M.; Sokoloski, E. A. *Organic Mass Spectrometry* 1989, 24, 632-636.

(250) Basile, A. S.; Pannell, L.; Jaouni, T.; Gammal, S. H.; Fales, H.; Jones, E. A.; Skolnick, P. *Proceedings of the National Academy of Sciences of the United States of America* 1990, 87, 5263-5267.

(251) Berkowitz, B. A.; Moriyama, T.; Fales, H. M.; Byrd, R. A.; Balaban, R. S. *Journal of Biological Chemistry* 1990, 265, 12417-12423.

(252) Cassels, F. J.; Fales, H. M.; London, J.; Carlson, R. W.; Van Halbeek, H. *Journal of Biological Chemistry* 1990, 265, 14127-14135.

(253) Eisner, T.; Eisner, M.; Jaouni, T.; Fales, H. M. *Naturwissenschaften* 1990, 77, 33.

(254) Fales, H. M.; Lloyd, H. A.; Ferretti, J. A.; Silverton, J. V.; Davis, D. G.; Kon, H. *Journal of the Chemical Society, Perkin Transactions 2: Physical Organic Chemistry* (1972-1999) 1990, 1005-1010.

(255) Fales, H. M.; Sokoloski, E. A.; Pannell, L. K.; Pu, Q. L.; Klayman, D. L.; Lin, A. J.; Brossi, A.; Kelley, J. A. *Analytical Chemistry* 1990, 62, 2494-2501.

(256) Kaufman, E. E.; Nelson, T.; Fales, H. M. *Biol. Oxid. Syst., [Proc. Symp.]* 1990, 2, 1059-1069.

(257) Mason, R. T.; Jones, T. H.; Fales, H. M.; Pannell, L. K.; Crews, D. *Journal of Chemical Ecology* 1990, 16, 2353-2369.

(258) Massiot, G.; Lavaud, C.; Delaude, C.; Van Binst, G.; Miller, S. P. F.; Fales, H. M. *Phytochemistry* 1990, 29, 3291-3298.

(259) Pannell, L.; Fales, H. *Proceedings of the Physics of Small Systems* 1990, 3, 33-50.

(260) Pannell, L. K.; Pu, Q. L.; Mason, R. T.; Fales, H. M. *Rapid Communications in Mass Spectrometry* 1990, 4, 103-107.
(261) Peden, D. B.; Hohman, R.; Brown, M. E.; Mason, R. T.; Berkebile, C.; Fales, H. M.; Kaliner, M. A. *Proceedings of the National Academy of Sciences of the United States of America* 1990, 87, 7638-7642.
(262) Rao, C. T.; Fales, H. M.; Pitha, J. *Pharmaceutical Research* 1990, 7, 612-615.
(263) Ryzlak, M.; Fales, H. M.; Russell, W. L.; Schaffner, C. P. *Alcoholism: Clinical and Experimental Research* 1990, 14, 490-495.
(264) Basile, A. S.; Pannell, L.; Jaouni, T.; Gammal, S. H.; Jones, E. A.; Fales, H. M.; Skolnick, P. *Fidia Research Foundation Symposium Series* 1991, 6, 305-323.
(265) Blum, M. S.; Everett, D. M.; Jones, T. H.; Fales, H. M. *ACS Symposium Series* 1991, 449, 14-26.
(266) Cahnmann, H. J.; Goncalves, E.; Ito, Y.; Fales, H. M.; Sokoloski, E. A. *Journal of Chromatography* 1991, 538, 165-175.
(267) Fales, H. M.; Pu, Q. L.; Mason, R. T.; Pannell, L. K. *International Journal of Mass Spectrometry and Ion Processes* 1991, 106, 273-281.
(268) Jaffe, H.; Raina, A. K.; Wagner, R. M.; Fales, H. M.; Kempe, T. G.; Keim, P.; Blacher, R. W.; Riley, C. T. *ACS Symposium Series* 1991, 453, 215-225.
(269) Jones, T. H.; Blum, M. S.; Fales, H. M.; Brandao, C. R. F.; Lattke, J. *Journal of Chemical Ecology* 1991, 17, 1897-1908.
(270) Mason, R. T.; Fales, H. M.; Eisner, M.; Eisner, T. *Die Naturwissenschaften* 1991, 78, 28-30.
(271) Murata, Y.; Yeh, H. J. C.; Pannell, L. K.; Jones, T. H.; Fales, H. M.; Mason, R. T. *Journal of Natural Products* 1991, 54, 233-240.
(272) Blum, M. S.; Fales, H. M.; Leadbetter, G.; Leonhardt, B. A.; Duffield, R. M. *Journal of Natural Toxins* 1992, 1, 57-63.
(273) Blum, M. S.; Foottit, R.; Fales, H. M. *Comparative Biochemistry and Physiology, Part C: Pharmacology, Toxicology & Endocrinology* 1992, 102C, 209-211.
(274) Cahnmann, H. J.; Goncalves, E.; Ito, Y.; Fales, H. M.; Sokoloski, E. A. *Analytical Biochemistry* 1992, 204, 344-350.
(275) Fales, H. M.; Jones, T. H.; Jaouni, T.; Blum, M. S.; Schmidt, J. O. *Journal of Chemical Ecology* 1992, 18, 847-854.
(276) Ito, Y.; Shibusawa, Y.; Fales, H. M.; Cahnmann, H. J. *Journal of Chromatography* 1992, 625, 177-181.
(277) Lee, Y. C.; Oeda, T.; Driscoll, W. J.; Fales, H. M.; Strott, C. A. *Journal of Biological Chemistry* 1992, 267, 10982-10987.
(278) Shinomiya, K.; Menet, J.-M.; Fales, H. M.; Ito, Y. *Journal of Chromatography* 1993, 644, 215-229.
(279) Liotta, A. S.; Kole, H. K.; Fales, H. M.; Roth, J.; Bernier, M. *Journal of Biological Chemistry* 1994, 269, 22996-23001.
(280) Ma, Y.; Ito, Y.; Sokolosky, E.; Fales, H. M. *Journal of Chromatography*, A 1994, 685, 259-262.
(281) Robrish, S. A.; Fales, H. M.; Gentry-Weeks, C.; Thompson, J. *Journal of Bacteriology* 1994, 176, 3250-3256.
(282) Weisz, A.; Scher, A. L.; Shinomiya, K.; Fales, H. M.; Ito, Y. *Journal of the American Chemical Society* 1994, 116, 704-708.

(283) Cao, J.; Fales, H. M.; Schaffner, C. P. *Proceedings of the Society for Experimental Biology and Medicine* 1995, 209, 195-204.

(284) Ijames, C. F.; Dutky, R. C.; Fales, H. M. *Journal of the American Society for Mass Spectrometry* 1995, 6, 1226-1231.

(285) Ito, Y.; Shinomiya, K.; Fales, H. M.; Weisz, A.; Scher, A. L. *ACS Symposium Series* 1995, 593, 156-183.

(286) Jaffe, K.; Blum, M. S.; Fales, H. M.; Mason, R. T.; Cabrera, A. *Journal of Chemical Ecology* 1995, 21, 379-384.

(287) Randazzo, P. A.; Terui, T.; Sturch, S.; Fales, H. M.; Ferrige, A. G.; Kahn, R. A. *Journal of Biological Chemistry* 1995, 270, 14809-14815.

(288) Davis, D. A.; Dorsey, K.; Wingfield, P. T.; Stahl, S. J.; Kaufman, J.; Fales, H. M.; Levine, R. L. *Biochemistry* 1996, 35, 2482-2488.

(289) Shinomiya, K.; Inokuchi, N.; Gnabre, J. N.; Muto, M.; Kabasawa, Y.; Fales, H. M.; Ito, Y. *Journal of Chromatography*, A 1996, 724, 179-184.

(290) Shinomiya, K.; Muto, M.; Kabasawa, Y.; Fales, H. M.; Ito, Y. *Journal of Liquid Chromatography & Related Technologies* 1996, 19, 415-425.

(291) Zhang, X. K.; Dutky, R. C.; Fales, H. M. *Analytical Chemistry* 1996, 68, 3288-3289.

(292) Koenig, S.; Fales, H. M. *BioTechniques* 1997, 23, 1010, 1012, 1014.

(293) Stebbins, J. W.; Jaffe, H.; Fales, H. M.; Moeller, J. R. *Biochemistry* 1997, 36, 2221-2226.

(294) Cassola, A. C.; Jaffe, H.; Fales, H. M.; Afeche, S. C.; Magnoli, F.; Cipolla-Neto, *J. Pfluegers Archiv* 1998, 436, 545-552.

(295) Degnore, J. P.; Konig, S.; Barrett, W. C.; Chock, P. B.; Fales, H. M. *Rapid Communications in Mass Spectrometry* 1998, 12, 1457-1462.

(296) Koenig, S.; Fales, H. M.; Haegele, K. D. *Analytical Chemistry* 1998, 70, 4453-4455.

(297) Konig, S.; Fales, H. M. *Journal of the American Society for Mass Spectrometry* 1998, 9, 814-822.

(298) Konig, S.; Weiler, S.; Fales, H. M. *BioTechniques* 1998, 24, 712, 714, 716.

(299) Qi, L.; Ma, Y.; Ito, Y.; Fales, H. M. *Journal of Liquid Chromatography & Related Technologies* 1998, 21, 83-92.

(300) Southan, G. J.; Srinivasan, A.; Keefer, L. K.; George, C.; Fales, *H. M. Chemical Communications* (Cambridge) 1998, 1191-1192.

(301) Xiang, B.; Ferretti, J.; Fales, H. M. *Analytical Chemistry* 1998, 70, 2188-2190.

(302) Zhang, X. K.; Fales, H. M. *Journal of the American Society for Mass Spectrometry* 1998, 9, 15-20.

(303) Barrett, W. C.; DeGnore, J. P.; Koenig, S.; Fales, H. M.; Keng, Y.-F.; Zhang, Z.-Y.; Yim, M. B.; Chock, P. B. *Biochemistry* 1999, 38, 6699-6705.

(304) Hanus, L. O.; Fales, H. M.; Spande, T. F.; Basile, A. S. *Analytical Biochemistry* 1999, 270, 159-166.

(305) Koenig, S.; Fales, H. M. *Journal of the American Society for Mass Spectrometry* 1999, 10, 273-276.

(306) Blum, M. S.; Fales, H. M.; Morse, R. A.; Underwood, B. A. *Journal of Chemical Ecology* 2000, 26, 801-807.

(307) Davis, D. A.; Newcomb, F. M.; Moskovitz, J.; Wingfield, P. T.; Stahl, S. J.; Kaufman, J.; Fales, H. M.; Levine, R. L.; Yarchoan, R. *Biochemical Journal* 2000, 346, 305-311.

(308) Driscoll, W. J.; Koenig, S.; Fales, H. M.; Pannell, L. K.; Eipper, B. A.; Mueller, G. P. *Biochemistry* 2000, 39, 8007-8016.
(309) Yang-Boja, E.; DeFilippes, F.; Fales, H. M. *Analytical Biochemistry* 2000, 285, 205-210.
(310) Boja, E. S.; Fales, H. M. *Analytical Chemistry* 2001, 73, 3576-3582.
(311) Ross, G. N.; Fales, H. M.; Lloyd, H. A.; Jones, T.; Sokoloski, E. A.; Marshall-Batty, K.; Blum, M. S. *Journal of Chemical Ecology* 2001, 27, 1219-1228.
(312) Wang, J.; Boja, E. S.; Tan, W.; Tekle, E.; Fales, H. M.; English, S.; Mieyal, J. J.; Chock, P. B. Journal of Biological Chemistry 2001, 276, 47763-47766.
(313) Weisz, A.; Andrzejewski, D.; Fales, H. M.; Mandelbaum, A. *Journal of Mass Spectrometry* 2001, 36, 1024-1030.
(314) Yang-Boja, E.; DeFilippes, F.; Fales, H. M. *Analytical Biochemistry* 2001, 289, 306.
(315) Davis, D. A.; Newcomb, F. M.; Moskovitz, J.; Fales, H. M.; Levine, R. L.; Yarchoan, R. *Methods in Enzymology* 2002, 348, 249-259.
(316) Davis, D. A.; Read-Connole, E.; Pearson, K.; Fales, H. M.; Newcomb, F. M.; Moskovitz, J.; Yarchoan, R. *Antimicrobial Agents and Chemotherapy* 2002, 46, 402-408.
(317) Pearson, K. M.; Pannell, L. K.; Fales, H. M. *Rapid Communications in Mass Spectrometry* 2002, 16, 149-159.
(318) Randazzo, P. A.; Fales, H. M. *Methods in Molecular Biology* (Totowa, NJ, United States) 2002, 189, 169-179.
(319) Weisz, A.; Andrzejewski, D.; Fales, H. M.; Mandelbaum, A. *Journal of Mass Spectrometry* 2002, 37, 1025-1033.
(320) Boja, E. S.; Hoodbhoy, T.; Fales, H. M.; Dean, J. *Journal of Biological Chemistry* 2003, 278, 34189-34202.
(321) Davis, D. A.; Brown, C. A.; Newcomb, F. M.; Boja, E. S.; Fales, H. M.; Kaufman, J.; Stahl, S. J.; Wingfield, P.; Yarchoan, R. *Journal of Virology* 2003, 77, 3319-3325.
(322) Nie, Z.; Boehm, M.; Boja, E. S.; Vass, W. C.; Bonifacino, J. S.; Fales, H. M.; Randazzo, P. A. *Developmental Cell* 2003, 5, 513-521.
(323) Zhao, M.; Gold, L.; Dorward, H.; Liang, L.-f.; Hoodbhoy, T.; Boja, E. S.; Fales, H. M.; Dean, *J. Molecular and Cellular Biology* 2003, 23, 8982-8991.
(324) Boja, E. S.; Sokoloski, E. A.; Fales, H. M. *Analytical Chemistry* 2004, 76, 3958-3970.
(325) De La Luz Sierra, M.; Yang, F.; Narazaki, M.; Salvucci, O.; Davis, D.; Yarchoan, R.; Zhang, H. H.; Fales, H.; Tosato, G. *Blood* 2004, 103, 2452-2459.
(326) Jeong, W.; Chang, T.-S.; Boja, E. S.; Fales, H. M.; Rhee, S. G. *Journal of Biological Chemistry* 2004, 279, 3151-3159.
(327) Joubert, F.; Fales, H. M.; Wen, H.; Combs, C. A.; Balaban, R. S. *Biophysical Journal* 2004, 86, 629-645.
(328) Sun, G.; Ma, Y.; Gao, X.; Konig, S.; Fales, H. M.; Kador, P. F. *Experimental Eye Research* 2004, 79, 919-926.
(329) Zhao, M.; Boja, E. S.; Hoodbhoy, T.; Nawrocki, J.; Kaufman, J. B.; Kresge, N.; Ghirlando, R.; Shiloach, J.; Pannell, L.; Levine, R. L.; Fales, H. M.; Dean, J. *Biochemistry* 2004, 43, 12090-12104.
(330) Che, M. M.; Boja, E. S.; Yoon, H.-Y.; Gruschus, J.; Jaffe, H.; Stauffer, S.; Schuck, P.; Fales, H. M.; Randazzo, P. A. *Cellular Signalling* 2005, 17, 1276-1288.

(331) Davis, D. A.; Singer, K. E.; De La Luz Sierra, M.; Narazaki, M.; Yang, F.; Fales, H. M.; Yarchoan, R.; Tosato, G. *Blood* 2005, 105, 4561-4568.
(332) Goren, M. B.; Sokoloski, E. A.; Fales, H. M. *Journal of Organic Chemistry* 2005, 70, 7429-7431.
(333) Zhao, H.; Joseph, J.; Fales, H. M.; Sokoloski, E. A.; Levine, R. L.; Vasquez-Vivar, J.; Kalyanaraman, B. *Proceedings of the National Academy of Sciences of the United States of America* 2005, 102, 5727-5732.
(334) Boja, E. S.; Hoodbhoy, T.; Garfield, M.; Fales, H. M. *Biochemistry* in press 2005.
(335) Ganesh, S.; Sharma, Y.; Billings, E.; Dayhoff, J.; Fales, H.; Kickler, T.; Nabel, E. *Cardiovascular Pathology* 2004, 13, 33.

In: Biomacromolecular Mass Spectrometry Yearbook. Volume 1 ISBN: 978-1-60876-167-8
Ed: Simone Koenig

Chapter 7

PHOSPHORYLATION SITE ANALYSIS OF MRP14 PURIFIED FROM GRANULOCYTES*

Martin Zeller, Thomas Vogl*, Johannes Roth* and Simone König[1]
Integrated Functional Genomics, Interdisciplinary Center for Clinical Research
*Institute of Experimental Dermatology
Medical Faculty, University of Münster, Germany

ABSTRACT

Myeloid-related protein MRP14 is a member of the S100 family of calcium-binding proteins and is known to be phosphorylated *in vivo*. With the combination of enzymatic dephosphorylation, reversed phase and metal ion affinity chromatography, and mass spectrometry methods the phosphorylation site T113 of human MRP14 was unambiguously determined.

Keywords: *protein phosphorylation, mass spectrometry, HPLC, IMAC*

INTRODUCTION

Myeloid-related protein (MRP14; MW 13 kDa) also known as S100A9 or calgranulin B, is a member of the S100 family of calcium-binding proteins. Members of this family are small proteins with two calcium-binding sites of the EF-hand type (helix-loop-helix calcium-

* A version of this chapter was also published in *Biomacromolecular Mass Spectrometry Volume 1, Number 2*, published by Nova Science Publishers, Inc. It was submitted for appropriate modifications in an effort to encourage wider dissemination of research.

1 Address correspondence to: PD Dr. Simone König, Integrated Functional Genomics, Röntgenstr. 21 48149 Münster, Germany, ph: xx-(0)251-8357164. fax: xx-(0)251-8355651, e-mail: koenigs@uni-muenster.de

binding domains) [1] and a tissue-specific expression pattern [2]. Even though the S100 protein family is the largest group of calcium-binding proteins, their functions are not clear. S100 proteins may play a dominant role during calcium-induced activation of phagocytes and there is evidence that these proteins are involved in various cellular functions, e.g. cell cycle progression, cellular differentiation, or modulation of cytoskeleton-membrane interactions [3-6]. MRP14 forms non-covalent complexes with MRP8 (S100A8, calgranulin A), another member of the S100 family, which is a prerequisite for biological functions [7-9].

In 1989 it was demonstrated that human MRP14 is phosphorylated *in vivo* [11]. Using ^{32}P-labelling, cyanogen bromide cleavage, and Edman sequencing T113 was indicated to be phosphorylated. Here, the modern analytical technique of mass spectrometry (MS) was used to unambiguously determine the phosphorylation site of hMRP14.

EXPERIMENTAL

Purification of MRP8 and MRP14 from human granulocytes by anion exchange chromatography has been described before [7, 8]. It included the separation of phosphorylated hMRP14 from the unphosphorylated species [10]. hMRP14 (Fig. 1) exists in two N-terminally acetylated isoforms where hMRP14* lacks the first four amino acids.

For tryptic digestion 2.5 µg trypsin sequencing grade (Roche Diagnostics, Mannheim, Germany) was dissolved in 50 µl 100 mM ammonium bicarbonate (pH 8; Sigma-Aldrich, Steinheim, Germany) and added to the protein in buffer (enzyme-substrate ratio 1:6). The digest was carried out at 37 °C overnight and stopped with 5 µl glacial acetic acid (HAc; Sigma). The intact protein and the protein digest were desalted on reversed phase (RP) material (ZipTip C4 and µC18, respectively, Millipore, Bedford, MA,USA) according to the manufacturers protocol.

For immobilized metal affinity chromtatography (IMAC) Co^{2+}-NTA-IMAC beads from TalonSpin columns (BD Biosciences Clontech, Palo Alto, CA, USA) were transferred into an Eppendorf tube as a slurry and washed twice with water (Sigma-Aldrich, Seelze, Germany). Co^{2+}-ions were removed from the coordination sites by repeated washing with 50 mM EDTA (Sigma). The beads were then equilibrated with 0.1 % HAC (pH 3.5) and charged with Fe^{3+}-ions by adding 200 mM $FeCl_3$ solution in 0.1 % HAc five times. Excess metal ions were removed by washing with 0.1 % HAc five times. Finally, the IMAC beads were equilibrated in storage solution consisting of methanol and 0.1 % HAc (1:1 v/v). Nanoscale bed volume columns were prepared by loading IMAC resin into a GELoader tip (Eppendorf, Hamburg, Germany) to pack a column of ~ 15-20 mm. A 10 ml syringe was used to force all liquid through the column by applying gentle air pressure. The IMAC column was washed three times each with 50 µl 0.1 % HAc, 50 µl 0.1 % HAc / ACN (3/1 v/v), and 50 µl 0.1 % HAc. The desalted phoshoprotein digest was lyophilised, redissolved in 40 µl 0.1 % HAc and slowly loaded onto the IMAC column. The IMAC column was then washed two times each with 50 µl 0.1 % HAc, 50 µl 0.1 % HAc / ACN (3/1 v/v), 50 µl 100 mM NaCl / HAc / ACN (74:1:25 v/v/v) to remove unspecifically bound peptides, and 50 µl 0.1 % HAc. Bound peptides were eluted from the nanoscale IMAC column with 40 µl 100 mM Na_2HPO_4 (Sigma), acidified with 5 µl glacial HAc and desalted with ZipTip-µC18 RP material.

```
1   MTCKMSQLER NIETIINTFH QYSVKLGHPD TLNQGEFKEL VRKDLQNFLK
51  LENKNEKVIE HIMEDLDTNA DKQLSFEEFI MLMARLTWAS HEKMHEGDEG
101 PGHHHKPGLG EGpTP
```

Figure 1. Amino acid sequence of hMRP14. The phosphopeptide generated by tryptic cleavage is underlined and the phosphorylation site is in bold and italicized.

Peptide and protein masses were determined with TofSpec-2E instrument (Micromass/Waters, Manchester, UK). α-cyano-4-hydoxycinnamic acid and sinapinic acid were used as matrices, respectively, and 1 µl of analyte was applied. Protein digests were separated using an Agilent HP1100 HPLC (high-performance liquid chromatography; Agilent, Palo Alto, CA, USA) with a 180 µm inner diameter LC Packings Pepmap C18 column (LC Packings, Netherlands) coupled to an $Esquire_{3000}$ 3D ion trap (Bruker Daltonics, Bremen, Germany). Single ion monitoring (SIM) experiments were carried out with an isolation window of ± 2 *m/z* units around the doubly charged mass of the phosphopeptide precursor which was determined by MALDI-TOF (matrix-assisted laser/desorption ionization time-of-flight) experiments.

Enzymatic dephosphorylation with alkaline phosphatase was performed directly after the acquisition of the MALDI peptide map of the IMAC enriched peptides. The spot was redissolved in 2 µl ACN and 100 mM NH_4HCO_3 (7/3 v/v) and transferred into a pre-equilibrated ZipTip-µC18 tip with 20 µl 100 mM NH_4HCO_3 and 1 unit alkaline phosphatase (Roche Diagnostics, Mannheim, Germany) loaded on top of the chromatographic resin. The tip was placed in an Eppendorf tube and incubated for 2 h at 37 °C. The chromatographic bed acted as a barrier and no drainage was observed. The reaction was then stopped with 2 µl formic acid and the solution was slowly pressed through the RP material and desalted. The bound peptides were eluted with 2 µl 80 % ACN containing 0.1 % TFA directly onto the MALDI target and mixed with matrix.

RESULTS AND DISCUSSION

The advantage of using MS for phosphorylation analysis is the capability of the method to detect the location of the modification on a specific amino acid residue by gas phase peptide sequencing. There are many different approaches both to prepare the sample for MS and the analysis itself [12]. Here we employed a strategy which was developed with the use of ion trap and MALDI-TOF instrumentation. In short, it is based on differential analysis of the intact protein using enzymatic dephosphorylation [13] followed by tryptic digestion, phosphopeptide-specific purification and comparative analysis on the peptide level.

Fig. 2 shows the MALDI-TOF spectrum of the C18 desalted intact hMRP14 before and after enzymatic dephosphorylation with alkaline phosphatase. The signal for both MRP isoforms shifts 80 Da to lower masses indicating the loss of one phosphogroup. Subsequent tryptic digestion of MRP14 and IMAC enrichment of the phosphopeptides allows to detect signals for the phosphopeptides [94-114] and [86-114] (one missed cleavage; Fig. 3) in MALDI-TOF. Methionine oxidation is observed as a second modification, known to be often introduced during handling. It is also evident in Fig. 3 that IMAC co-purifies acidic peptides which is a drawback of the method. In MALDI-TOF, peaks were observed ~90 Da below the peptide signal for both phosphopeptides including the Met-ox species. Those could be attributed to the fragmentation of metastable ions and the mass shift is instrument-specific. Therefore, this effect could be used as a phosphopeptide indicator. Enzymatic dephosphorylation confirmed the assignment of the phosphopeptides by the loss of the phosphogroup and an 80 Da-shift in MS (Fig. 4).

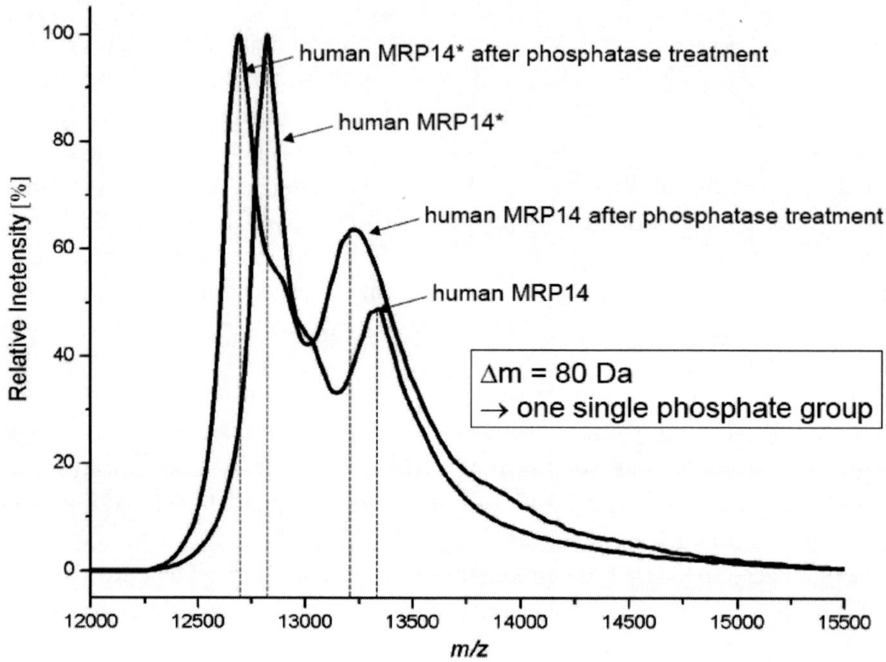

Figure 2. MALDI-TOF spectrum of intact hMRP14 before and after enzymatic dephosphorylation.

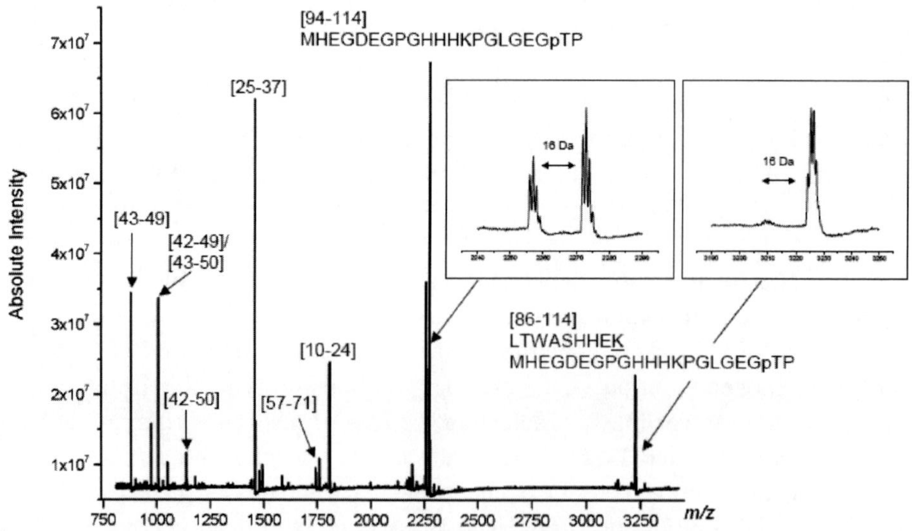

Figure 3. MALDI-TOF spectrum of the IMAC enriched fraction of hMRP14 digest.

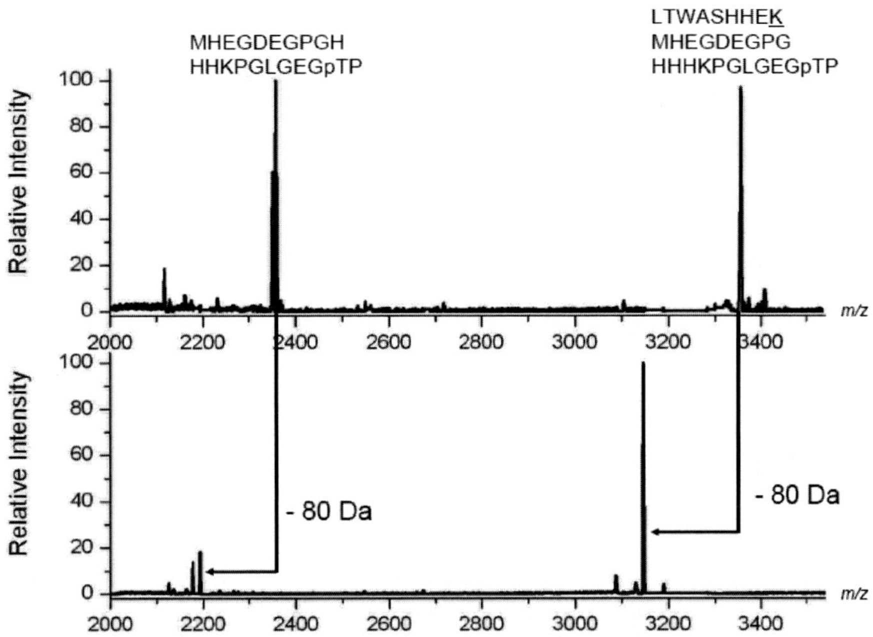

Figure 4. MALDI-TOF spectrum of the IMAC enriched fraction of hMRP14 digest before and after enzymatic dephosphorylation.

For the determination of the modified amino acid residue, SIM-HPLC-MS/MS was carried out. The mass in the SIM experiment was set to the mass of the doubly charged Met-ox phosphopeptide [94-114] (*m/z* 1136 Da, mass window 4 Da, Fig. 5). In the MS/MS spectrum generated (Fig. 6), the phosphorylation site can be assigned to T113 by the presence of b_{19} and b_{20} ions. In that, tandem MS allowed to unambiguously identify the hMRP14 phosphorylation site.

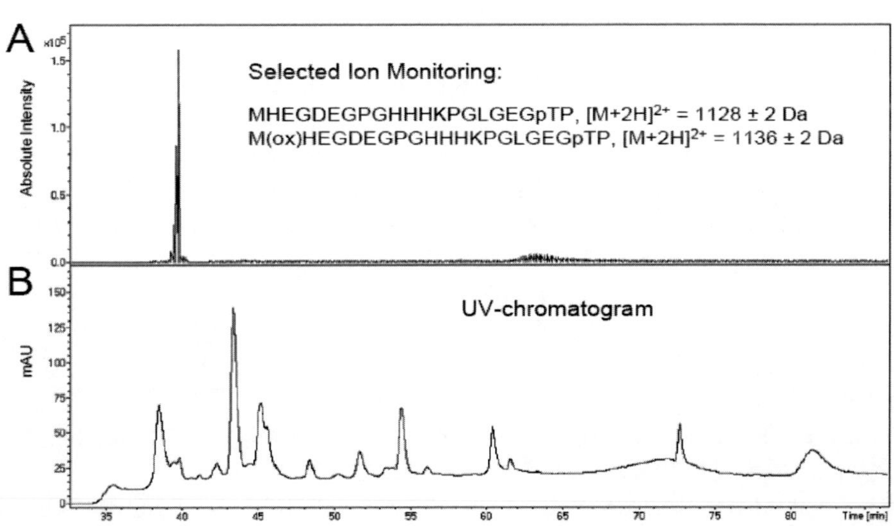

Figure 5. SIM-HPLC-MS/MS; (A) Extracted ion chromatogram, (B) UV chromatogram.

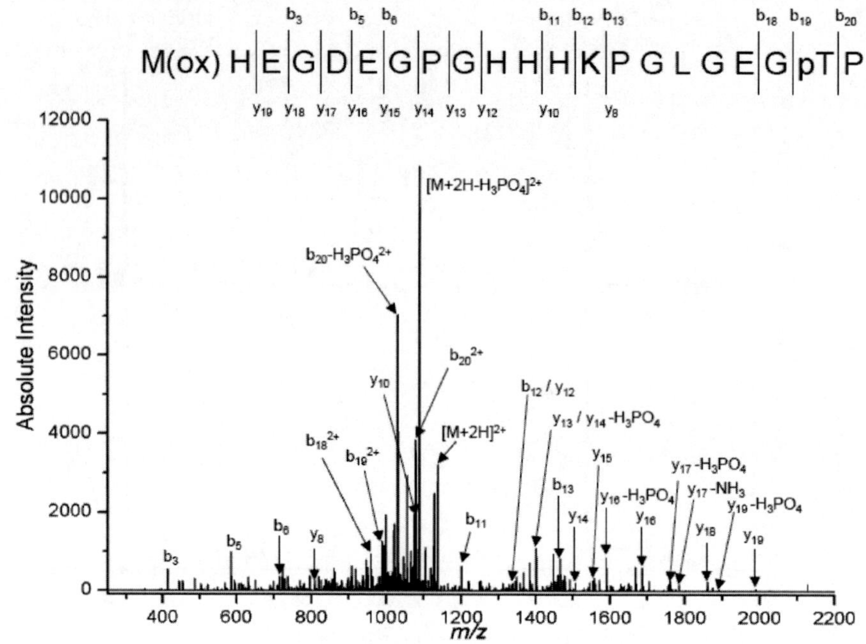

Figure 6. MS/MS spectrum of hMRP14 phosphopeptide [94-114].

Conclusion

With the combination of enzymatic dephosphorylation, IMAC and C18 separation and different MS methods it was possible to unambiguously determine T113 as the only *in vivo* phosphorylation site of MRP14 purified from human granulocytes.

References

[1] Kretsinger, R. H., *Annu. Rev. Biochem.*, 1976, *45*, 239-266.

[2] Roth, J.; Vogl, T.; Sorg, C.; Sunderkötter, C. *Trends Immunol.* 2003, *24*, 155-158.

[3] Niki, I.; Yokokura, H.; Sudo, T.; Kato, M.; Hidaka, H. *J. Biochem.* 1996, *120*, 685-698.

[4] Schafer, B. W.; Heizmann, C. W. *Trends Biochem. Sci.* 1996, *21*, 134-140.

[5] Kligman, D.; Hilt, D. C. *Trends Biochem. Sci.* 1988, *13*, 437-443.

[6] Nacken, W.; Roth, J.; Sorg, C.; Kerkhoff, C. *Microsc. Res. Tech.* 2003, *60*, 569-580.

[7] Vogl, T.; Roth, J.; Sorg, C.; Hillenkamp, F.; Strupat, K. *J. Am. Soc. Mass Spectrom.* 1999, *10*, 1124-1130.

[8] Strupat, K.; Rogniaux, H.; Van Dorsselaer, A.; Roth, J.; Vogl, T. *J. Am. Soc. Mass Spectrom.* 2000, *11*, 780-788.

[9] Vogl, T.; Ludwig, S.; Goebeler, M.; Strey, A.; Thorey, I. S.; Reichelt, R.; Foell, D.; Gerke, V.; Manitz, M. P.; Nacken, W.; Werner, S.; Sorg, C.; Roth, J. *Blood* 2004, *104*, 4260-4268.

[10] Van den Bos, C.; Rammes, A.; Vogl, T.; Boynton, R.; Zaia, J.; Sorg, C.; Roth, J. *Protein Expr. Purif.* 1998, *13*, 313-318.
[11] Edgeworth, J.; Freemont, P.; Hogg, N. *Nature* 1989, *342*, 189-192.
[12] Zeller, M.; König, S. *Anal. Bioanal. Chem.* 2004, *378*, 898-909, Erratum *379*, 318.
[13] Zeller, M.; König, S. Proc. *51th Am. Soc. Mass Spectrom. Montreal, Canada*, 2003.

In: Biomacromolecular Mass Spectrometry Yearbook. Volume 1 ISBN: 978-1-60876-167-8
Ed: Simone Koenig

Chapter 8

THE CRUX OF PROTEOME ANALYSIS*

Simone König[1]
Integrated Functional Genomics, Interdisciplinary Center for Clinical Research, Medical Faculty, University of Muenster, Roentgenstr. 21, 48149 Muenster

The dynamic range of protein expression has been recognized as a main difficulty for analyses decades ago with the advent of two-dimensional gel electrophoresis (2D-PAGE) and later in particular with the option of mass spectrometric protein identification. The general observation is that the abundant proteins are seldom of interest and that the proteins which matter are hard to catch.

The problem has been discussed by several authors. Corthals et al. (1) stated that one needs at least 10^9 cells to perform comprehensive proteome analysis and suspected up to 12 orders of magnitude in dynamic range. Anderson and Anderson (2) demonstrated that the two clinically useful proteins albumin and interleukin 6 differ in plasma abundance by a factor of 10^{10} and that a technology allowing analysis of the single molecule of interest would require a dynamic range of 10^{17}. In addition, the 1 gene - 1 protein illusion is long gone and protein scientists have to struggle with splice variants, isoforms, modifications and glycosylation microheterogeneity of one predicted sequence. Not unexpectedly, the correlation between mRNA and protein levels was shown to be insufficient to predict protein expression levels from quantitative mRNA data (3; see Table 1).

As a technology platform in clinical research we provide protein separation and identification techniques and we teach the importance of sample pre-fractionation and sub-proteome analysis. However, in daily practice we are hardly heard. Most of the time, scientist wish a general proteome scan and end up battling actin in heart projects, albumin in serum analysis, lactoglobulin in milk, and enolase, tubulin, ATPase and glyceraldehyde 3-phosphate dehydrogenase in most systems as we observe routinely. For instance, metabolic enzymes

* A version of this chapter was also published in *Biomacromolecular Mass Spectrometry Volume 1, Number 2*, published by Nova Science Publishers, Inc. It was submitted for appropriate modifications in an effort to encourage wider dissemination of research.

[1] Address correspondence to:PD Dr. Simone König, Integrated Functional Genomics, Röntgenstr. 21, 48149 Münster, Germany, ph: +49-(0)251-8357164, fax: +49-(0)251-8357255, e-mail: koenigs@uni-muenster.de

were indicated to be regulated in a differential protein analysis of transient focal ischemia (Fig. 1; with M. Besselmann and O. Schmidt, University of Muenster). Even though they are possibly affected during stroke, they show up in many other biological systems as well simply due to the fact that they are present at the concentration levels accessible to this kind of experiment. It also appears that some enzymes are present at over-abundance at not too well defined concentrations. The question arises what proteins are expected to be visible in proteome overviews? A PAGE-based study on yeast (3; Table 1) provides insights to that effect. They reflect our observations in mammalian systems.

Table 1. Abundant proteins in yeast sorted by copy number modified after ref. 3. Protein names were obtained from the Comprehensive Yeast Genome Database public webserver (Munich Information Center for Protein Sequences, http://mips.gsf.de/genre/proj/yeast/).

gene or entry name	Protein name	Protein abundance (10^3 copies/ cell)	mRNA abundance (copies/cell)
TDH2	Glyceraldehyde-3-phosphate dehydrogenase 2	863.5	473.0
ADH1	Alcohol dehydrogenase I	746.1	260.0
ENO2	Enolase II (2-phosphoglycerate dehydratase)	635.5	289.1
TEF2	Translation elongation factor eEF1 α-A chain, cytosolic	558.5	282.0
FBA1	Fructose-bisphosphate aldolase	427.2	183.6
PGK1	Phosphoglycerate kinase	315.2	165.7
PDC1	Pyruvate decarboxylase, isozyme 1	303.2	200.7
GPM1	Phosphoglycerate mutase	231.4	169.4
PYK1	Pyruvate kinase	225.3	101.8
TSA1	Thiol-specific antioxidant	110.6	40.1
BEL1	40S small subunit ribosomal protein	103.2	81.0
RPS3	Ribosomal protein S3.e	96.8	
PSA1	Mannose-1-phosphate guanyltransferase	96.4	27.5
AHP1	Alkylhydroperoxide reductase	94.4	9.7
SSA2	Heat shock protein of HSP70 family, cytosolic	77.1	18.6
ILV5	Ketol-acid reducto-isomerase	76.0	4.5
WTM1	Transcriptional modulator	74.5	13.4
GPP1	DL-glycerol phosphatase	70.2	11.2
TDH1	Glyceraldehyde-3-phosphate dehydrogenase 1	69.8	32.7
OMP2	Voltage dependent anion-selective channel (YVDAC1) of the mitochondrial outer membrane	67.4	41.6
PDI1	Protein disulfide-isomerase precursor	66.2	14.1
EFB1	Translation elongation factor eEF1beta	66.1	23.8
SSB1	Heat shock protein of HSP70 family	64.5	79.5
IPP1	Inorganic pyrophosphatase, cytoplasmic	63.1	3.7
TMA19	Protein putative involved in cytoplasmic ribosome function	61.2	88.4

Table 1 continued

gene or entry name	Protein name	Protein abundance (10^3 copies/ cell)	mRNA abundance (copies/cell)
ADH2	Alcohol dehydrogenase II	58.9	260.0
URA1	Dihydrorotate dehydrogenase	49.5	8.9
TAL1	Transaldolase	44.8	3.7
MET17	O-acetylhomoserine sulfhydrylase	38.1	29.0
ALD6	Aldehyde dehydrogenase, cytosolic	37.7	2.2
ENO1	Enolase I (2-phosphoglycerate dehydratase)	35.4	0.7
VMA2	H^+-ATPase V1 domain 60 KD subunit, vacuolar	33.7	8.9
IDP2	Isocitrate dehydrogenase, cytosolic	32.4	
YNL010W	Protein of unknown function localised to cytoplasm and nucleus	31.6	3.7
BAT1	Branched chain amino acid aminotransferase, mitochondrial	30.9	11.2
PAB1	mRNA polyadenylate-binding protein	30.4	1.5
YFR044C	Protein of unknown function localised to cytoplasm and mitochondria	30.2	6.7
SUP45	Translational release factor	29.6	11.9
IDH2	Isocitrate dehydrogenase (NAD^+) subunit 2, mitochondrial	29.4	6.7
GSP1	GTP-binding protein of the ras superfamily	26.3	5.2
SBA1	Hsp90 Associated Co-chaperone	29.2	10.4
HXK2	Hexokinase II	26.5	7.4
URA5	Orotate phosphoribosyltransferase	25.4	6.0
ADO1	Adenosine kinase	27.5	17.1
ERG10	Acetyl-CoA C-acetyltransferase, cytosolic	24.1	4.5
GLK1	Aldohexose specific glucokinase	22.6	6.0
HIS1	ATP phosphoribosyltransferase	22.4	4.5
TOM40	Forms the hydrophilic channel of the mitochondrial import pore for preproteins	22.3	2.2
CYS4	Cystathionine beta-synthase	22.2	
ATP1	F1F0-ATPase complex, F1 alpha subunit	21.6	2.2
THR4	Threonine synthase (o-p-homoserine p-lyase)	21.4	3.7
PRB1	Protease B, vacuolar	21.2	1.5
ICL1	Isocitrate lyase	20.1	
SHM2	Serine hydroxymethyltransferase, cytoplasmic	19.7	7.4
BAT2	Branched chain amino acid aminotransferase, cytosolic	19.0	8.9
LPD1	Dihydrolipoamide dehydrogenase precursor	18.9	2.2

Table 1 continued

gene or entry name	Protein name	Protein abundance (10^3 copies/ cell)	mRNA abundance (copies/cell)
CYB2	Lactate dehydrogenase cytochrome b2	18.9	
RPS5	Ribosomal protein S5.e	18.6	
FSH1	Member of family of serine hydrolases	18.4	2.2
EFT1	Translation elongation factor eEF2	17.9	41.6
ADK1	Adenylate kinase, cytosolic	17.4	16.4
GUK1	Guanylate kinase	16.5	3.7
DLD3	D-lactate dehydrogenase	16.3	3.0
LYS9	Saccharopine dehydrogenase (NADP+, L-glutamate forming)	16.2	3.7
DYS1	Deoxyhypusine synthase	15.8	5.2
CAR2	Ornithine aminotransferase	15.4	
CPR1	Cyclophilin (peptidylprolyl isomerase)	15.2	61.7
SPE3	Putrescine aminopropyltransferase (spermidine synthase)	15.1	6.7
YNL134C	Putative alcohol dehydrogenase	14.9	1.5
BMH1	14-3-3 protein involved in rapamycin-sensitive signalling	14.7	28.2
YMR226C	NADP(+)-dependent dehydrogenase acting on 3-hydroxy acids	14.5	2.2
ARO8	Aromatic amino acid aminotransferase I	14.3	3.0
ILV6	Acetolactate synthase, regulatory subunit	13.9	3.0
YST1	40S ribosomal protein p40 homolog A	13.6	52.8
PRO2	γ-glutamyl phosphate reductase	13.6	5.2
YBR025C	Protein of unknown function localised to cytoplasm	13.1	6.0
SOD2	Superoxide dismutase (Mn) precursor, mitochondrial	12.6	2.2
AAT2	Aspartate aminotransferase, cytosolic	11.7	6.0
TUB2	β-tubulin	11.2	7.4
ATP7	F1F0-ATPase complex, FO D subunit	11.0	
SEC14	Phosphatidylinositol (PI)/phosphatidylcholine (PC) transfer protein	10.9	6.0
MET6	5-methyltetrahydropteroyltriglutamate--homocysteine methyltransferase	10.9	

Table 1 continued

gene or entry name	Protein name	Protein abundance (10^3 copies/ cell)	mRNA abundance (copies/cell)
VMA4	H^+-ATPase V1 domain 27 KD subunit, vacuolar	10.5	3.7
SER1	Phosphoserine transaminase	10.5	1.5
MRP8	Ribosomal protein, mitochondrial	9.3	
ADE1	Phosphoribosylamidoimidazole-succinocarboxamide synthase	8.7	5.2
APA1	ATP adenylyltransferase I	8.7	0.7
TFS1	Cdc25-dependent nutrient- and ammonia-response cell-cycle regulator	8.1	0.7
IDP1	Isocitrate dehydrogenase ($NADP^+$), mitochondrial	7.7	0.7
SSB2	Heat shock protein of HSP70 family, cytosolic	7.0	
PRE8	20S proteasome subunit Y7 (α2)	6.9	0.7
CYS3	Cystathionine γ-lyase	6.7	8.9
STI1	Stress-induced protein	6.7	0.7
SRV2	Adenylate cyclase-associated protein, 70kDa	6.5	
YDL124W	α -keto amide (ester) reductase, NADPH-dependent	6.4	4.5
ADE13	Adenylosuccinate lyase	6.3	1.5
RPE1	D-ribulose-5-phosphate 3-epimerase	5.8	0.7
HOR2	DL-glycerol phosphatase	5.7	0.7
ZWF1	Glucose-6-phosphate dehydrogenase	5.6	0.7
GRS1	Glycine-tRNA ligase	5.5	7.4
SAR1	GTP-binding protein of the ARF family	5.4	10.4
ACH1	Acetyl-CoA hydrolase	5.4	1.5
ILV3	Dihydroxy-acid dehydratase	5.3	4.5
CAR1	Arginase	5.2	3.0
DPM1	Dolichyl-phosphate β-D-mannosyltransferase	5.0	11.2
ADE3	C1-tetrahydrofolate synthase (trifunctional enzyme),cytoplasmic	4.8	5.2
RVB1	RUVB-like protein	4.8	2.2
GYP6	GTPase-activating protein	4.4	0.7
PUP2	20S proteasome subunit (α5)	4.4	0.7
PDC5	Pyruvate decarboxylase, isozyme 2	4.3	
YEL047C	Soluble fumarate reductase, cytoplasmic	3.8	1.5
YER067W	Protein of unknown function localised to cytoplasm and nucleus	3.7	6.7

Table 1 continued

gene or entry name	Protein name	Protein abundance (10^3 copies/ cell)	mRNA abundance (copies/cell)
PRE4	20S proteasome subunit (β7)	3.4	3.7
LEU4	2-isopropylmalalate synthase	3.1	3.0
TIF1	Translation initiation factor 4A	2.9	39.4
MAE1	Malic enzyme	2.8	3.7
MCM3	Replication initiation protein	2.7	
COR1	Ubiquinol-cytochrome-c reductase 44K core protein	2.5	0.7
FRS2	Phenylalanine-tRNA ligase α chain, cytosolic	2.3	0.7
ERG6	S-adenosyl-methionine δ-24-sterol-c-methyltransferase	2.2	14.1
PGM2	Phosphoglucomutase, major isoform	2.2	3.0
CCT8	Component of chaperonin-containing T-complex	2.2	1.5
QCR2	Ubiquinol-cytochrome-c reductase 40KD chain II		2.2
HSP26	Heat shock protein		0.7
TPI1	Triose-phosphate isomerase		

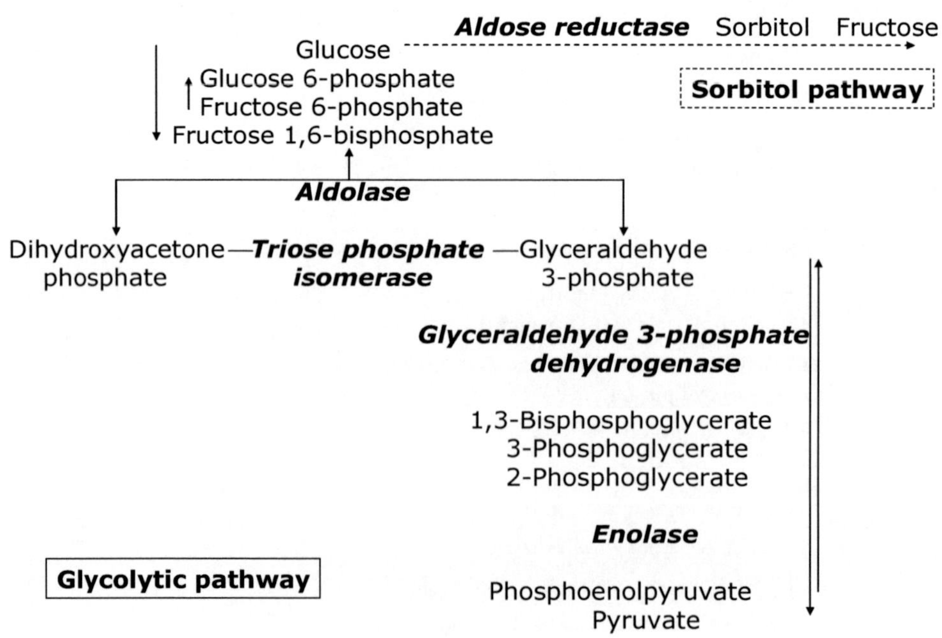

Figure 1. Schematic of glycolytic (7) and sorbitol (8) pathways with an emphasis on the enzymes indicated in differential analysis of transient focal ischemia.

So, are we all doing superficial analyses? Yes and no. A general proteome scan is a valid starting point for any analysis to define the experimental problem. The technology of differential gel electrophoresis (DIGE) allows to obtain significant information on that level (4-6), although it is an expensive technique by itself. Reasons for the reluctance to dig deeper into proteomes are mainly funds and/or sample availability. Extensive sub-proteome analysis is time and man-hour, and conclusively also analyte-consuming and this is often underestimated. Therefore, it is not too surprising that proteomics is most successful in bundled efforts with a solid financial basis.

References

[1] Corthals, G.L., Wasinger, V.C., Hochstrasser, D. F., Sanchez, J.-C. 2000 *Electrophoresis* 21: 1104-1115.

[2] Anderson, N.L. and Anderson N.G. 2002 *Molecular and Cellular Proteomics* 1.11: 845-867.

[3] Gygi, S.P., Rochon, Y., Franza, B.R., and Aebersold, R. 1999 *Molecular and Cellular Biology* 19, 3: 1720-1730.

[4] Alban A, David SO, Björkesten L, Andersson C, Sloge E, Lewis S, Currie I. 2003 *Proteomics* 3:36-44.

[5] Marouga R, David S, Hawkins E. 2005 *Anal. Bioanal. Chem.* 382: 669-78.

[6] Yan JX, Devenish AT, Wait R, Stone T, Lewis S, Fowler S. 2002 *Proteomics* 2: 1682-1698.

[7] Stryer, L. *Biochemistry*, 4 ed.; W.H. Freeman and Co.: New York, 1995.

[8] Yabe-Nishimura, C. 1998 *Pharmacological Reviews* 50: 21-34.

In: Biomacromolecular Mass Spectrometry Yearbook. Volume 1 ISBN: 978-1-60876-167-8
Ed: Simone Koenig

Chapter 9

MÜNSTER CONFERENCE ON SINGLE CELL ANALYSIS

PROGRESS IN RESEARCH AND TECHNOLOGY MÜNSTER, DECEMBER 1-2, 2005

Organizers: Interdisciplinary Center for Clinical Research, Core Facility Integrated Functional Genomics, Münster, Germany; Kurt Sieberns, Martin Eisenacher, Simone König

Chairs: Heinrich F. Arlinghaus, Klaus Dreisewerd, Cornelia Denz, Simone König

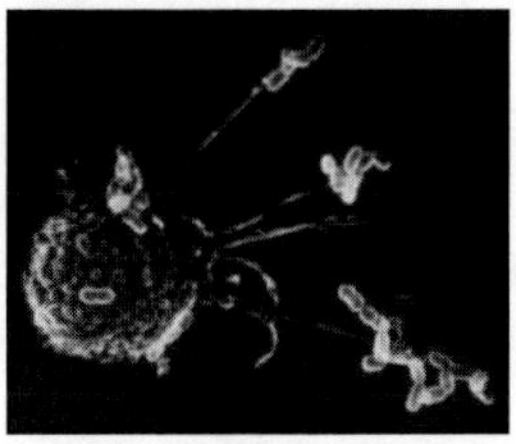

SPEAKER ABSTRACTS

DOWNSCALING ANALYSIS: SINGLE CELLS AND MOLECULES

Simone König

The Münster Conference Series on Single Cell Analysis brings together scientists of very different fields – physicists, physicians, biologists, chemists – with the common goal of studying events on a cellular, sub-cellular, and molecular level. The impact of mass spectrometry in that area is increasing and, therefore, the conference will be monitored in this journal.

This 2nd conference underscored that the TOF-SIMS (time-of-flight secondary-ion mass spectrometry) technique grows in importance in the biological field. J. Hoheisel and H. Arlinghaus collaborate to develop an alternative microarray detection method which avoids labelling and amplification. Their efforts for protein microarrays have been recognized on an

European level and the project is part of the Framework 6 MolTools Project as is J. Kochs area of research as detailed in his abstract.

While mass spectrometric imaging and profiling continue to be improved and developed as illustrated by one manufacturer as well as the neuroscientists Ka Wan Li and J. Bergquist, for other analytical demands mass spectrometry is not yet a suitable detector. Although capillary electrophoresis is a recognized separation technique in conjunction with mass spectrometry, on the single cell level researchers such as E. Arriaga still have to rely on fluorescence detection.

Events in a single cell are not representative for cell populations and cell-to-cell variation needs to be considered. Nevertheless, insights on this level drive our understanding as well as our technology and take bioinformatics along with it. It is fantastic to watch this process.

Quantitative Single Cell Molecular Biology

James Eberwine, Kevin Miyashiro, Lindy Barrett, Theresa Kannanayakal, Jeanine Jochems, Tiina Peritz, Kalle Kilk, Emelia Eriksdottir, Ulo Langel, Jai-Yoon Sul, Hajime Takano, Phillip Hayden, Elizabeth Van Bockstaele and Jason Glanzer

University of Pennsylvania, University of Stockholm, Thomas Jefferson University

There are an estimated 1012 neurons and glia in the mammalian brain, the interactions of which give rise to the functional central nervous system. To examine specific neuronal systems we have been developing methodologies that permit single cell resolution and quantitation of mRNA and protein abundances. These approaches including single cell aRNA amplification, single cell PCR and IDAT (immunodetection amplified by RNA polymerase) have recently been complimented with approaches that permit analysis of interactions between proteins and RNA. Using a combination of these methodologies, single dendrite transfection, APRA (antibody positioned RNA amplification) and PAIR (PNA-assisted identification of RNA binding proteins), we have been able to prove that protein synthesis is differentially regulated in distinct subcellular regions and have been able to show that RNA binding proteins differentially bind to the same RNA dependent upon particular neuronal stimulation paradigms. Application of these methodologies to the characterization of live cell physiologies as well as fixed cell biochemical states will be presented. Finally the use of single cells for functional genomics analysis, using novel methodologies, will also be discussed.

HETEROGENEITY IN THE MITOCHONDRIA OF SINGLE CELLS REVEALED BY INDIVIDUAL ORGANELLE CAPILLARY ELECTROPHORETIC ANALYSIS

Edgar A. Arriaga
University of Minnesota, Minneapolis

Capillary electrophoresis with laser-induced fluorescence detection (CE-LIF) has been previously utilized to investigate properties of whole single cells. In these studies, the information at the subcellular level is lost because the entire cell contents are solubilized and mixed. In this presentation, we describe a novel approach to analyze by CE-LIF individual mitochondria after their release from a single mammalian cell. The results of this analysis are data clusters that reveal the heterogeneity in the organelle population of the analyzed cell. Data clusters that contain the number of detected organelles per cell, their individual electrophoretic mobilities, and their individual organelle properties measured via fluorescent reporters will be illustrated with studies using the human osteosarcoma 143B and leukemia CCRF-CEM cell lines. The implications of detecting the anti-cancer agent doxorubicin in individual mitochondria from single CCRF-CEM cell lines will also be discussed. The single cell data clusters of individual organelle CE-LIF measurements are a new approach to more comprehensively describe single cell properties.

GENE EXPRESSION PROFILING OF MICRODISSECTED HODGKIN AND REED/STERNBERG (HRS) CELLS

Ralf Küppers
University of Duisburg-Essen Medical School, Essen

In classical Hogkin lymphoma (cHL) the neoplastic HRS cells represent <1% of the lymph node cellularity. Attempts to reveal the largely unknown pathogenesis of cHL through gene expression profiling have so far been restricted to HL cell lines. However, these lines most likely do not retain all important features of primary HRS cells. To generate gene expression profiles from primary HRS cells, ~1000 HRS cells were laser-microdissected from H&E-stained frozen sections of cHL biopsies. After two rounds of *in vitro* transcription, RNA was hybridized to Affymetrix HG-U133 Plus 2.0 chips. Expression profiles were also generated from similar cell numbers of HL cell lines, microdissected tumour cells from various other B cell lymphomas and lymphocyte-predominant HL (LPHL), and normal mature B-cell subsets that were MACS/FACS-sorted from tonsil or peripheral blood of healthy donors. Unsupervised hierarchical clustering of the first 55 samples grouped the 22 normal B-cell samples separately from the 33 tumour samples. The further branching of the dendrogram showed that each of the four B-cell subsets tended to form discrete clusters, and that, among tumour samples, cell lines grouped apart from primary cases. The latter were further split in two sub-branches: one with diffuse large B cell lymphomas and follicular lymphomas, and the other mainly comprising HLs. A preliminary supervised comparison of primary HRS with HL cell lines showed a highly differential expression ($\geq$ 4fold change) of ~1200 genes, including many involved in intercellular signalling, chemotaxis, and immune/inflammatory response. These preliminary results suggest that expression profiles can be reliably generated from small numbers of microdissected cells, and that primary HRS cells and HL cell lines differ in a number of biological features.

COMBINED GENOME AND TRANSCRIPTOME ANALYSIS OF SINGLE DISSEMINATED CANCER CELLS

Christoph A. Klein
Ludwig Maximilian University of Munich

Single disseminated cancer cells reside in various organs after so-called "curative" surgery and may eventually grow out to lethal metastases. Adjuvant therapies are currently tested to prevent metastatic outgrowth, however detailed knowledge about the target cells is lacking. Unfortunately, the analysis of the progenitor cells of metastases is hampered by their extreme rareness, because they are detected in bone marrow or lymph nodes at a frequency of 10^{-5} to 10^{-6}. Therefore, we started to develop techniques for the study of single cells and to investigate the early stages of systemic tumor progression. Thus far we succeeded in establishing protocols for global analysis of the genome and transcriptome of a single cell. While genome analysis was initially established for comparative genomic hybridization on metaphase chromosomes, recent development now enables the high-resolution analysis using BAC arrays. Global gene expression analysis of single cells using high-densitiy oligonucleotide arrays has been achieved by several modifications of our previously published protocol. The first results obtained from samples of cancer patients already have an impact on current models about systemic cancer progression. We show that dissemination is a very early event in the genomic and morphologic development of a tumor and provide novel strategies to search for efficient target structures of adjuvant therapies.

TOWARD SINGLE-MOLECULE DETECTION ON DNA-, RNA- AND PROTEIN-LEVEL

Jörg Hoheisel
German Center for Cancer Research (DKFZ), Heidelberg

In the DKFZ Division of Functional Genome Analysis, we are developing technologies for the identification, description and evaluation of cellular functions and their regulation by producing and processing biological information on a genomic scale. One emphasis in our efforts is work on DNA-, protein- and peptide-microarrays. Many chemical and biophysical issues are being addressed as part of this work in an attempt to understand the underlying procedural aspects, thereby eventually establishing superior analysis procedures. Concerning the analysis of human material, systems are being developed toward early diagnosis, prognosis and evaluation of the success of disease treatment with accentuation on cancer.

An important issue is the sensitivity of the analyses not only for the purpose of detecting small numbers of molecules but also for achieving a good measurement dynamics. Based on self-assembly processes, we designed a system that allows a very sensitive detection of chip-based analyses, avoiding both labelling and amplification prior to microarray hybridisation. In another approach, secondary ion mass spectrometry is being applied for detecting binding events directly, again avoiding any amplification or labelling. Last, also in protein-protein binding assays, we established conditions, which led to a significant improvement of detection limits.

SINGLE CELL AND SINGLE MOLECULE ANALYSIS IN CELL AND TISSUE PREPARATIONS

Jørn Koch
Aarhus Sygehus

Molecules do not function in cells by simply existing, but rather by interacting with and affecting other molecules. It is therefore important to know where a molecule is, when it is there, and what other molecules are in the neighbourhood. A rather simple example of this is an mRNA molecule with a mutation disabling its export into the cytoplasm. While such a molecule would exist, possibly in great abundance since the cell might increase its production in response to a lack of effect, it would get no further than the cellular nucleus, and therefore would be unable to mediate the synthesis of the appropriate protein. It is similarly important to know the context in which a cell exists. To control and modulate the functions of the individual cells, any multicellular organism relies on communication between cells. Therefore, no cell exists in splendid isolation, but is constantly influenced by its neighbours. To understand what goes on inside a cell, it may therefore be necessary to know what other cells are present and what they do. A prime example of this is found in embryology, where one single starting cell eventually develops into a great variety of vastly different cells depending upon what organ or tissue they end up in. Many pathological processes can be understood in this context as well. This is not least the case for cancer, in which cancer cells, among other things, rely on ignoring appropriate signalling and on producing their own signalling to fool the organism into not eradicating them.

For the analysis of this, we are developing new detection reactions with probes capable of resolving minor molecular details in bio-molecules and amplification reactions capable of visualising single probe molecule *in situ*. The technology uses oligonucleotide probes to identify targets, if necessary at single nucleotide resolution, and localised DNA synthesis at sites binding the probe as the amplification medium. More particularly, various formats of circular hybridisation probes are employed for target identification, whereupon the target molecule primes an "endless" DNA synthesis in a rolling circle format copying the probe over and over again to produce a long tandem array of DNA extending from the hybridisation target and covalently linked to it. The circular probes can resolve single nucleotide differences in the target molecules, and the rolling circle DNA synthesis produces enough copies of the probe to enable single probe molecules to show up under the microscope.

Analysis of a Simple Neural Circuit by Single Cell Profiling

Ka Wan Li
Vrije University, Amsterdam

Neuropeptides form a large group of signalling molecules with multiple effects on the target cells. Given that even closely apposed neurons may have distinct peptide contents and serve different functions, it becomes clear that the analysis of peptides at the single cell level is the first step toward the understanding of the functioning of the neuronal circuit of interest. In the present study, we utilized the fact that functionally connected neurons share a common nerve innervating the same or overlapping targets, and therefore, they can be retrograde-labelled from the nerve. Next, these labelled cells can be isolated for subsequent analysis. To demonstrate proof of principle, we analyzed single stained neurons from the parietal ganglion of several individual freshwater snail. Direct single cell MALDI-TOF-MS peptide profiling reveals the presence of several putative peptides. From a single labelled neuron, the peptides were structurally characterized by MALDI time-of-flight/time-of-flight MS/MS. Single cell peptide profiling is (semi)quantitative; the relative signal intensities of the distinct peptides displayed on the mass spectrum correspond closely to the expected ratio of these peptides expressed in the cell. Currently, hundreds of single neurons have been analyzed, and reveals that closely apposed neurons often contain overlapping sets of peptides suggesting that these cells may have similar functions. LC-MS/MS analysis of peptides contained in the nerve confirms the transport of these peptide messengers to the target organ.

Combined Optical Trapping and Chemical Analysis of Single Cells Within a Capillary

Dave Barrett
University of Nottingham

An approach is described which has potential for combined trapping and chemical analysis of a single cell. Optical tweezers have been used to facilitate the rapid trapping of cells within a narrow glass capillary. A single cell can be isolated and held using a infrared Nd: YAG laser beam focussed through the capillary at a point just above the inner surface. Preliminary studies with polystyrene microspheres (radius 6.8 μm) established the basic parameters required for successful optical trapping. Individual yeast cells (*Saccharomyces Cerevisae*) could be isolated from an injected 100 nL suspension of many thousands of cells, the process taking 2-3 minutes. The relationship between capillary flow rate and the power of the optical trap was measured to optimise the conditions for isolation of single cells. A single cell could be readily held in the optical trap for periods of tens of minutes or more and it was established that cell viability was not compromised by monitoring yeast budding rates.

This process does not require special micro-manipulation skills; the cells can simply be injected directly into the capillary. This allows rapid cell isolation even for inexperienced users. One a cell is trapped within the capillary, it offers the possibility of directly analysing release of biomolecules from the cell or the entire cell contents after *in situ* lysis using integrated capillary electrophoresis with high sensitivity detection using laser-induced fluorescence or mass spectrometry.

ANALYTICAL TOOLS FOR THE SINGLE CELL

Jonas Bergquist
Uppsala University

Mammalian tissues are composed of very heterogeneous assortments of different cell types and their biologically consequential analysis necessitates the preparation of pure samples of the cells of interest. A variety of microdissection techniques have been used in attempts to overcome the heterogeneity of tissues. Initially, these involved scraping off selected parts the tissue of interest from a histological section, very slow, tedious, and highly operator dependent approaches. In recent years new methods for single cell analysis have improved our possibilities for in depth analysis of what is happening in cells, and why. Clearly cytomics (and its sub-disciplines proteomics and peptidomics), if defined as the global analysis of cell function (or protein and peptide content respectively) at the cellular level, becomes an important addition to the available tools in cell biology. To fully appreciate e.g. single-cell proteomics, consider that a typical mammalian cell is a mere 10 μm in diameter, has a volume of about 500 fL, and contains about 50 pg (2 femtomoles) of total protein. In this presentation mass spectrometric approaches to obtain "fingerprints" of a single-cell proteome or peptidome will be presented. We have developed a methodology able to study on-target grown single cells using MALDI-TOF MS technology that enables studies of variation in protein expression among individual cells, cell-cell interactions, and to follow on-target manipulations of cells including e.g. stimulation with growth factors. Yet another methodological approach utilises a highly sensitive Fourier transform ion cyclotron resonance mass spectrometry (FTICR MS) system for protein analysis in minimal amounts of spinal cord from patients with neurodegenerative disease. This approach also includes the technique of laser microdissection in combination with pressure catapulting (LMPC) for the dissection of samples and specific neurons. This proteomic approach promises to allow the identification of qualitative or quantitative differences in proteins between normal and diseased cells. As well as providing insights into pathogenesis, such information might yield new screening or therapeutic targets.

INTRACELLULAR DYNAMICS OF SINGLE MOLECULES

Ulrich Kubitscheck
University of Bonn

The nuclei of mammalian cells are highly organized structures, in which complex tasks like DNA transcription and replication occur in dedicated nuclear domains. We examine the structure and dynamics of such molecular assemblies at the single-molecule level with a time resolution of ≥ 2 ms and spatial precision of ≥ 20 nm using state-of-the-art fluorescence microscopy. New results on the dynamics of the large pore complex in the nuclear envelope, on functional splicing factors, and on non-functional tracer molecules within the cell nucleus will be presented. These examples show that the light-microscopic imaging, tracking and nanolocalization of single protein molecules allows fascinating insights into the intracellular pathways of single protein factors, the structure and dynamics of supramolecular arrangements and the elucidation of biomolecular interactions at the single cell level.

MULTIPARAMETER FLUORESCENCE SPECTROSCOPY AND IMAGING WITH SINGLE-MOLECULE SENSITIVITY

Claus A. M. Seidel
Heinrich-Heine-University, Düsseldorf

Using a confocal fluorescence microscope the newly developed multiparameter fluorescence detection (MFD) enables us to simultaneously collect all fluorescence information [1-4]. MFD is applied to perform single-molecule fluorescence-resonance-energy-transfer (FRET) studies on biological systems labelled with a fluorescent donor and acceptor dye. Thus, it is possible to circumvent the classical pitfalls of the FRET method in ensemble measurements (incomplete labelling, uncertainty about assumptions concerning dye anisotropy and quantum yield). Direct access to the time trajectories of the FRET efficiency allows one to reveal the dynamics of the system in real time. Work on HIV-1 Reverse Transcriptase: DNA/DNA complexes and the SNARE-protein Syntaxin and on various DNAs will be presented showing that MFD has developed to a powerful tool in single-molecule biophysics [5-8]. Moreover it will be shown that the MFD principles can also be applied to imaging.

[1]. Kühnemuth, R. & Seidel, C. A. M. (2001) *Single Mol.* 2, 251-254.

[2]. Eggeling, C., Berger, S., Brand, L., Fries, J. R., Schaffer, J., Volkmer, A. & Seidel, C. A. M. (2001) *J. Biotechnol.* 86, 163-180.

[3]. Eggeling, C., Schaffer, J., Seidel, C. A. M., Korte, J., Brehm, G., Schneider, S. & Schrof, W. (2001) *J. Phys. Chem. A* 105, 3673-3679.

[4]. Felekyan S., Kühnemuth R., Kudryavtsev V., Sandhagen C., Becker W. & Seidel C. A. M.; (2005). *Rev. Sci. Instr.* (2005) in press.

[5]. Widengren, J., Schweinberger, E., Berger, S. & Seidel, C. A. M. (2001) *J. Phys. Chem. A* 105, 6851-6866.

[6]. Rothwell, P. J., Berger, S., Kensch, O., Felekyan, S., Antonik, M., Wöhrl, B. M., Restle, T., Goody, R. S. & Seidel, C. A. M. (2003) *Proc. Natl. Acad. Sci. U. S. A.* 100, 1655-1660.

[7]. Margittai, M., Schweinberger, E., Widengren, J., Fasshauer, D., Schroder, G., Haustein, E., Felekyan, S., König, M., Grubmüller, H., Jahn, R., Seidel, C. A. M. (2003) *Proc. Natl. Acad. Sci. USA*. 100, 15516-15521.

[8]. Diez, M., Zimmermann, B., Börsch, M., König, M., Schweinberger, E., Steigmiller, S., Reuter, R., Felekyan, S., Kudryavtsev, V. Seidel, C. A. M., Gräber, P (2004) *Nat. Struct. Mol. Biol.* 11,135-141.

DIGITAL HOLOGRAPHIC MICROSCOPY - A NEW TECHNOLOGY FOR LIVING SINGLE CELL ANALYSIS

Gert von Bally
Westphalian-Wilhelms-University, Münster

Marker free, in vivo screening methods actually gain increasing importance, as well as stand alone analysis technique as in combination with fluorescence labelled methods as reference. Especially nanoholographic methods like digital holographic microscopy provide simultaneously a label free, non destructive, non contactive, full field quantitative online 3D-imaging of living cells with an axial resolution of several nanometers. Therefore, nanointerferometric holographic techniques open new possibilities for the analysis of morphology, form and volume, specifically the analysis of new parameters in dynamic processes like volume changes, micromovements and (localisation of) elasticity differences at supracellular, cellular and subcellular level. In this way, important new application domains in dynamic living cell analysis in the fields of biophotonics, life sciences and pharmacy are opened. At the Laboratory of Biophysics, University of Muenster, modular holographic, nanointerferometric microscopy systems in combination with conventional fluorescence microscopes, laser scanning microscopes and laser micromanipulation systems were developed in cooperation with partners from industry, biology and medicine. Investigations of migration and reaction to drug trafficking of living tumor cells demonstrate a new way of functional imaging by holographic analysis of dynamic processes down to the nanometer scale.

GENE EXPRESSION PROFILING OF SINGLE AND FEW CELLS SORTED BY MACS, FACS AND MICRODISSECTION

Andreas Bosio
Miltenyi Biotec GmbH, Cologne

The molecular analysis of single or few cells is a fascinating topic which is gaining increasing awareness in the scientific community. The demands and possible applications are increasing as both, technologies for cell sorting via MACS, FACS and microdissection, as well as tools for molecular analysis like microarrays, have reached a high level.

Stem cells, grafted cells or certain neurons have individual characteristics and must therefore sometimes be analysed as a single cell. Furthermore, multiparametric sorting of an ever increasing number of different cell subsets via MACS or FACS results in decreasing numbers of cells for downstream analysis.

We have established a single cell amplification procedure (SuperAmp) combining bead based mRNA extraction and conversion, global PCR amplification and Klenow fragment labelling. Our amplification protocol circumvents two main experimental obstacles of published protocols: i) all cDNA fragments have a comparable length avoiding a PCR bias due to different transcript length ii) PCR amplification is performed with one single primer avoiding PCR bias due to different primer annealing conditions. To gain sensitivity, all reactions are performed in a volume of 10 - 25 μl.

In more than 200 set up experiments we have varied and optimised all possible parameters. The resulting protocol shows a robust amplification down to a single cell. Regression coefficients of samples independently amplified and subsequently hybridised to the same array as well as Pearson correlation coefficients of repeated hybridisations of different cells hybridised against each other are in the range of 0.2 for a single cell, > 0.6 for 10 cells and > 0.9 for 100 cells.

The amplification procedure has been used to perform repeated analysis of 12 different T cell subsets sorted by FACS from the same donor allowing to dissect and compare differentiation pathways of T cells.

Furthermore, analysis of four different Natural Killer cell subsets sorted by novel magnetic cell sorting (MACS) methods from the same donor have illuminated underlying molecular mechanisms of NK cell identity and reactivity.

The analysis of laser microdissected samples from snap frozen and paraffin embedded tissue has also shown very promising results which are currently repeated and evaluated.

HIGH-SPEED MALDI-TOF IMAGING: THE FUTURE OF BIOMARKER DISCOVERY AND PROTEOMICS?

Wolfgang Albrecht, Martin Schürenberg, Sören-Oliver Deininger, Arne Fütterer and Detlev Suckau

Bruker Daltonics, Bremen

Biomarkers detected by clinical proteomics approach-es are expected to increase the sensitivity and specificity of diagnostic assays for various diseases. However, the direct analysis of serum/plasma did not provide reliable biomarkers so far that were of diagnostic or prognostic value, as the actual analytical sensitivity is not sufficient to deal with typical biomarker con-centrations. The con-centration of such biomarkers in diseased tissue, such as cancer, is expected to be a lot higher and more amenable to mass spectrometric detection. Therefore, tissue as a direct platform for marker dis-covery is becoming very interesting and its direct analysis by MALDI imaging.

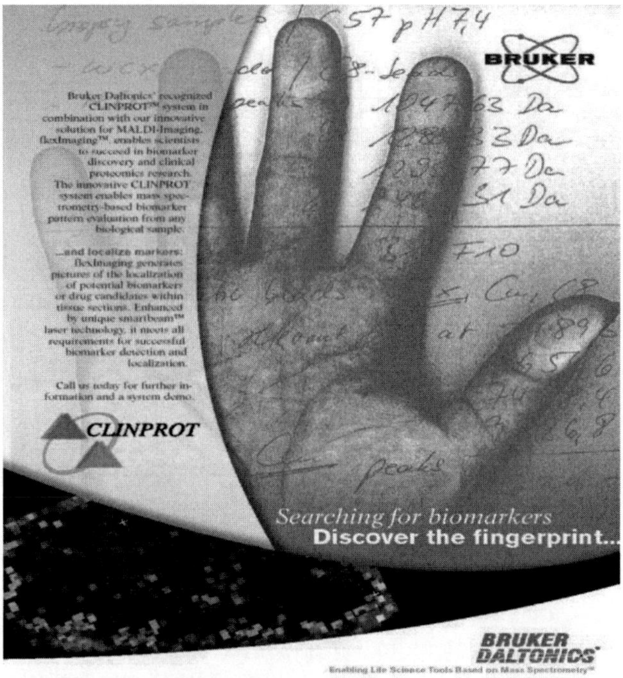

MALDI imaging allows the characterization of the protein composition of a number of cells covering approx. 500-5000 µm² - at best 20 µm² - on the MALDI sample carrier. Typically, thin tissue slices from a microtome are deposited on an electrically conductive microscope slide, photographed and then covered by MALDI matrix. The prepared tissue on the microscope slide is then scanned by MALDI with a 200 Hz Smartbeam or a 50 Hz N_2 laser and the spectra are displayed as a heatmap for each selected molecular weight. These maps allow to determine the special distribution of proteins and other biomolecules in the mass range of 500-20,000 Da.

However, the identification of detected biomarkers still requires the isolation of regions of interest from the tissue (e.g., by laser capture micro dissection) and the subsequent isolation/digestion of the markers for subsequent MS analysis.

First examples from recent work will be discussed, covering technical aspects as well as applications in cancer diagnostics and fundamental research.

New Horizons by the Use of Parallel Qualitative and Quantitative Proteomics Without Isotopic Labelling

Marc Kipping
Waters GmbH, Eschborn

Proteins are the key players in living processes of the cell. Mass spectrometry proved to be an ideal technique to identify and characterize cellular proteins with high sensitivity, therefore proteomics was mainly based on qualitative mass spectrometric techniques in the last years. In order to understand the complex interactions from thousands of proteins within a cell it becomes more and more important to get information about quantitative changes of protein expression levels in different cell states or under distinct cellular conditions. So far approaches have been used which utilise stable isotope labelling of samples with separation and subsequent analysis by mass spectrometry. Despite the ability of these techniques to quantitatively compare protein levels between samples and across conditions, complexity of samples, sample amount, speed of analysis and costs are limiting factors. To improve these shortcomings Waters has introduced a label-free technique. The Protein Expression System is an exact mass LC-MS strategy, where quantification is achieved via normalisation of the LC-MS datasets and comparison of the observed tryptic peptide intensities across samples. This completely new approach allows to open up new possibilities in proteomic research.

LASER-BASED ISOLATION OF SINGLE LIVING CELLS – NEW APPROACHES IN CELL AND MOLECULAR BIOLOGY

Yilman Niyaz, M. Stich, A. Buchstaller, R. Gangnus, R. Burgemeister

P.A.L.M Microlaser Technologies AG, Bernried

The understanding of the molecular mechanisms of cellular interactions depends on identification, isolation and finally characterization of a single cell or a population of cells and thereafter of their specific subsets of biomolecules. Thus, the generation of homogeneous samples is a very crucial step for high-resolution downstream applications.

Laser Microdissection and Pressure Catapulting (LMPC) is a well-known method to isolate and collect selected cells from complex human/animal or plant tissues, resulting in an eminent increase in the specificity of downstream molecular analyses like PCR and RT-PCR amplification of DNA or RNA, respectively, and protein analysis via, e.g., MALDI/SELDI spectrometry.

With laser microdissection, focused laser light is utilized to excise selected tissue areas or single cells of a specimen down to a resolution of subcellular components like organelles or chromosomes, respectively. After microdissection, the separated sample is directly transferred against gravity into an appropriate collection device by using the same laser (Laser Pressure Catapulting, LPC). As this process is working entirely without any mechanical contact, LMPC enables pure and homogeneous sample retrieval from e.g., paraffin and cryosections, cell smears and virtually any other preparation on microscope slides.

An important innovation is PALMs laser driven isolation of cultured cells. The work with selected living cells is extremely facilitated with this new approach and opens a wide field of new applications and research possibilities. Small groups or individual cultured cells can be separated by morphology or label for direct molecular analysis, clonal expansion or re-cultivation (e.g., stem cells).

Here we present the proof-of-principle for generation of pure stem cell cultures from a heterogeneous, mixed starting culture. Two different murine stem cell lines, either expressing a specific marker (CD-34) or not, were co-cultured. CD-34 expressing cells were then separated by LMPC from those not expressing this marker. These isolated cell types were individually re-cultivated. It could be shown, that generation of pure stem cell lines is feasible with LMPC assisted isolation.

The applied UV-A laser does not affect the viability of the specimens: There is no impairment of viability or biological information, even after several cycles of catapulting and recultivation, as is proven by, e.g., Comparative Genomic Hybridization (CGH) or Multiplex Fluorescence In-Situ Hybridization (M-FISH) analysis.

The efficacy of this method was also recently shown by LMPC-mediated isolation of transgenic murine embryonic stem cells, which were successfully used to generate chimeric mice.

In: Biomacromolecular Mass Spectrometry Yearbook. Volume 1 ISBN: 978-1-60876-167-8
Ed: Simone Koenig

Chapter 10

MASS SPECTROMETRY APPLICATIONS IN VIROLOGY*

Alexander Hahn **and Frank Neipel***
Institute for Clinical and Molecular Virology,
University of Erlangen, Nürnberg, Germany

ABSTRACT

Mass spectrometry has found a growing number of applications in many disciplines of the life sciences. This review will focus on the field of molecular and clinical virology. Examples for established as well as potential applications for mass spectrometry techniques will be given from different fields of virology, among them receptor identification, identification of cellular interaction partners, post translational modification analysis, virus particle structure and clinical applications.

The rapid advancement in the field of mass spectrometry (MS) and proteomics has found its way into virology as documented by recent publications. MS is providing valuable new tools for the identification of cellular receptors and other interaction partners, post-translational modifications of viral and cellular proteins, or components of the virus particle itself. It provides new possibilities for the genotyping of viruses and might soon play an important role in the discovery and clinical monitoring of markers for disease progression and therapeutical response. It may also be used for the diagnostic identification of viral infections. This review gives a short overview over applications for MS in the field of virology (see figure 1 for synopsis).

---

* A version of this chapter was also published in *Biomacromolecular Mass Spectrometry Volume 1, Number 3*, published by Nova Science Publications, Inc. It was submitted for appropriate modifications in an effort to encourage wider dissemination of research.

* Corresponding Author: Alexander Hahn; Institut fuer Klinische und Molekulare Virologie; Schlossgarten 4, 91054 Erlangen; Ph. 09131 8526496; Fax: 09131 8526493; Email: arhahn@viro.med.uni-erlangen.de

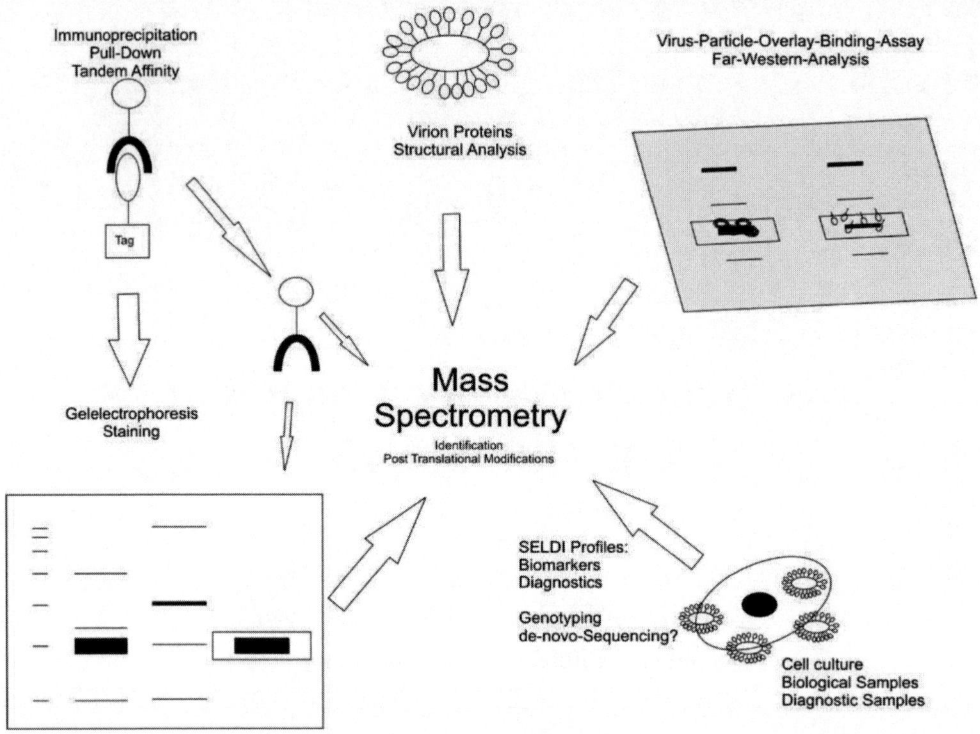

Figure 1. Overview of established and potential applications for MS in a virological context.

VIRUS HOST-INTERACTIONS

The increase in availability of high precision mass spectrometers for MALDI-TOF and ESI-MS/MS together with the growth of protein databases now makes it possible to identify proteins even in complex mixtures. Protein amounts roughly corresponding to a silver stained spot or gel band are sufficient. This allows for a new range of strategies to identify protein-protein interactions.

The first step in any virus infection is attachment to and entry of the host cell. For this purpose, viruses engage one or more receptors on the cell surface. Subsequently, viruses highjack and exploit the cellular machinery to accomplish either lytic replication or silent persistence. Retroviruses even become an integral part of the cellular genome. As a consequence of this parasitic behaviour and their almost complete dependence on cellular functions, viruses have to interact very closely and in a defined manner with cellular proteins to influence cellular processes. Identification of these interaction partners provides valuable information about the viral life cycle and pathogenesis. It can even give cues for drug development. Cellular drug targets are of particular interest, as they are not subject to the high mutation rate of many viruses. Thus, development of drug resistance is less likely.

Interactions at the Plasma Membrane and Receptors

The classical approach to identify virus receptors employs recombinant DNA technology. If both permissive and non-permissive cells are available, cDNA libraries can be generated from permissive cells and are transfected into non-permissive cells. The next step is the generation of - usually - retro- or lentiviral particles pseudotyped with viral surface proteins [1] or fusion competent cells carrying viral surface proteins [2]. In case non-permissive cells are rendered infectable by expression of a protein from transfected cDNA, a selection marker is transferred to these cells by infection or fusion. Alternatively, cells carrying cDNA-clones that enable them to bind recombinant viral protein can be expanded [3-5]. Another approach is to test antibodies with known specificity against putative receptors for effects on viral infection [6]. This approach is quite straight-forward and may yield results more quickly. However, it is less reliable as it requires an “informed-guessing” of the receptor.

The possibility to directly identify proteins by mass spectrometry provides promising new ways for receptor identification. It is particularly useful where neither clear-cut permissive nor clear-cut non-permissive cells are available, and/or when biological properties make guessing of the receptor difficult.

An elegant approach is to generate a panel of monoclonal antibodies using permissive cells or tissues as immunogen and screen this panel for the ability to block infection [7]. Once such a clone has been identified, immunoaffinity procedures should deliver the right antigen without too much effort [8]. Affinity procedures can also be used to enrich and purify cellular receptors out of complex lysates by using the viral surface protein as bait. In addition, this allows for the identification of surface molecules engaged by the virus in processes other than fusion or entry. These affinity procedures can encompass methods quite similar to ‘classical’ immunoprecipitation, e.g. using so-called ’immuno-adhesins’: fusion-constructs comprising viral surface protein(-fragments) and the Fc-part of immunoglobulins, a strategy which lead very quickly to the identification of ACE2 as the receptor for SARS-Coronavirus [9-11]. Another common approach is affinity chromatography. Usually, peptide-tags are used to immobilize the bait protein to a carrier matrix [12]. A combination of these techniques may be used to improve specificity (see table 1 for common tags). Briefly, after incubation of the bait with cellular lysate and several washing steps either the whole complex is eluted or the bait protein stays immobilized and anything bound is washed-off. These methods have in common that they require the viral protein to be recombinantly expressed. The latter frequently proves difficult e.g. due to unusual codon usage, improper post-translational processing, or toxicity of the protein for the producing cells.

Viral-Particle-Overlay-Binding-Assay (VPOBA) and Far-Western-Blot techniques seem promising for the identification of interaction partners, especially for those not amenable to immunoprecipitation. Cellular proteins are separated by electrophoresis and transferred onto a membrane. The membrane is then incubated with purified virus or proteins and these are subsequently detected on the membrane. Once an interacting protein band is found in VPOBA or Far-Western analysis, gel or membrane pieces of the corresponding molecular weight range can be prepared on a larger scale and are analyzed by LC-MS/MS [13].

Table 1. Common Protein Tags

Tag / Fused Protein	Description	Elution of Bait (tagged) Protein	Conditions under which Bait (tagged) Protein remains immobilized	Advantages	Disadvantages
IgG Fc	Fc-fusionprotein, very similar to an antibody, binds to Protein A matrices	Harsh conditions	High salt, partly denaturing	Easy to purify, very versatile, dimerization, relatively cheap, secretory pathway	Unspecific and harsh elution may damage proteins during purification, dimerization, secretory pathway
Peptide Tags: Myc, Flag, HA…	Short, usually non-interfering tag, binds to antibody matrices	Peptide elution	High salt	Easy to purify, mild and specific elution	Peptides very expensive, sometimes incomplete elution
Strep-Tag	Mimics biotin, binds to modified streptavidin	Biotin elution		Easy to purify, mild and specific elution	Rather expensive
His-Tag	Binds to immobilized metal-chelate via six or more histidines	Imidazole, low pH, chelating agents	High salt, maybe denaturing agents	Cheap, easy to purify	Sometimes low or no affinity for metall-chelate matrix, rather unspecific elution and binding
Calmodulin binding peptide	Binds to calmodulin matrices	Chelating agents		Specific elution	Calmodulin binding proteins may copurify
Glutathione-S-Transferase (GST)	Binds to Glutathione-matrices	Glutathione	High salt, denaturing	Mild and specific elution	?
Protein A	Binds to Ig-matrices	Harsh conditions	High salt, partly denaturing, cleavage	Easy to purify	Cleavage between ProtA and protein required, otherwise unspecific elution
TEV (tobacco etch virus) protease site	Cleaved by TEV-protease, used for elution by cleaving between protein and affinity tag	Proteolytic digest		Mild and specific elution	Site might be present in proteins, some incubation time necessary

Intracellular Interactions

In contrast to transmembrane proteins, most intracellular proteins are readily soluble and amenable to pulldowns or immunoprecipitation. Thus, techniques similar to those mentioned above for membrane-bound interactions can be applied. However, their usefulness may often be limited by the large number of interactions. As a consequence, single-step affinity procedures may be less suitable here as they frequently result in the identification of highly abundant intracellular proteins. Invention of the double-tag or tandem-affinity strategy [14;15] provides a more powerful tool to identify protein-protein interactions or to characterize protein complexes. Two independent affinity steps minimize false positive interactions. The double-tagged bait protein is preferably introduced into cells by retroviral [16] or lentiviral vectors. This directly points at the exceptional potential of this technique for the field of virology. Most viruses now can be genetically modified by means of reverse genetics [17], bacmid [18], or plasmid technology. Introducing a double-tagged viral protein this way would allow for the isolation of interacting proteins in the context of the viral lifecycle, thus avoiding more artificial circumstances like over-expression of single proteins in non-permissive cell lines. It is even conceivable that this technique in combination with viruses could be used to purify protein complexes from infected tissues of experimental animals.

Proteins interacting with viral nucleic acid can be prepared by affinity procedures using immobilized DNA or RNA probes to enrich interacting proteins. Again, isolated proteins can subsequently be identified by mass spectrometry [19;20].

Virus Particles and Structure

It is possible to identify proteins out of virus preparations by MS. This enables researchers to characterize the protein components of complete viral particles [21], the so-called virions. In a bottom-up approach, viral particles are directly digested and all peptides are analyzed directly by one- or two dimensional LC-MS [22]. Alternatively, virion proteins are pre-fractionated and resolved by SDS-PAGE before MS-analysis [23]. Recently, these techniques did not only lead to the identification of viral proteins not known before to be part of the virion [24;25], but also of surprisingly numerous cellular components incorporated into viral particles as well. Especially enveloped viruses regularly include several structural proteins including actin and myosin, as well as a number of heat shock proteins and cyclophilin A. Interestingly, they also incorporate regulatory host proteins, such as members of the ras-family. The latter could be identified in virions of Kaposi-sarcoma associated herpesvirus and vaccinia virus [26;27].

As MS is able to identify disulfide bridges [28], chemical crosslinks and other protein-modifications [29] by differing ion masses as well as absolute molecular weight of intact proteins, it should be possible to determine stochiometry [30], orientation, or neighbourhood of viral structural proteins in viral particles.

Orientation of proteins on the viral surface for instance can be examined by limited digestion of viral particles, thereby generating peptides from viral proteins exposed to the solvent and accessible to proteolytic treatment [31;32].

Post-Translational Modifications

As with cellular proteins, post-translational modifications are essential for the function of many viral proteins. For example, viral proteins may reach their fully functional state by post translational modification, or viral proteins influence the function of cellular proteins by modifying these cellular proteins at the post-translational level. For instance, certain spike proteins of coronaviruses are palmitoylated [33] and the HIV Nef protein fulfills some of its functions only when myristoylated [34]. Myristoylation has been shown to increase the affinity of the hepatitis B virus surface antigen for its cellular receptor [35]. Modifications such as phosphorylation, ubiquitination or sumoylation [36;37] seem to play major roles in the functionality of many viral proteins and their cellular interaction partners. Detection of these modifications by MS becomes increasingly feasible [38] and may provide completely new insights into the fine tuning of viral infection. Employing MS-techniques, phosphorylation status has recently been discovered as a key switch in RNA-packaging and nucleocapsid maturation of hepadnaviruses [39]. HHV-8 D-like cyclin has been shown to influence cellular cyclin-dependent kinases 4 and 6 towards a novel substrate specificity [40], just to name a few examples. Post translational modifications are of particular interest, as they point to unique cellular targets for drug development, for example kinase or phosphatase inhibitors. The suppression of protein modification pathways by drugs might prove very effective [41], especially combined with topical drug application.

Clinical Applications

MALDI-TOF can be used for genotyping of certain virus strains in combination with PCR [42]. As this technique is generally able to determine small mass differences of oligonucleotide samples, for example PCR products, it could be of general use for the genotyping of viruses and the detection of certain drug resistance variants. SELDI (surface enhanced laser assisted desorption/ionization) [43] MS seems promising for viral diagnostics as well as for the discovery of new markers for disease progression and treatment response in biological samples such as serum. A very rapid SELDI-MS based screening method for SARS in patients with pneumonia has been reported [44] and putative markers for disease progression and treatment response have been identified for hepatitis B [45], hepatitis C [46], and HIV [47]. The most striking difference of SELDI-MS compared to more conventional techniques lies in the complete independence from immunological or nucleic acid methods. Even previous knowledge about the nature of the disease may not be required. Marker peptides are identified by statistical comparison of their intensities in samples from different patient groups only.

Perspectives

As virus often accumulates in considerable amounts in the supernatant of infected cells, it is conceivable that it may be feasible to identify viruses from such cell culture supernatants by MS [48]. It might even be possible to identify peptide sequences from unknown pathogens, in case they are present in cell culture supernatants in sufficient quantity, very similar to the identification of biomarkers. With growing availability of mass spectrometers in combination with SELDI or different forms of HPLC combined with automated sample preparation, the number of applications for MS will continue to grow in the near future.

REFERENCES

Tailor, C. S.; Nouri, A.; Zhao, Y.; Takeuchi, Y.; Kabat, D. *J.Virol.* 1999, *73*, 4470-74.

Feng, Y.; Broder, C. C.; Kennedy, P. E.; Berger, E. A. *Science.* 1996, *272*, 872-77.

Shimojima, M.; Miyazawa, T.; Sakurai, Y.; Nishimura, Y.; Tohya, Y.; Matsuura, Y.; Akashi, H. *Analytical Biochemistry.* 2003, *315*, 138-40.

Jeffers, S. A.; Tusell, S. M.; Gillim-Ross, L.; Hemmila, E. M.; Achenbach, J. E.; Babcock, G. J.; Thomas, W. D., Jr.; Thackray, L. B.; Young, M. D.; Mason, R. J.; Ambrosino, D. M.; Wentworth, D. E.; DeMartini, J. C.; Holmes, K. V. *Proceedings of the National Academy of Sciences* 2004, *101*, 15748-53.

Shimojima, M.; Miyazawa, T.; Ikeda, Y.; McMonagle, E. L.; Haining, H.; Akashi, H.; Takeuchi, Y.; Hosie, M. J.; Willett, B. J. *Science.* 2004, *303*, 1192-95.

Nemerow, G. R.; Wolfert, R.; McNaughton, M. E.; Cooper, N. R. *J. Virol.* 1985, *55*, 347-51.

Naniche, D.; Wild, T. F.; Rabourdin-Combe, C.; Gerlier, D. *J. Gen. Virol.* 1992, *73 (Pt 10)*, 2617-24.

Naniche, D.; Varior-Krishnan, G.; Cervoni, F.; Wild, T. F.; Rossi, B.; Rabourdin-Combe, C.; Gerlier, D. *J. Virol.* 1993, *67*, 6025-32.

Li, W.; Moore, M. J.; Vasilieva, N.; Sui, J.; Wong, S. K.; Berne, M. A.; Somasundaran, M.; Sullivan, J. L.; Luzuriaga, K.; Greenough, T. C.; Choe, H.; Farzan, M. *Nature.* 2003, *426*, 450-54.

Chai, N.; Bates, P. *PNAS.* 2006, *103*, 5531-36.

Negrete, O. A.; Levroney, E. L.; Aguilar, H. C.; Bertolotti-Ciarlet, A.; Nazarian, R.; Tajyar, S.; Lee, B. *Nature.* 2005, *436*, 401-05.

Reyes-Del Valle, J.; Chavez-Salinas, S.; Medina, F.; Del Angel, R. M. *J. Virol.* 2005, *79*, 4557-67.

Thepparit, C.; Smith, D. R. *J. Virol.* 2004, *78*, 12647-56.

Puig, O.; Caspary, F.; Rigaut, G.; Rutz, B.; Bouveret, E.; Bragado-Nilsson, E.; Wilm, M.; Seraphin, B. *Methods.* 2001, *24*, 218-29.

Rigaut, G.; Shevchenko, A.; Rutz, B.; Wilm, M.; Mann, M.; Seraphin, B. *Nat Biotechnol.* 1999, *17*, 1030-32.

DeMasi, J.; Huh, K. W.; Nakatani, Y.; Munger, K.; Howley, P. M. *PNAS.* 2005, *102*, 11486-91.

Enami, M.; Luytjes, W.; Krystal, M.; Palese, P. *PNAS.* 1990, *87*, 3802-05.

Luckow, V. A.; Lee, S. C.; Barry, G. F.; Olins, P. O. *J. Virol.* 1993, *67*, 4566-79.

Choi, K.; Kim, J. H.; Li, X.; Paek, K. Y.; Ha, S. H.; Ryu, S. H.; Wimmer, E.; Jang, S. K. *Nucl.Acids Res.* 2004, *32*, 1308-17.

Si, H.; Verma, S. C.; Robertson, E. S. *J. Virol.* 2006, *80*, 9017-30.

Varnum, S. M.; Streblow, D. N.; Monroe, M. E.; Smith, P.; Auberry, K. J.; Pasa-Tolic, L.; Wang, D.; Camp, D. G., II; Rodland, K.; Wiley, S.; Britt, W.; Shenk, T.; Smith, R. D.; Nelson, J. A. *J. Virol.* 2004, *78*, 10960-66.

Chung, C. S.; Chen, C. H.; Ho, M. Y.; Huang, C. Y.; Liao, C. L.; Chang, W. *J. Virol.* 2006, *80*, 2127-40.

Yoder, J. D.; Chen, T. S.; Gagnier, C. R.; Vemulapalli, S.; Maier, C. S.; Hruby, D. E. *Virol. J.* 2006, *3*, 10.

Varnum, S. M.; Streblow, D. N.; Monroe, M. E.; Smith, P.; Auberry, K. J.; Pasa-Tolic, L.; Wang, D.; Camp, D. G., II; Rodland, K.; Wiley, S.; Britt, W.; Shenk, T.; Smith, R. D.; Nelson, J. A. *J. Virol.* 2004, *78*, 10960-66.

Chung, C. S.; Chen, C. H.; Ho, M. Y.; Huang, C. Y.; Liao, C. L.; Chang, W. *J. Virol.* 2006, *80*, 2127-40.

Chung, C. S.; Chen, C. H.; Ho, M. Y.; Huang, C. Y.; Liao, C. L.; Chang, W. *J. Virol.* 2006, *80*, 2127-40.

Zhu, F. X.; Chong, J. M.; Wu, L.; Yuan, Y. *J. Virol.* 2005, *79*, 800-11.

Schnaible, V.; Wefing, S.; Resemann, A.; Suckau, D.; Bucker, A.; Wolf-Kummeth, S.; Hoffmann, D. *Anal. Chem.* 2002, *74*, 4980-88.

S.Wefing, V.Schnaible, and D.Hoffmann. http://www.searchxlinks.de. 2001. Ref. Type: *Generic.*

Loo, J. A.; Holler, T. P.; Foltin, S. K.; McConnell, P.; Banotai, C. A.; Horne, N. M.; Mueller, W. T.; Stevenson, T. I.; Mack, D. P. *Proteins.* 1998, *Suppl. 2*, 28-37.

Phinney, B. S.; Blackburn, K.; Brown, D. T. *J. Virol.* 2000, *74*, 5667-78.

Bothner, B.; Dong, X. F.; Bibbs, L.; Johnson, J. E.; Siuzdak, G. *J..Biol..Chem.* 1998, *273*, 673-76.

Thorp, E. B.; Boscarino, J. A.; Logan, H. L.; Goletz, J. T.; Gallagher, T. M. *J. Virol.* 2006, *80*, 1280-89.

Harris, M. P.; Neil, J. C. *J. Mol. Biol.* 1994, *241*, 136-42.

De Falco, S.; Ruvo, M.; Verdoliva, A.; Scarallo, A.; Raimondo, D.; Raucci, A.; Fassina, G. *J. Pept. Res.* 2001, *57*, 390-400.

Lamsoul, I.; Lodewick, J.; Lebrun, S.; Brasseur, R.; Burny, A.; Gaynor, R. B.; Bex, F. *Mol.Cell. Biol.* 2005, *25*, 10391-406.

Izumiya, Y.; Ellison, T. J.; Yeh, E. T. H.; Jung, J. U.; Luciw, P. A.; Kung, H. J. *J. Virol.* 2005, *79*, 9912-25.

Wilkins, M. R.; Gasteiger, E.; Gooley, A. A.; Herbert, B. R.; Molloy, M. P.; Binz, P. A.; Ou, K.; Sanchez, J. C.; Bairoch, A.; Williams, K. L.; Hochstrasser, D. F. *J. Mol. Biol.* 1999, *289*, 645-57.

Perlman, D. H.; Berg, E. A.; O'Connor, P. B.; Costello, C. E.; Hu, J. *PNAS.* 2005, *102*, 9020-25.

Cuomo, M. E.; Knebel, A.; Platt, G.; Morrice, N.; Cohen, P.; Mittnacht, S. *J.Biol.Chem.* 2005, *280*, 35844-58.

Root, C. N.; Wills, E. G.; McNair, L. L.; Whittaker, G. R. *J. Gen. Virol.* 2000, *81*, 2697-705.

Hong, S. P.; Kim, N. K.; Hwang, S. G.; Chung, H. J.; Kim, S.; Han, J. H.; Kim, H. T.; Rim, K. S.; Kang, M. S.; Yoo, W.; Kim, S. O. *J. Hepatol.* 2004, *40*, 837-44.

Kuwata, H.; Yip, T. T.; Yip, C. L.; Tomita, M.; Hutchens, T. W. *Biochem. Biophys. Res. Commun.* 1998, *245*, 764-73.

Pang, R. T. K.; Poon, T. C. W.; Chan, K. C. A.; Lee, N. L. S.; Chiu, R. W. K.; Tong, Y. K.; Wong, R. M. Y.; Chim, S. S. C.; Ngai, S. M.; Sung, J. J. Y.; Lo, Y. M. D. *Clin. Chem.* 2006, *52*, 421-29.

Zhu, X. D.; Zhang, W. H.; Li, C. L.; Xu, Y.; Liang, W. J.; Tien, P. *World J. Gastroenterol.* 2004, *10*, 2327-29.

Paradis, V.; Asselah, T.; Dargere, D.; Ripault, M. P.; Martinot, M.; Boyer, N.; Valla, D.; Marcellin, P.; Bedossa, P. *Gastroenterology.* 2006, *130*, 2189-97.

Sun, B.; Rempel, H. C.; Pulliam, L. *AIDS.* 2004, *18*, 1009-12.

Krokhin, O.; Li, Y.; Andonov, A.; Feldmann, H.; Flick, R.; Jones, S.; Stroeher, U.; Bastien, N.; Dasuri, K. V.; Cheng, K.; Simonsen, J. N.; Perreault, H.; Wilkins, J.; Ens, W.; Plummer, F.; Standing, K. G. *Mol.Cell Proteomics.* 2003, *2*, 346-56.

In: Biomacromolecular Mass Spectrometry Yearbook. Volume 1 ISBN: 978-1-60876-167-8
Ed: Simone Koenig

Chapter 11

PROTEOMICS OF NOCARDIA AFRICANA (SD769) RECENTLY ISOLATED FROM PATIENTS WITH PULMONARY INFECTION IN SUDAN*

***Mogahid M. El Hassan*[*1]*, Kanury V. S. Rao*[2]*, Zaved Siddique*[2]*, Dinesh S. Kumar*[2]*, Rashmi Tickoo*[2]*, Nageeb S. Saeed*[3]*, Moawia M. Mukhtar*[4] *and Mohamed E.Hamid*[5]**

[1] College of Medical Laboratory Sciences,
Sudan University of Science and Technology, Khartoum, Sudan
[2] Immunology Group, International Center for Genetic
Engineering and Biotechnology (ICGEB) New Delhi, India
[3] General Directorate of Laboratories and Medical Researches,
Federal Ministry of Health, Khartoum, Sudan
[4] General Directorate of Institute of Endemic Diseases,
University of Khartoum, Khartoum, Sudan
[5] Department of Microbiology, College of Medicine,
King Khalid University, Abha, PO. Box 641,
Kingdom of Saudi Arabia

ABSTRACT

Nocardia africana is a pulmonary pathogen which has been recently isolated from human in Sudan. There is great need for proteomic investigations but they are sparse at present. First attempts on *N. africana SD 769* (DSM44491) visualized 74 protein of molecular weights between 10 -250 kDa and isoelectric points between 4–7. Due to the limited database information, identified proteins were assigned by homology to sequences from the other

* A version of this chapter was also published in *Biomacromolecular Mass Spectrometry Volume 1, Number 3*, published by Nova Science Publications, Inc. It was submitted for appropriate modifications in an effort to encourage wider dissemination of research.

* Corresponding author: Mogahid M. El Hassan, e-mail: greensudan70@yahoo.com

bacterial species such as *Staphylococcus aureus, Oceanobacillus iheyensis, N. farcinica* and *Yersinia pestis.*

Keywords: Nocardia, proteomics

Nocardiae are Gram positive aerobic actinomycetes, which are predominantly saprophytic [1,2], but also include species forming parasitic association with animals and plants [3,4]. They occur in a wide range of man made and natural habitat including activated sewage sludge, soil, water and tissues of plants and animals including human [5]. Nocardiae are known to cause a variety of localized and disseminated suppurative infections in humans and animals [6,7]. The most commonly reported pathogenic species are *Nocardia asteroides, N. farcinica* and *N. nova*, followed in importance by *N. brasiliensis, N. otitidiscaviarum, N. pseudobrasiliensis* and *N. transvalensis*. Infections caused by *Nocardia* species are infrequent but challenging to clinicians. In recent years, the number of case reports has been increasing, and this can be attributed to the improvements in diagnostic capabilities and the higher clinical index of suspicion accompanying the increased prevalence of immune-suppressed patients [7,8]. Although nocardiosis had been reported from most regions of the world it is well established that nocardiae can be easily overlooked under routine culture and smear examinations. Moreover, its incidence in many countries remains unknown.

Nocardiosis is usually considered to be a late-presenting immunity acquired infection [9], but there is growing evidence that the disease is transmissible [10]. Many cases with *N. asteroides* have been reported from liver [11] and renal transplant units [9,12]. In another outbreak it was suggested [13] that nosocomial air-born transmission, possibly in the operating room environment, was the cause of a cluster of *N. farcinica* post-operative wound infections in patients undergoing cardiac and other vascular surgeries at a university hospital. In France, it has been estimated that up to 250 new cases of nocardiosis occur each year [14].

It was also found that nocardiosis, although infrequent, is an important cause of morbidity and mortality in patients with cancer (immuno-compromised). It had pleomorphic clinical manifestations and it can be seen in breakthrough infections. Cytomegalovirus co-infection influences the outcome of patients with cancer and nocardiosis [15].

Primary pulmonary nocardiosis may be sub clinical or pneumonic; it may be chronic or acute with secondary involvement of other organs, mainly the brain. In non-tropical countries, most infections are caused by *N. asteroides, N. farcinica* and *N. nova*, relatively few *by N. brasiliensis, N. otitidiscaviarum* and *N. transvalenses. N. farcinica* shows a greater degree of virulence than *N. asteroides* [16]. Co-infection by *N. asteroides* and *Mycobacterium tuberculosis* was diagnosed in sporadic cases, which were treated by combined therapy for both diseases [17].

Nocardia africana, a new species of *Nocardia* was isolated from clinical specimens (sputum) obtained from patients suffering from pulmonary infection admitted at Chest Unit of Khartoum Teaching Hospital in Sudan. The isolates showed morphological features of actinomycetes and they have been assigned to the genus *Nocardia* according to their morphological criteria and chemical properties. The strains were considered to merit species status and were designated as *N. africana* sp. nov. [18] (figures 1 & 2).

The techniques of proteomics have also moved into microbiology. Two-dimensional gel electrophoresis of pulse-labeled proteins was used for global analysis of protein synthesis and turnover in *Escherichia coli* [19]. Comparative proteomics of the human pathogen *Campylobacter jejuni* revealed an important first step in characterizing strain differences potentially responsible for different disease outcomes associated with this organism [20]. Investigation of wild type and phoB mutants of *Vibrio cholerae* determined clearly the effect of the phosphate limitation [21]. Through proteomics, analysis of protein phosphorylation in *Corynebacterium glutamicum* was possible [20] and also the analysis of *Bacillus anthraces* cell membrane proteins [23]. COX- and AOX-dependent respiratory chains in the filamentous fungus *Podospora anserina* were supramolecularly organized using proteomic analysis [22]. *Rickettsia prowazekii*, the etiologic agent of epidemic typhus, has virulent and avirulent strains and their proteomes have been analyzed [24]. Several studies subjected *M. tuberculosis* to comprehensive proteomic analysis. Culture supernatant proteins of virulent *M. tuberculosis* H37Rv and attenuated *M. bovis* BCG were analyzed [25]. In another study, proteomics revealed open reading frames in *M. tuberculosis* H37Rv not predicted by genomics [26].

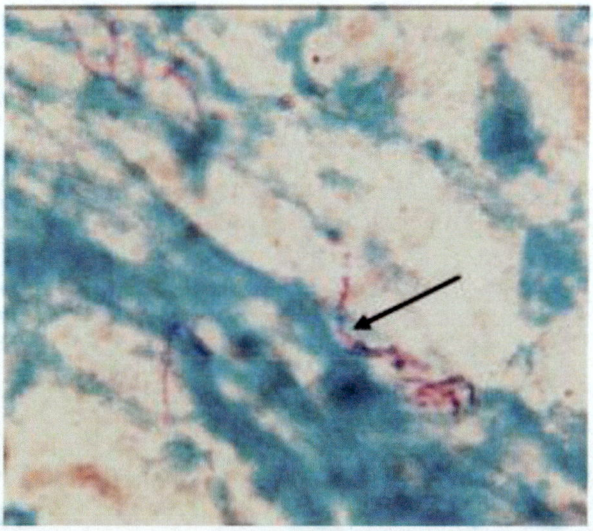

Figure 1. Modified Zeihl-Neelsen-stained smear from sputum sample taken from patient infected with *N. africana* (notice the branching filamentous acid fast organism).

This clearly showed that the analysis of proteins may generate new knowledge and thereby alter the approaches to disease diagnosis, control and treatment. Definitive diagnosis of nocardiosis depends on the isolation and identification of the causative organism from clinical materials. These procedures are not straightforward; hence the true incidence of the disease is masked, a problem which is compounded by poor documentation. Generally, *Nocardia* cases are difficult to diagnose [27]. So, regarding pulmonary nocardiosis, physicians have to be alert to suspect nocardial infection if pulmonary symptoms are clear, especially, when patients with respiratory infection fail to respond to anti-tubercular therapy [18].

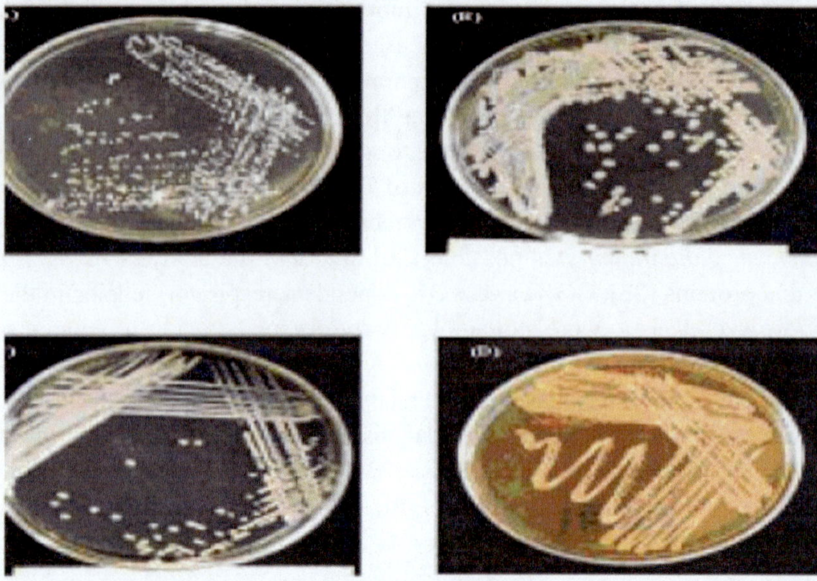

Figure 2. Growth of *N. africana* SD 769 (A); *N. asteroidis* N 317 (B); *N. farcinica*SD 828(*C*); *N. farcinica* SD 1819 (D); on glucose yeast extract agar medium. 7 days old cultures. The growth of *N. africana* is visible in white pigments.

A first study on *N. africana* SD769 (DSM44491) showed 74 proteins on a silver-stained gel (figure 3). Some proteins were identified in an in-solution digest using Q-STAR instrument (table 1), but a careful proteome analysis of a Nocardia strain is still lacking. Although some efforts are in progress, they are compromised by the difficult database situation. Nevertheless, even protein assignments based on homology as shown here are valuable and advance the knowledge with respect to disease treatment.

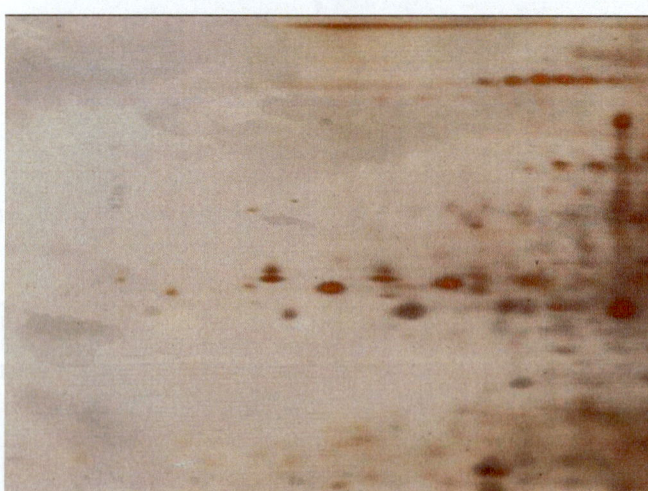

Figure 3. 2D 12% gel of *N. africana* SD 769. 70 μg applied, pH4-7, MW10-250, silver-stain.

Table 1. Proteins identified in *N. africana* SD769 (DSM44491) by Q-STAR peptide sequencing

SwissProt Accession #	Protein name	Peptides sequenced	Taxonomy
Q6GB54	Enolase	LGANAILGVSIAVA AAADLLGQPLYK WGTEIFHNLK VNQIGTLTETFDAIEMAQK	*Staphylococcus aureus*
Q6GBT9	Elongation factor Tu	LLDYAEAGDNIGALLR TVGSGVVTEIIK	*Staphylococcus aureus*
Q8ENP5	Enolase	VQLVGDDLFVTNTNK	*Oceanobacillus iheyensis*
Q5YQ30	Enolase	LGANALLGVSLAVAR TMMSHRSGETEDTTIADLAVAVGSGQIK	*Nocardia farcinica*
Q8ZBY6	Acyl-coenzyme A dehydrogenase	GAPDWK ISAEEAALLTKAENSR	*Yersinia pestis*

ACKNOWLEDGMENTS

The authors are grateful to Adil Mahgoub, Mr. Modathir M. Ahmed and Dr. Amanwel for their help during the preparation of this project. The study was supported by the University of Sudan for Science and Technology, Deutscher Akademischer Austausch Dienst (DAAD) and ICGEB (M.E.) and DFID Higher Education (M.E.H). The article was edited by S. König.

REFERENCES

[1] Orchard, V. A. (1981). The ecology of *Nocardia* and related taxa. *Zentralbl. Bakteriol. Suppl.* 11: 167-80

[2] Goodfellow, M., Williams, S.T. (1983) Ecology of actinomycetes. *Ann. Rev. Microbiol.* 37: 189-216

[3] Goodfellow, M., (1992) The family Nocardiaceae. The prokaryotes; Vol 2. 2nd Ed., Eds. Balows, A., Truper, H. G. *et al.*, Springer-Verlag New York, 1188-213

[4] Beaman, B.L., Beaman, L., (1994) *Nocardia* species: host-parasite relationships. *Clin. Microbiol. Rev.* 7: 203-64

[5] Weber, L., Yium, J., Hawkins, S.. (2002) Intracranial *Nocardia* dissemination during minocycline therapy. *Transp. J. Infect. Dis.*; 4(2): 108-12

[6] Goodfellow M. Nocardia and related genera. In A. Balows, B. I. Duerden (Eds) *Topley and Wilson's Microbiology and Microbial Infections* 9th Ed., Vol. 2. Edward Arnold, London, 463-489, 1998

[7] Laurent, F., Mick, V., Boiron. P. (1999) Nocardia infections: clinical and biological aspects. *Ann. Biol. Clin. (Paris) 2000* 57: 545-554

[8] Corti, M. E., Villafane-Fioti, M.F. Nocardiosis: a review. (2003) *Int. J. Infect. Dis.* 7: 243-250

[9] Houang, E. T. and Lovett, I. S. (1980) *Nocardia asteroids* infection, a transmissible disease, *J. Hosp. Infect.*, 1:31-40

[10] McNeil, M. M., Brown, J. M. The medically important aerobic actinomycetes: Epidemiology and microbiology. (1994) *Clin. Microbiol. Rev.* 7: 357-417

[11] Sahathevan, M. and Harvey, F. A. H. (1991) Epidemiology, bacteriology and control of an outbreak of *Nocardia asteroides* infection in a liver unit. *J. Hosp. Infect.* 18, Suppl A: 478-80

[12] Baddour, L. M., Baselski, V. S. (1986) Nocardiosis in recipients of renal transplant: evidence for nosocomial acquisition, *Am. J. Infect. Control*; 14:214-19

[13] Schaal, K. P., (1991), Medical and microbiological problems arising from air born infections in hospitals. *J. Hosp. Infect*, 18, Suppl, A: 451—9

[14] Boiron, P., Provost, F. (1990) Characterization *of Nocardia, Rhodococcus* and *Gordona* species by *in-vitro* susceptibility testing. *Zentralbl. Bakteriol.* 274 (2): 203-13

[15] Torres, H. A., Reddy, B.T., Raad, I.I., Tarrand, J., Bodey, G.P., Hanna, H.A., Rolston, K.V., Kontoyiannis, D.P. (2002) Nocardiosis in cancer patients. *Medicine (Baltimore)*; 81(5)

[16] Schaal, K.P. and Lee H.-J. (1992), Actinomycetes infections in humans: a review. *Gene*, 115: 201-11.

[17] Shimokubo, T., Ashitani, J., Ihi, T., Higa, T., Kumamoto, K. (2002) A case of mixed infection by *Nocardia asteroids* and *Mycobacterium tuberculosis*. *Nihon Kokyuki Gakkai Zasshi* 40(8): 703-7

[18] Hamid, M. E., Maldonado, L., Sharaf Eldin, G. S., Mohammed, M. F., Saeed, N. S., Goodfellow, M. (2001) *Nocardia africana* sp. nov, a new pathogen isolated from patients with pulmonary infections. *Clin. Microbiol.* 39(2) 625-30

[19] Weichart, D., Querfurth, N., Dreger, M., Hengge-Aronis, R. (2003) Global role for ClpP-containing proteases in stationary-phase adaptation of Escherichia coli. *J. Bacteriol.* 185, 1: 115-125

[20] Brubacher, J. L., DeWitte-Orr, S. J., Zorzitto, J. R., Playle, R. C., Bols, N. C. (2003) Redox-active metals in commercial preparations of lipopolysaccharide: implications for studies of cellular responses to bacterial products. *Cell Microbiol.* 5, 4: 233-243

[21] Manta, P., Terzis, G., Papadimitriou, C., Kontou, C., Vassilopoulos, D. Emerin expression in tubular aggregates. (2004) *Acta Neuropathol.* (Berl.) 107, 6: 546-552

[22] Chitlaru, T., Ariel, N., Zvi, A., Lion, M., Velan, B., Shafferman, A., Elhanany, E. (2004) Identification of chromosomally encoded membranal polypeptides of Bacillus anthracis by a proteomic analysis: prevalence of proteins containing S-layer homology domains. *Proteomics* 4, 3: 677-691

[23] Krause, R., von Mering, C., Bork, P. (2003) A comprehensive set of protein complexes in yeast: mining large scale protein-protein interaction screens. *Bioinformatics.* 19, 15: 1901-1908

[24] Chelius, D., Zhang, T., Wang, G., Shen, R. F. (2003) Global protein identification and quantification technology using two-dimensional liquid chromatography nanospray mass spectrometry. *Anal. Chem.* 75, 23:6658-6665

[25] Mattow, J., Jungblut, P. R., Schaible, U. E., Mollenkopf, H. J., Lamer, S., Zimny-Arndt, U., Hagens, K., Müller, E. C., Kaufmann, S. H. (2001) Identification of proteins from *Mycobacterium tuberculosis* missing in attenuated *Mycobacterium bovis* BCG strains. *Electrophoresis* 22(14):2936-46.

[26] Jungblut, P. R., Müller, E. C., Mattow, J., and Kaufmann, S. H. E. (2001) Proteomics reveals open reading frames in *Mycobacterium tuberculosis* H37Rv not predicted by genomics *Infect. and Imm.* 69 (9): 5905-5907

[27] Garcia-Benitez, V., Garcia-Hidalgo, L., Archer-Dubon, C. and Orozco-Topete, R. (2001) Acute primary superficial nocardiosis due to *Nocardia brasiliensis*: a case report in an immunocompromised patient. *Eur. J. Epidemiol.* 17(11): 1019-22

In: Biomacromolecular Mass Spectrometry Yearbook. Volume 1 ISBN: 978-1-60876-167-8
Ed: Simone Koenig

Chapter 12

PROTEIN KINASE B AUTOPHOSPHORYLATION AFFECTS ACTIVITY*

Martin Zeller, Bodo Levkau[1] and Simone König*

Integrated Functional Genomics, Interdisciplinary Center for Clinical Research;
Medical Faculty, University of Münster, Germany
[1] Institute for Pathophysiology, University of Duisburg-Essen, Germany

ABSTRACT

Protein kinase B (PKB) plays an important role in various signal transduction pathways. PKB in its active state is phosphorylated on Thr-308 and Ser-473 and phosphorylates various target proteins. Mass spectrometric data suggests that PKB also has a high potential for autophosphorylation at random sites on expense of regulatory sites affecting protein activity.

Keywords: protein kinase B, autophosphorylation, mass spectrometry, IMAC.

INTRODUCTION

Protein kinase B (PKB, also termed Akt or RAC) is a serine/threonine kinase of the second-messenger or AGC (protein kinase A/protein kinase G/protein kinase c-like) subfamily of protein kinases [1, 2]. This group of kinases shares similarities within their catalytic structure and in their activation mechanism. PKB has been of great interest in the last decade for medical and pharmaceutical research due to its role in diverse cellular processes such as glucose uptake, cell-cycle progression, cancer, apoptosis and transcriptional regulation [3-8]. Three isoforms of PKB exist in mammals (α, β, γ; Akt1, 2, 3). They are

---

* A version of this chapter was also published in *Biomacromolecular Mass Spectrometry Volume 1, Number 3*, published by Nova Science Publications, Inc. It was submitted for appropriate modifications in an effort to encourage wider dissemination of research.

* Address correspondence to: PD Dr. Simone König; Integrated Functional Genomics; Röntgenstr. 21; 48149 Münster, Germany; ph: xx-(0)251-8357164; fax: xx-(0)251-8355651; e-mail: koenigs@uni-muenster.de

composed of three functionally distinct regions: an N-terminal pleckstrin homology domain, a central catalytic domain, and a C-terminal hydrophobic motif. The catalytic domain is structurally similar to other protein kinases of the AGC family [9].

PKB was shown to be stimulated by the phosphorylation of two regulatory residues, Thr-308 of the activation segment and Ser-473 of the hydrophobic motif which was demonstrated by mutation of these amino acids and measurement of the activity of wild-type and mutant PKB *in vitro* upon activation by insulin or insulin-like growth factor-1 [10]. Phosphorylation of Thr-308 is catalysed by 3-phosphoinositide-dependent kinase [11-13] and causes a charge-induced change in the conformation of PKB which allows substrate binding [10]. Mutation of Thr-308 to Ala reduces the PKB activity to a large extent [10]. Phosphorylation of Ser-473 of the hydrophobic motif is tightly coupled to that of the activation loop. Activation of PKB by Ser-473 phosphorylation occurs via a disorder-to-order transition of the αC-helix with concomitant restructuring of the activation segment carrying Thr-308 phosphorylation and reconfiguration of the kinase bilobal structure [14]. Monophosphorylated PKB on Thr-308 is approximately 10 % as active as the fully phosphorylated PKB [14] so that the autophosphorylation at Ser-473 regulates PKB. Autophosphorylation at the hydrophobic motif has also been found in other members of the AGC subfamily such as protein kinase C [15, 16]. The importance of Thr-308 and Ser-473 phosphorylation for activation of PKB is now well accepted and antibodies for the detection of these sites are commercially available. Nevertheless, Tyr-474 has been identified as an additional phosphorylation site by NH_2-terminal sequencing and phosphate release analysis that appears to be required for full activation of the kinase [17]. Substitution of Tyr-474 with Phe abolished Tyr phosphorylation of Akt and resulted in up to 55 % inhibition. Several other phosphorylation sites have been reported in addition (table 1, figure 1) whose role in the activation process of PKB is not clear. For instance, epidermal growth factor (EGF) phosphorylated Tyr-315 and Tyr-326. The present research was initiated by the lack of activity of a commercial PKB leading to an investigation of its phosphorylation using mass spectrometry based methods [25].

Table 1. Reported phosphorylation sites in PKB

Phosphorylation site	Reference	Method of determination
Thr-308	[10]	mutation, monitoring of PKB activity
Ser-473	[10]	mutation, monitoring of PKB activity
Tyr-474	[17]	phosphoamino acid analysis, Edman degradation
Ser-124	[10, 17]	phosphoamino acid analysis, Edman degradation
Ser-129	[17]	phosphoamino acid analysis, Edman degradation
Thr-450	[10, 17]	phosphoamino acid analysis, Edman degradation
Tyr-315	[18]	mutation, monitoring of PKB activity
Tyr-326	[18]	mutation, monitoring of PKB activity

1	MSDVAIVKEG	WLHKRGEYIK	TWRPRYFLLK	NDGTFIGYKE	RPQDVDQREA
51	PLNNFSVAQC	QLMKTERPRP	NTFIIRCLQW	TTVIERTFHV	ETPEEREEWT
101	TAIQTVADGL	KKQEEEEMDF	RSG***S***PSDN***S***G	AEEMEVSLAK	PKHRVTMNEF
151	EYLKLLGKGT	FGKVILVKEK	ATGRYYAMKI	LKKEVIVAKD	EVAHTLTENR
201	VLQNSRHPFL	TALKYSFQTH	DRLCFVMEYA	NGGELFFHLS	RERVFSEDRA
251	RFYGAEIVSA	LDYLHSEKNV	VYRDLKLENL	MLDKDGHIKI	TDFGLCKEGI
301	KDGATMK**T**FC	GTPE***Y***LAPEV	LEDND***Y***GRAV	DWWGLGVVMY	EMMCGRLPFY
351	NQDHEKLFEL	ILMEEIRFPR	TLGPEAKSLL	SGLLKKDPKQ	RLGGGSEDAK
401	EIMQHRFFAG	IVWQHVYEKK	LSPPFKPQVT	SETDTRYFDE	EFTAQMITI***T***
451	PPDQDDSMEC	VDSERRPHFP	QF**S*****Y***SASSTA		

Figure 1. Sequence of PKB. Sites required for activation (Thr-308, Ser-473) are in bold and underlined; other known phosphorylation sites are in bold and italic. Phosphopeptides AA371-386 and 122-142 assigned here are underlined.

EXPERIMENTAL

Enzymatic digestion was performed by dissolving the lyophilised PKB in in 20 μl 100 mM ammonium bicarbonate (pH 8.0; Sigma-Aldrich, Steinheim, Germany) containing 1 μg trypsin sequencing grade (Roche Diagnostics, Mannheim, Germany). The digest was carried out overnight at 37 °C and stopped with 2 μl glacial acetic acid (HAc); Sigma). The protein digest was desalted on reversed phase (RP) material (ZipTip C4 and μC18, respectively, Millipore, Bedford, MA,USA) according to the manufacturers protocol.

For immobilized metal affinity chromatography (IMAC) Co^{2+}-NTA-IMAC beads from TalonSpin columns (BD Biosciences Clontech, Palo Alto, CA, USA) were transferred into an Eppendorf tube as a slurry and washed twice with water (for gradient elution, Sigma-Aldrich, Seelze, Germany). Co^{2+}-ions were removed from the coordination sites by repeated washing with 50 mM EDTA (Sigma). The beads were then equilibrated with 0.1 % Hac (pH 3.5) and charged with Fe^{3+}-ions by adding 200 mM $FeCl_3$ solution in 0.1 % HAc five times. Excess metal ions were removed by washing with 0.1 % HAc five times. Finally, the IMAC beads were equilibrated in storage solution consisting of methanol and 0.1 % HAc (1:1 v/v). Nanoscale bed volume IMAC columns were prepared by loading IMAC resin into a GELoader tip (Eppendorf, Hamburg, Germany) to pack a column of ~ 15-20 mm. A 10 ml syringe was used to force all liquid through the column by applying gentle air pressure. The IMAC column was washed three times each with 50 μl 0.1 % HAc, 50 μl 0.1 % HAc and acetonitrile (ACN; 3:1 v/v), and 50 μl 0.1 % HAc. The desalted phoshoprotein digest was lyophilised, redissolved in 40 μl 0.1 % HAc and slowly loaded onto the IMAC column. The IMAC column was then washed two times each with 50 μl 0.1 % HAc, 50 μl HAc 0.1 % and ACN (3:1 v/v), 50 μl 100 mM NaCl in water, HAc and ACN (74:1:25 v/v) to remove unspecifically bound peptides and 50 μl 0.1 % HAc. Bound peptides were eluted from the nanoscale IMAC column with 40 μl 100 mM Na_2HPO_4 (Sigma), acidified with 5 μl glacial HAc and desalted with ZipTip-μC18 RP material.

Peptide masses were determined with TofSpec-2E instrument (Micromass/Waters, Manchester, UK). Dihydroxybenzoic acid (DHB) was used as a matrix and 1 μl of analyte was applied.

Enzymatic dephosphorylation with alkaline phosphatase was performed directly after the acquisition of the MALDI-TOF (matrix-assisted laser desorption/ionisation time-of-flight) peptide map of the IMAC enriched peptides. The spot was redissolved in 2 µl ACN and 100 mM NH_4HCO_3 (7/3 v/v) and transferred into a pre-equilibrated ZipTip-µC18 tip with 20 µl 100 mM NH_4HCO_3 and 1 unit alkaline phosphatase (Roche Diagnostics, Mannheim, Germany) loaded on top of the chromatographic resin. The tip was placed in an Eppendorf tube and incubated for 2 h at 37 °C. The chromatographic bed acted as a barrier and no drainage was observed. The reaction was then stopped with 2 µl formic acid and the solution was slowly pressed through the RP material and desalted. The bound peptides were eluted with 2 µl 80 % ACN containing 0.1 % TFA directly onto the MALDI target and mixed with matrix.

RESULTS AND DISCUSSION

Mass spectrometric phosphorylation analysis of PKB was based on the enrichment of phosphopeptides with IMAC and their confirmation by enzymatic dephosphorylation. In this way, phosphopeptides are unambiguously identified. Figure 2 and 3 show the MALDI-TOF spectra of two IMAC enriched phosphopeptides before and after dephosphorylation. There is an intense signal for the mono-phosphorylated peptide amino acids (AA) 371-386 and a minor signal for double phosphorylation (figure 2a). In addition, a characteristic metastable ion approximately 90 Da below the peak for the mono-phosphorylated peptide AA371-386 is indicative of phosphoserine or phosphothreonine containing peptides in our instrument [25]. Metastable ions are formed with sufficient excitation to dissociate spontaneously during their flight from the ion source to the detector (IUPAC definition). The fragment ions have lost kinetic energy during dissociation and since the instrument is not calibrated for their analysis the observed mass shift depends on the instrument geometry. A mass difference of 94 Da is reported by other authors [19]. After enzymatic dephosphorylation the peak for the doubly phosphorylated peptide disappears (figure 2b). The peak for the mono-phosphorylated peptide is reduced in intensity and a new peak shifted by 80 Da towards lower *m/z*-values appears. It should be noted that the reduction in the abundances is not only caused by stoichiometric dephosphorylation but also by sample loss due to handling. The sequence of the peptide contains one Thr and two Ser residues, two of which are at least phosphorylated. The peptide AA371-386 is located in the central catalytic domain of the protein.

For peptide AA122-142 (figure 3a) two intense peaks for the doubly and triply phosphorylated peptide and a minor peak for the mono-phosphorylated peptide are present. Metastable ions are observed as minor peaks. Dephosphorylation causes the peaks for the doubly and triply phosphorylated peptides to disappear and generates a new peak for the unphosphorylated peptide. Some of the singly phosphorylated peptide is still present indicating incomplete dephosphorylation. Ser-124 and Ser-129 have been identified as phosphorylation sites in this peptide earlier [17], here, a third site is indicated.

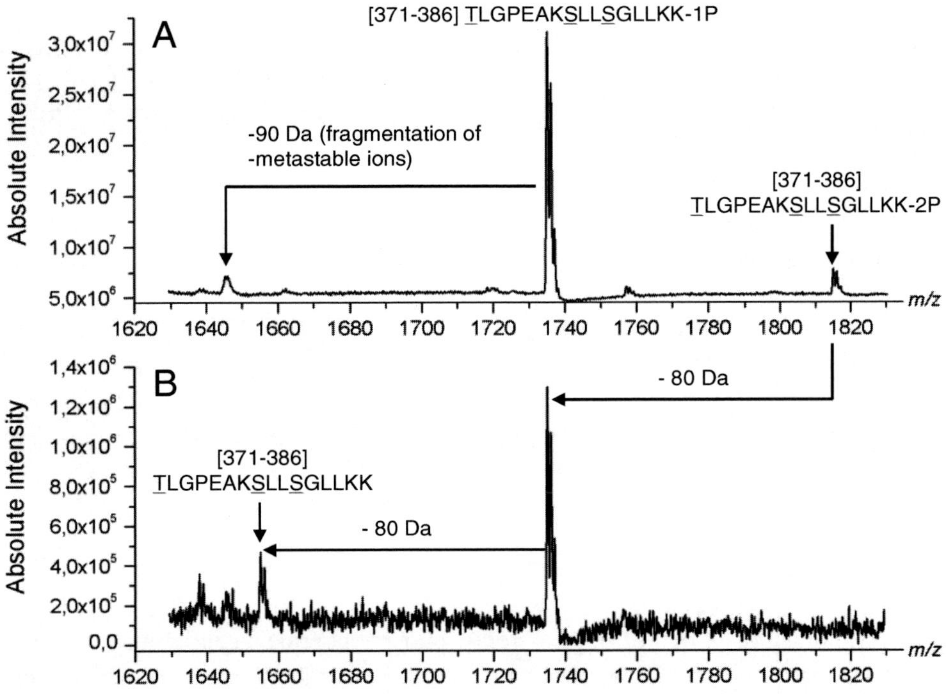

Figure 2. MALDI-TOF spectrum of IMAC enriched PKB before and after enzymatic dephosphorylation, peptide AA371-386. Potential phosphorylation sites are underlined.

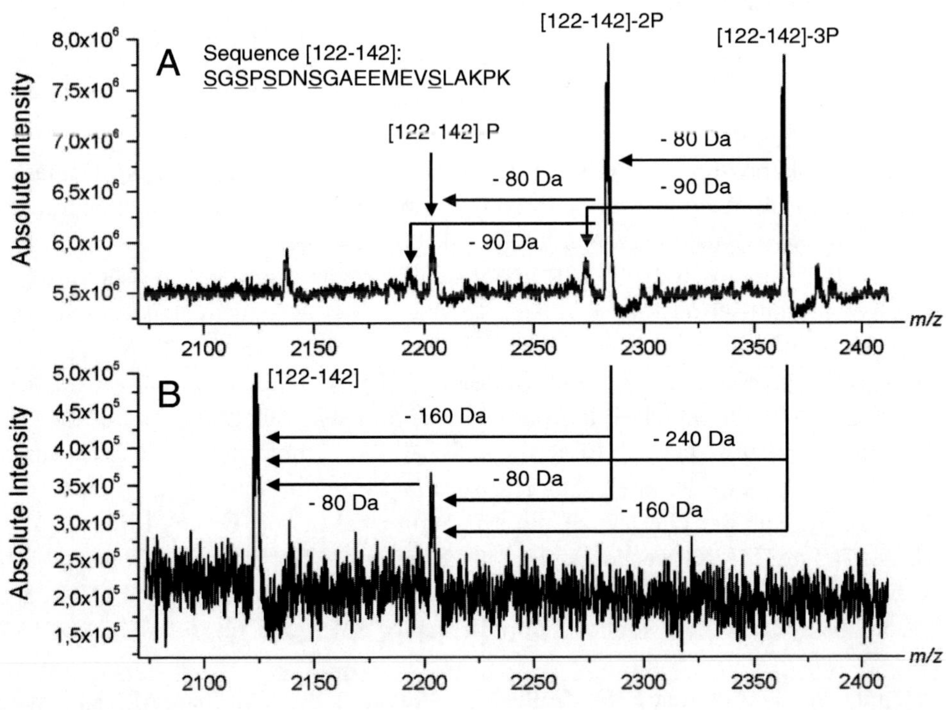

Figure 3. MALDI-TOF spectrum of IMAC enriched PKB before and after enzymatic dephosphorylation, peptide AA122-142. Potential phosphorylation sites are underlined.

Peptide and protein library screenings as well as studies on synthetic peptides revealed a consensus motif R-X_1-R-X_2-X_3-S/T-X_4 for PKB phosphorylation sites, where X_1 is any amino acid, X_2 and X_3 are small residues other than Gly and X_4 is a bulky hydrophobic residue [20, 21]. T/S-P-X-D was found as another consensus motif for PKC [22]. Phosphorylation on the hydrophobic region on the other hand requires a F-X-X-F-S/T-Y motif which is conserved within a large proportion of AGC kinases [23, 24]. In the phosphopeptide AA371-386 none of these motifs are present while phosphorylation of Ser-124 is supported by the presence of the consensus sequence motif S-P-X-D. Again, there is no motif for the phosphorylation of Ser-129. Therefore, experimental evidence indicates that these motifs are not necessarily mandatory and that particularly autophosphorylation can access additional sites. It appears that storage might have substantial influence on this process and in that on the activity of PKB. No evidence for phosphorylation of the known sites Thr-308 and Ser-473 was found which suggests that autophosphorylation on expense of the regulatory sites destroyed the PKB activity.

CONCLUSION

The combination of IMAC for phosphopeptide enrichment, enzymatic dephosphorylation, and mass spectrometry allowed the assignment of three additional phosphorylation sites in two phosphopeptides not known to be required for activity in commercial inactive PKB. It was shown that PKB is prone to autophosphorylation which accessed random sites with no consensus motifs on expense of regulatory phosphorylation sites quenching PKB activity.

REFERENCES

[1] Jones, P., Jakubowicz, T., Pitossi, F., Maurer, F., and Hemmings, B., Molecular cloning and identification of a serine/threonine protein kinase of the second-messenger subfamily, *Proc. Natl. Acad. Sci. USA* 1991, 88, 4171-4175.

[2] Coffer, P. J., and Woodgett, J. R., Molecular cloning and characterisation of a novel putative protein-serine kinase related to the cAMP-dependent and protein kinase C families, *Eur. J. Biochem.* 1991, 201, 475-481.

[3] Brazil, D. P., Yang, Z. Z., and Hemmings, B. A., Advances in protein kinase B signalling: AKTion on multiple fronts, *Trends Biochem. Sci.* 2004, 29, 233-242.

[4] Hanada, M., Feng, J., and Hemmings, B. A., Structure, regulation and function of PKB/AKT--a major therapeutic target, *Biochim. Biophys. Acta* 2004, 1697, 3-16.

[5] Nofer, J.-R., van der Giet, M., Tolle, M., Wolinska, I., von Wnuck Lipinski, K., Baba, H. A., Tietge, U. J., Godecke, A., Ishii, I., Kleuser, B., Schafers, M., Fobker, M., Zidek, W., Assmann, G., Chun, J., and Levkau, B., HDL induces NO-dependent vasorelaxation via the lysophospholipid receptor S1P3, *J. Clin. Invest.* 2004, 113, 569-581.

[6] Schmitz, K. J., Otterbach, F., Callies, R., Levkau, B., Holscher, M., Hoffmann, O., Grabellus, F., Kimmig, R., Schmid, K. W., and Baba, H. A., Prognostic relevance of

activated Akt kinase in node-negative breast cancer: a clinicopathological study of 99 cases, *Mod. Pathol.* 2004, 17, 15-21.

[7] Schönherr, E., Levkau, B., Schaefer, L., Kresse, H., and Walsh, K., Decorin-mediated signal transduction in endothelial cells. Involvment of Akt/Protein kinase B in up-regulation of p21WAF1/CIP1 but not p27/KIP1, *J. Biol. Chem.* 2001, 276, 40687-40692.

[8] Nofer, J.-R., Levkau, B., Wolinska, I., Junker, R., Fobker, M., von Eckardstein, A., Seedorf, U., and Assmann, G., Suppression of endothelial cell apoptosis by high density lipoproteins (HDL) and HDL-associated lysosphingolipids, *J. Biol. Chem.* 2001, 276, 34480-34485.

[9] Yang, J., Cron, P., Good, V. M., Thompson, V., Hemmings, B. A., and Barford, D., Crystal structure of an activated Akt/protein kinase B ternary complex with GSK3-peptide and AMP-PNP, *Nat. Struct. Biol.* 2002, 9, 940-944.

[10] Alessi, D. R., Andjelkovic, M., Caudwell, B., Cron, P., Morrice, N., Cohen, P., and Hemmings, B. A., Mechanism of activation of protein kinase B by insulin and IGF-1, *EMBO J.* 1996, 15, 6541-6551.

[11] Alessi, D. R., James, S. R., Downes, C. P., Holmes, A. B., Gaffney, P. R., Reese, C. B., and Cohen, P., Characterization of a 3-phosphoinositide-dependent protein kinase which phosphorylates and activates protein kinase Balpha, *Curr. Biol.* 1997, 7, 261-269.

[12] Stokoe, D., Stephens, L. R., Copeland, T., Gaffney, P. R., Reese, C. B., Painter, G. F., Holmes, A. B., McCormick, F., and Hawkins, P. T., Dual role of phosphatidylinositol-3,4,5-trisphosphate in the activation of protein kinase B, *Science* 1997, 277, 567-570.

[13] Stephens, L., Anderson, K., Stokoe, D., Erdjument-Bromage, H., Painter, G. F., Holmes, A. B., Gaffney, P. R., Reese, C. B., McCormick, F., Tempst, P., Coadwell, J., and Hawkins, P. T., Protein kinase B kinases that mediate phosphatidylinositol 3,4,5-trisphosphate-dependent activation of protein kinase B, *Science* 1998, 279, 710-714.

[14] Yang, J., Cron, P., Thompson, V., Good, V. M., Hess, D., Hemmings, B. A., and Barford, D., Molecular mechanism for the regulation of protein kinase B/Akt by hydrophobic motif phosphorylation, *Mol. Cell.* 2002, 9, 1227-1240.

[15] Dutil, E. M., Toker, A., and Newton, A. C., Regulation of conventional protein kinase C isozymes by phosphoinositide-dependent kinase 1 (PDK-1), *Curr. Biol.* 1998, 8, 1366-1375.

[16] Behn-Krappa, A., and Newton, A. C., The hydrophobic phosphorylation motif of conventional protein kinase C is regulated by autophosphorylation, *Curr. Biol.* 1999, 9, 728-737.

[17] Conus, N. M., Hannan, K. M., Cristiano, B. E., Hemmings, B. A., and Pearson, R. B., Direct Identification of tyrosine 474 as a regulatory phosphorylation site for the akt protein kinase, *J. Biol. Chem.* 2002, 277, 38021-38028.

[18] Chen, R., Kim, O., Yang, J., Sato, K., Eisenmann, K. M., McCarthy, J., Chen, H., and Qiu, Y., Regulation of Akt/PKB activation by tyrosine phosphorylation, *J. Biol. Chem.* 2001, 276, 31858-31862.

[19] Müller, D. R., Schindler, P., Coulot, M., Voshol, H., and Oostrum, J. v., Mass spectrometric characterization of stathmin isoforms separated by 2D PAGE, *J. Mass Spectrom.* 1999, 34, 336-345.

[20] Alessi, D. R., Caudwell, F. B., Andjelkovic, M., Hemmings, B. A., and Cohen, P., Molecular basis for the substrate specificity of protein kinase B; comparison with MAPKAP kinase-1 and p70 S6 kinase, *FEBS Lett.* 1996, 399, 333-338.

[21] Obata, T., Yaffe, M. B., Leparc, G. G., Piro, E. T., Maegawa, H., Kashiwagi, A., Kikkawa, R., and Cantley, L. C., Peptide and protein library screening defines optimal substrate motifs for AKT/PKB, *J. Biol. Chem.* 2000, 275, 36108-36115.

[22] Edwards, A. S., Faux, M. C., Scott, J. D., and Newton, A. C., Carboxyl-terminal phosphorylation regulates the function and subcellular localization of protein kinase C beta II, *J. Biol. Chem.* 1999, 274, 6461-6468.

[23] Keranen, L. M., Dutil, E. M., and Newton, A. C., Protein kinase C is regulated in vivo by three functionally distinct phosphorylations, *Curr. Biol.* 1995, 5, 1394-1403.

[24] Pearson, R. B., Dennis, P. B., Han, J. W., Williamson, N. A., Kozma, S. C., Wettenhall, R. E., and Thomas, G., The principal target of rapamycin-induced p70s6k inactivation is a novel phosphorylation site within a conserved hydrophobic domain, *EMBO J.* 1994, 14, 5279-5287.

[25] Zeller, M and König, S., Impact of chromatography and mass spectrometry on the detection of protein phosphorylation sites. Analytical and Bioanalytical Chemistry 2004, 378, 898-909; *Erratum ABC* 2004, 379, 318

In: Biomacromolecular Mass Spectrometry Yearbook. Volume 1 ISBN: 978-1-60876-167-8
Ed: Simone Koenig

Chapter 13

MASS SPECTROMETRIC MONITORING OF MEMBRANE PROCESSING: NEUROPEPTIDE DIGESTION BY SKIN CELLS*

Simone König* and Thomas E. Scholzen[1]

Integrated Functional Genomics, Interdisciplinary Center for Clinical Research,
[1] Ludwig Boltzmann Institute, Department of Dermatology,
Medical Faculty, University of Münster, Germany

ABSTRACT

A method is described which allows the generation of bioactive endothelial cell membrane preparations for subsequent processing of proopiomelanocortin peptides. Semi-quantitative mass spectrometry is used as a detector for monitoring of the products of membrane protease (neutral endopeptidase and angiotensin-converting enzyme) action.

Keywords: Mass spectrometry, endothelial membranes, POMC, ACTH, MSH, ACE, NEP.

Mass spectrometry (MS) is a valuable detector for cell surface activity as was shown during investigations of the processing of proopiomelanocortin (POMC) peptides by dermal microvascular endothelial cell (EC) extracellular peptidases neutral endopeptidase (NEP) and angiotensin-converting enzyme (ACE) [1-7]. Methodological advances in that respect involve the preparation and incubation of bioactive cell membranes and semi-quantitative MALDI (matrix-assisted laser desorption ionization)-MS. Both steps present a number of difficulties

---

* A version of this chapter was also published in *Biomacromolecular Mass Spectrometry Volume 1, Number 3*, published by Nova Science Publications, Inc. It was submitted for appropriate modifications in an effort to encourage wider dissemination of research.

* Address correspondence to: PD Dr. Simone König; Integrated Functional Genomics; Röntgenstr. 21; 48149 Münster, Germany; ph: +49-(0)251-8357164; fax: +49-(0)251-8357255; e-mail: koenigs@uni-muenster.de

and here, we provide a successful procedure which covers all aspects from cell culture to measurement of membrane protease activity and MS.

CELL CULTURE

HDMEC were obtained from PromoCell (Heidelberg, Germany). HDMEC and the endothelial cell line HMEC-1 were grown in EC basal media (EBM-MV Kit system, PromoCell), and otherwise treated as described in ref. [8]. For the analysis of HDMEC or HMEC-1 cell surface expression, cells were detached from culture dishes using Accutase (PAA Laboratories, Linz, Austria) and stained with mouse anti human CD10 mAb conjugated to fluorescein isothiocyanate (FITC; clone B-E3, Acris GmbH, Hiddenhausen, Germany). For the detection of ACE, an anti human CD143 mAb (clone 9B9, Acris GmbH, Hiddenhausen, Germany) followed by FITC-conjugated anti mouse IgG Ab (Sigma-Aldrich, Taufkirchen, Germany) were used. Flow cytometry was performed on an EPICS XL Flow cytometer (Beckman-Coulter, Miami, FL).

For preparation of crude membranes, cells were grown in 175 cm^2 culture flasks to subconfluency (80–90 %), washed with PBS, and detached with Accutase (PAA Laboratories). They were collected by centrifugation (300 xg), washed with PBS, and resuspended in 0.05 M Tris/HCl pH 7.4. To collect the membranes, cells were lyzed by sonication (10 pulses 5 sec) and subjected to 100,000 xg centrifugation (30 min, 4 °C). Membrane pellets were washed twice with ice cold 0.05 M Tris-HCl pH 7.4 and resuspended in 0.05 M Tris-HCl pH 7.4. The protein content was determined using the method of Lowry and was adjusted to 1 mg/ml.

NEP activity was determined using the artificial substrate glutaryl-Ala-Ala-Phe-4-methoxy-β-naphtylamide (Glu-MNA) that is converted to the fluorescent product 4-Methoxy-2-naphtylamin (MNA; Sigma-Aldrich) in the presence of aminopeptidase M (AMP; Calbiochem/Merck Bioscience GmbH, Schwalbach, Germany). Cell fragment, membrane or supernatant samples adjusted to a defined protein concentration were preincubated with 1 μg AMP for 10 min at 37 °C in a total volume of 240 μl 50 mM Tris-HCl pH 7.5. Glu-MNA (Sigma-Aldrich; from a 1 mM stock solution in 90 % dimethylformamide (DMF), 10% 50 mM Tris-HCl pH 7.4) was added to the incubation mix in a final concentration of 40 μM, and samples were incubated for exactly 30 min at 37 °C in the absence or presence of 1 μM NEP inhibitor (Phosphoramidon or Thiorphan in ethanol; Sigma-Aldrich) to determine the amount of MNA specifically generated by NEP. After centrifugation (5 min/14,000 xg at 4 °C), dithiothreitol (2 mM DTT in H_2O; Sigma-Aldrich) was added in a final concentration of 1 mM DTT. Fluorescence of MNA was detected at 425 nm emission wavelength in a Fluoromax-2 Fluorimeter with Datamax 2.2 Software (Instruments S.A. Inc., Edison, NJ) and an integrated GRAMS/386 data processing and database software (Galactic Industries Corp., Salem, NH) using an excitation wavelength of 340 nm. NEP enzyme activity was calculated from a standard curve of 0–3000 pmol MNA (1 mM stock solution in 90 % DMF/10 % 50 mM Tris-HCl pH 7.5) in 50 mM Tris-HCl pH 7.5 containing 1 mM DTT, which was linear in the range of 0–3000 pmol MNA and expressed in μU with 1 μU = 1 pmol MNA/h/μg protein.

POMC Peptide Processing by EC Membranes

2.5 nmol adrenocorticotropin hormone $ACTH_{1-39}$, $ACTH_{1-24}$, or melanocyte-stimulating hormone α-MSH_{1-13} (all from Bachem AG, Bubendorf, Switzerland) were incubated with 1–100 μg membrane protein in 250 μl 0.05 M Tris-HCl pH 7.4 at 37 C for various time points (1–240 min). To determine specificity of degradation, 1 μM of the NEP and ACE inhibitors Phosphoramidon or Thiorphan and Captopril, respectively, (Sigma-Aldrich) or various other protease inhibitors [7] were added to the membranes 5 min prior to the addition of peptides. For incubation with recombinant human NEP, peptides were incubated with 10 pmol (about 1.0 μg) rhNEP (generously provided by Nigel W. Bunnett, UCSF San Francisco) in 0.05 M Tris-HCl pH 7.4. To terminate the reactions, samples were placed on ice and/or NEP or ACE inhibitors, respectively, were added in a final concentration of 10 μM.

Peptide Purification and MALDI-MS

Contamination with glycol-derived or other polymeric components presents a major difficulty in preparing biomolecular analytes for MS and created substantial problems here. These compounds are used for preservation or anti-freezing purposes as well as for sample solubilization and they are hard to remove using common reversed-phase solid-phase extraction. Residual polymers or molecules prone to extensive gas phase adduct formation affect the quality of mass spectrometric data, because they ionize easily and can effectively suppress all analyte information (figure 1). In our procedures contaminants originated either from protease inhibitor cocktail (Complete, Roche) or from Amicon Microcon® filters devices (Millipore Corporation, Bedford, MA, USA) which were used for separation of peptides from higher molecular components. While the first problem was solved by the use of other inhibitors, the second could only be approached by extensive washing of the filters prior to separation. This appeared to be the critical step. Soluble peptides were separated from cell membranes or recombinant proteins by ultrafiltration at 4 °C through Amicon centrifugal filter devices with a 10 kDa cut-off, which were prepared by 2 washes with 500 μl H_2O/15 acetonitrile according to the manufacturer's protocol. This allowed successful analysis in most cases although occasionally residual contaminant signals were detected (figure 2) indicating the need for even more washing. Samples were desalted into a volume of 10 μl using C_{18} ZipTips following the manufacturers protocol (Millipore). Peptide maps were generated with TofSpec-2E instrument (Micromass/Waters, Manchester, UK) in reflectron mode using α-cyano-4-hydoxycinnamic acid as matrix. 1 μl of analyte was applied.

Above procedure was standardized and routinely used for the measurement of POMC peptide degradation products.

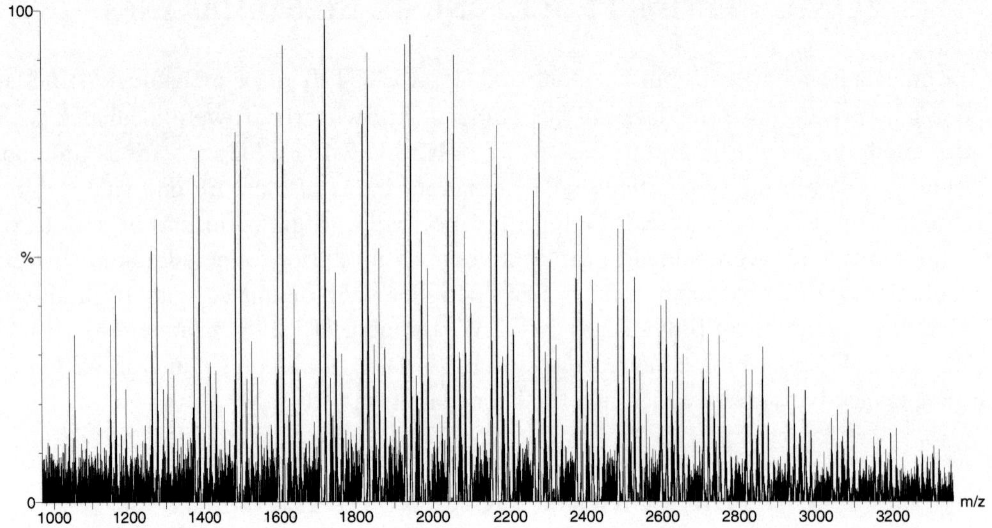

Figure 1. MALDI-MS of membrane preparation containing protease inhibitor Complete.

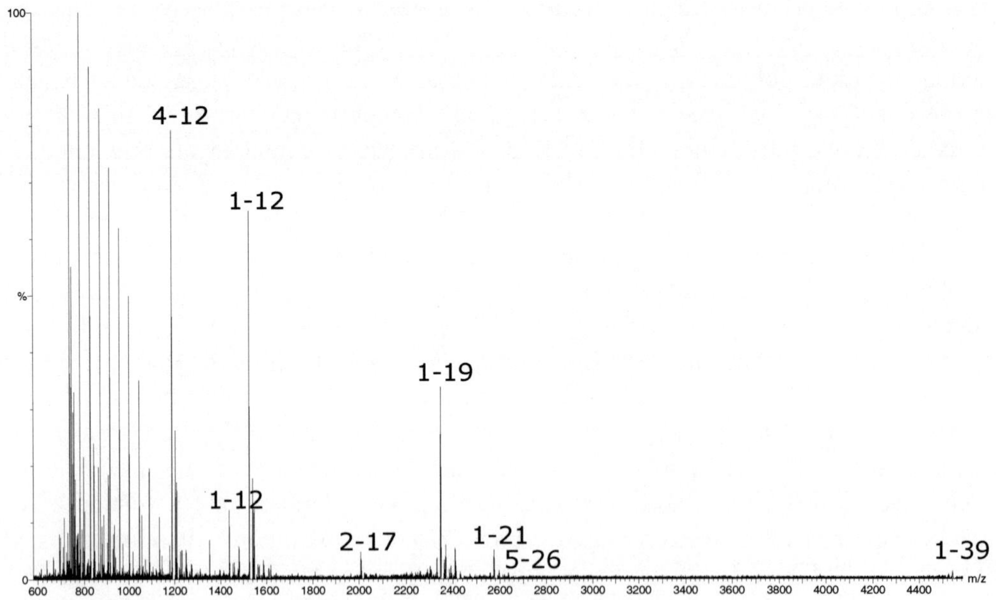

Figure 2. MALDI-MS of 30 min NEP incubation of $ACTH_{1-39}$ showing residual glycol in the range m/z 600-1200. Peptide peaks were labelled according to the fragment amino acids of $ACTH_{1-39}$.

ACKNOWLEDGMENTS

The authors thank A.-M. Mehlich and M. Fastrich for technical assistance and the Innovative Medical Research Münster for funding.

REFERENCES

[1] S. König, M. Fastrich, T.A. Luger, and T.E. Scholzen 52nd ASMS, Nashville, TN, May 2004

[2] T.E. Scholzen, S. König, M. Fastrich, M. Böhm, and T.A. Luger. *Exp. Dermatol.* 13, 9 (2004) 587-588

[3] T.E. Scholzen, S. König, M. Fastrich, Böhm, M., and T.A. Luger. *Exp. Dermatology.* 14, 2 (2004) 158

[4] T.E. Scholzen, S. König, M. Fastrich, and T.A. Luger. *J. Investigative Dermatology.* 122, (2004) 3 A3 018

[5] T.E. Scholzen, S. König, M. Fastrich, M., Böhm, T., Brzoska, T.A., Luger, *J. Investigative Dermatology.* 125 (2005) 273 A47

[6] T.E. Scholzen, T.A. Luger, S. König (2006) *J. Investigative Dermatology,* (2007) in print

[7] S. König, T.A. Luger, T.E. *Scholzen Exp. Dermatol.* 15, 10 (2006) 751-761

[8] T.E. Scholzen, D.-H. Kalden, T. Brzoska, T., M. Fastrich, T. Schwarz, M. Schiller, M. Böhm, C.A. Armstrong, J.C. Ansel, T.A. Luger, T. A. *J. Invest. Dermatol.* 115 (2000) 1021-1028

In: Biomacromolecular Mass Spectrometry Yearbook. Volume 1 ISBN: 978-1-60876-167-8
Ed: Simone Koenig

Chapter 14

DOWNSCALING ANALYSIS: 3RD MÜNSTER CONFERENCE ON SINGLE CELL AND MOLECULE TECHNOLOGIES; DECEMBER 4-5, 2006

Organizers: Kurt Sieberns, Rita Naskar and Simone König

Interdisciplinary Center for Clinical Research, Core Facility Integrated Functional Genomics, Münster, Germany

Chairs: Franz Hillenkamp, Stephan Ludwig, Michael Schäfers, Frank-Ulrich Müller, Simone König

INTRODUCTION: PROTEOMICS, GENOMICS, CYTOMICS...

Simone König

Integrated Functional Genomics, Röntgenstr. 21, 48149 Münster,
e-mail: koenigs@uni-muenster.de

The Münster Conference Series on Single Cell Analysis brings together scientists of very different fields – physicists, physicians, biologists, chemists – with the common goal of studying events on a cellular, sub-cellular, and molecular level. The impact of mass spectrometry in that area is increasing and, therefore, the conference is monitored in this journal.

This 3rd conference started with a keynote on neuropeptide imaging with mass spectrometry (MS) given by Jonathan Sweedler (University of Illinois). This particular technique was further evaluated in a special session where the status of the current instrumental developments was given along with recent experiments in infrared orthogonal time-of-flight (TOF) by Klaus Dreisewerd (Münster). In the discussion the opinion was agreed on that imaging MS is valuable in cases of localized high concentrations of particular peptides such is the case in neurons. There, information would be lost using classical homogenization techniques. In addition, imaging MS is expected to have considerable use for tracing of small molecules in pharmaceutical oriented dose studies. High molecule tissue profiling, however, although possible, seems to provide little further understanding.

TOF-SIMS (secondary-ion MS) has advanced more into the biological field due to the use of cluster primary ions (e.g. Bi3+, C60+) which limit sample destruction upon ion impact (Daniel Breitenstein, Münster). Lipids can be observed with the technique and other smaller molecules m/z <2000 to some extend while larger biomolecules are fragmented. As a special guest the conference featured Thomas Stephan (Münster) who became famous for his involvement in the NASA Stardust mission studying cometary dust using TOF-SIMS.

The conference presented further interesting topics such as cytomics which is based on the differential analysis of molecular cell phenotypes as obtained by the analysis of multitudes of single cells by multiparameter flow or image cytometry followed by bioinformatics knowledge extraction (Günther Valet). Single object tracking was discussed as a tool to analyze virus entry and cell biological transport phenomena (Mario Schelhaas, Zürich). Klaus Gerwert introduced his technique of monitoring proteins in action by time-resolved FTIR spectroscopy. Defined tissue extraction and RNA preparation were topics of other talks. All together the conference touched several important subjects and was a great success.

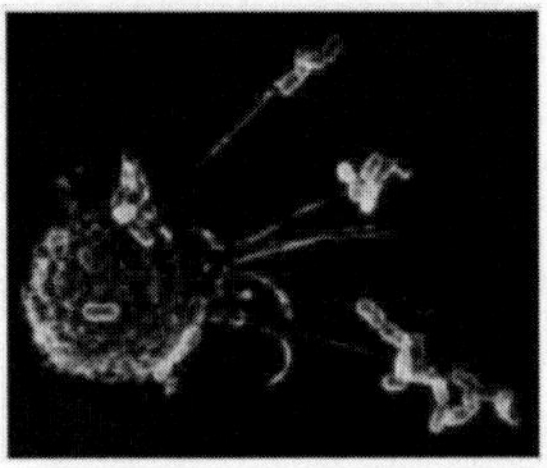

SPEAKER ABSTRACTS

"TECHNIQUES FOR PROBING THE BRAIN'S CHEMISTRY ONE CELL AT A TIME"

Jonathan V. Sweedler

Department of Chemistry, University of Illinois, 71 Roger Adams Lab,
600 South Mathews Avenue, Urbana, IL 61801

Understanding the functioning of networks of neurons is hampered by a lack of knowledge of the full complement of neuro-transmitters and neuromodulatory compounds used in most neuronal systems. As neurotransmitters range from gaseous molecules such as NO to large peptides that are only bioactive with particular posttranslational modifications, a variety of capillary separations and sample limited mass spectrometric approaches have been developed. Several capillary electrophoretic (CE) and capillary liquid chromatographic (cLC) methods allow neurotransmitters to be measured at individual neurons and neuronal release sites. Mass spectrometric approaches ranging from single cell mass spectrometry (MS), single bead solid phase extraction and MS-based imaging are described.

Besides new technology, several applications of single cell separations are highlighted: novel indolamine neurochemistry, determining the role of d-Asp in the brain, and single cell peptide measurements. Specifically, new serotonin-related compounds and literally hundreds of new neuropeptides have been characterized in well-defined neuronal networks, and in several cases, the functional roles of these molecules described.

"FIRE AND ICE – FIRST RESULTS FROM THE SPACE MISSION STARDUST"

Thomas Stephan

Institute of Planetology, University of Münster, Wilhelm-Klemm-Str. 10, 48149 Münster, Germany

Comets are probably the most primitive objects in our solar system. Since the formation of the solar system some 4.6 billion years ago, these objects spent most of their lifetime in the outer, cold regions of the solar system, in the Kuiper belt or in the Oort cloud. Due to gravitational perturbations, they eventually get into the inner regions of the solar system, where they form remarkable tails through the interaction with solar wind and solar photons.

Although comets are believed to be ideally suited to study the early stages of our solar system and its origin, only little was known about these celestial bodies prior to NASA's space mission Stardust, the first mission to return solid samples from a specific astronomical body other than the moon.

Stardust returned its precious freight to Earth on January 15, 2006, after almost seven years of travel in interplanetary space. During the passage through the coma of comet Wild 2 in January 2004, Stardust collected cometary dust at a velocity of 6.12 km/s. During the cruise phase before the cometary encounter, a sample of contemporary interstellar dust was also col-lected. The primary collection medium for both cometary and interstellar dust was low-density silica aerogel, arranged in 132 individual cells with a total area of 1039 cm2 for each collection. In addition, ~153 cm2 of aluminum foil was exposed to each dust source. Aerogel, a solid with extremely low densities of 5–50 mg/cm3 was used to gently decelerate the im-pinging dust grains, while particles cannot survive unaltered an impact at ~6 km/s on Al foil, but generate small craters. However, in most cases, residual material can be found associated with the craters that allow an identification of the respective impactors.

After Stardust's return to Earth and recovery of the capsule in the Utah desert, different types of samples, entire particle tracks in aerogel, individual particles extracted from aerogel, or impact craters on Al foil, were distributed to selected scientists all over the world. During a so-called preliminary examination phase, lasting for six months, these scientists addressed some of the key questions relating to the elemental, isotopic, organic, and mineralogical composition, spectroscopic characteristics, and size distribution of Wild 2 dust.

First results showed that during formation of Wild 2 in the cold regions of the solar system, high-temperature minerals, which probably formed close to the early sun, were incorporated. Such large-scale mixing in the early solar system was not expected before.

"CYTOMICS AND THE RESOLUTION OF ORGANISMAL BIOCOMPLEXITY"

Günter K. Valet
Max-Planck-Institut fuer Biochemie, Am Klopferspitz 18a,
D-82152 Martinsried, Germany

The high biocomplexity of organisms is caused to a significant degree by the functional specialization of cells as well as by the organisation of various cell types in form of tissues and organs.

Molecular disregulations in cells or cellular systems may cause disease in humans, prompting for the characterisation of molecular pathways for example during cell proliferation, differentiation, signaling or apoptosis in a *bottom-up* direction that is from the genome over the proteome up to the cell system and organism levels to find the molecular causes of a disease. Molecular pathways can be mathematically modeled in a *hypothesis-driven* way. The number of components in the model can be gradually increased or model cell systems can be perturbed, followed by molecular network analysis as in *systems biology*. The bottom-up approach is complex and it is uncertain whether clinically useful biomarker patterns can be obtained to predict therapy outcome in individual patients (*personalised, individualised medicine*) in a foreseeable time.

Alternatively, available molecular patient data can be used *top-down* in a comparatively *hypothesis-free, data-driven* way (*inductive discovery* science). Data may be obtained for example by the *differential* analysis of molecular cell phenotypes as obtained by the analysis of multitudes of single cells by multiparameter flow or image cytometry followed by bioinformatic knowledge extraction from the multiparametric data (*cytomics*). Since diseases are caused by molecular aberration in cells as the elementary function units of organisms, *molecular cell phenotypes* of diseased or disease associated cells in affected body compartments correlate with the molecular disease processes as consequence of the sum of *genotypic* disposition and *exposure of the organism to* disease inducing conditions.

Differential cell phenotype analysis may concern therapy response versus non-response, disease progression versus stationary state or survivor versus non-survivor patients. Additional patient specific clinical chemistry or clinical parameters can be included into the differential cell analysis data to initially enrich the discriminatory parameters from the highest possible number of molecular parameters. Discriminatory parameters are enriched as standardised and hospital independent *biomarker profiles* for diagnosis or *therapy dependent predictions* about disease progression of individual patients in stratified patient groups (*predictive medicine by cytomics*) while non-discriminatory parameters are eliminated by *algorithmic* "data sieving". Such biomarker profiles may serve as starting points for retrograde analysis that is *molecular reverse engineering* of underlying disease pathways by systems biology (-> *drug discovery*).

No detailed *a-priori* knowledge on specific disease processes is initially required for the analysis of the entire observable cellular heterogeneity of affected tissues downwards to the organelle and molecular network levels. Due to its mainly *inductive* nature, the cytometric approach may open so far *hidden* knowledge spaces, *inaccessible* to traditional *deductive hypothesis*. The systematic analysis of bioparameter profiles may furthermore permit to discriminate between preferentially genotypic or exposure induced disease at the individual patient level, thus facilitating early *preventive* measures in case of mainly exposure induced disease. The validity of this approach will be shown for two clinical multicenter leukemia studies.

"Combined Cell Sorting and Few Cell Gene Expression Profiling for the Characterisation of GABAergic Interneurons and Neural Precursors"

Andreas Bosio
Miltenyi Biotec GmbH, MACSmolecular Business Unit,
Stöckheimer Weg 1, D-50829 Cologne, Germany

Gamma-aminobutyric acid (GABA)ergic neurons play a crucial role in information processing as they regulate the activity of other neurons in the central nervous system. Accordingly, the massive loss of GABAergic neurons is central to neurodegenerative diseases like Huntington disease. To extend the knowledge about GABAergic neurons and neural precursors we have isolated and characterized several populations of GABAergic neurons differing in their spatiotemporal distribution as well as neuronal precursor defined by the expression of certain surface markers. A two-step, mild enzymatic digestion protocol (Neural Tissue Dissociation Kit) was developed to produce a single cell suspensions of brain tissue, focusing on high yield of living cells and conservation of antibody epitopes for subsequent FACS and MACS applications. GABAergic neurons were isolated from different brain regions of glutamic acid decarboxylase–green fluorescence protein (GAD67-GFP) knock-in mice by microdissection or fluorescence activated cell sorting (FACS). Using surface markers like A2B5 (glial precursors), CD24 (neuronal precursors, ependymal cells) and PSA-NCAM (neuronal and oligodendroglial precursors), different populations of neural precursors were isolated by magnetic cell sorting (MACS).

As the number of isolated cells was partly very low, we have used a new single cell amplification procedure (SuperAmp) combining bead based mRNA extraction and reverse transcription, global PCR amplification and Klenow fragment labelling to generate gene expression profiles.

"Single Object Tracking as Tool to Analyze Virus Entry and Cell Biological Transport Phenomenon"

M. Schelhaas, H. Ewers, I. Sbalzarini, P. Day, J. Schiller and A. Helenius
Swiss Federal Institute of Technology Zurich, ETH Hoenggerberg, HPM E10.1
Schafmattstrasse 18, 8093 Zürich

Viruses are excellent objects to study not only viral pathogenesis but also cell biological processes. Viruses use a 'Trojan horse' strategy to enter their host cells. Entry, i.e. the delivery of the viral genome and accessory proteins to the cell interior, is a highly dynamic process and involves binding to cell surface, diffusion and/or transport on the cell surface, internalisation - often by endocytosis -, and intracellular transport. Here, we used high resolution time-lapse microscopy to follow single viruses during entry. As tool to analyze this data quantitatively we developed single object tracking software that we applied during various steps of virus entry. To highlight the use and limitations of this approach examples on diffusion and active transport of viruses will be presented.

"AQUAPORIN 2 ACTIVITY IN RENAL EPITHELIAL CELLS MONITORED WITH DIGITAL HOLOGRAPHIC MICROSCOPY"

Jens Klokkers[1], Patrik Langehanenberg[2], Bayram Edemir[1], Björn Kemper[2], Gert von Bally[2] and Eberhard Schlatter[2]

[1] University Clinic Münster, Experimentelle Nephrologie, Domagkstr. 3a, D-48149 Münster

[2] Laboratory for Biophysics, Robert-Koch-Str. 45, D-48129 Münster

Water permeability across biological membranes is supported by aquaporins in the plasma membrane. In the kidney the final step in urine concentration occurs in the collecting duct. In this segment aquaporin-2 (AQP2) in the luminal membrane of principal cells is the final target of vasopressin (AVP). AVP binds to the vasopressin 2 receptor activating the adenylate cyclase via cAMP, which leads to activation of PKA. Phosphorylation of AQP2 by PKA leads to translocation of AQP2 from intracellular vesicles to the luminal membrane. To further study AQP2 regulation we present a method to dynamically monitor regulation of water permeability in living cells.

In this work we examined the influence of AVP on water permeability of primary cultured principal cells of rat inner medullary collecting ducts (IMCD). IMCD-cells endogenously expressing AQP2 were grown to confluency in cell culture medium. For quantitative phase contrast imaging of living IMCD cells digital holographic microscopy was applied. As this marker-free technique is non-scanning it permits the quantification of the swelling process with a high spatial and temporal resolution (up to video repetition rate). In time-lapse measurements dynamic changes in the cell thickness and shape were investigated. A hypoosmolar change of the growth medium from 600 to 200 mosm/kg was used to stimulate water influx across the cell membranes. Cell thickness was plotted with time followed by regression analysis. Control cells were compared to cells stimulated with AVP (0.5 μM, 30 min). With digital holographic microscopy the changes in cell thickness as a result of changes of water permeability of the membrane could be recognized in IMCD cells. Kinetics of changes of the cell thickness was described by sigmoidal curve fitting allowing determination of the time constant τ and maximal changes in optical length (OPL).

τ after this hypoosmolar shock in control cells was 101 ± 4 s and ΔOPL was approximately 55%. τ and ΔOPL decreased in these first measurements after AVP by approximately 5% and 29%, respectively (n=4 each).

In conclusion, the digital holographic microscopy is a suitable method to measure changes in cell thickness in living renal cells to indicate changes in water permeability with combined high temporal and spatial resolution.

"FOLLOWING PRIMARY HYPERPARATHYROIDISM FROM SURGERY TO PROTEOMIC ANALYSIS"

Payam Minoofar[1], Martin Eisenacher[2], Dave S.B. Hoon[1], Christian Stephan[2] and Frederick R. Singer[1]

[1] John Wayne Cancer Institute, Santa Monica, California, USA.
[2] Medizinisches Proteom-Center, Ruhr Universität Bochum, Germany

Primary hyperparathyroidism is a serious disorder of bone metabolism that often requires surgical treatment. The main causes of primary hyperparathyroidism are a single adenoma (85%) and hyperplasia of all four glands (10-12%) These two forms of the disorder are frequently indistinguishable prior to surgery. The diagnosis is made at the time of surgery, and identifying the particular form prior to surgery remains a challenge. A proteomic study was therefore undertaken to determine if any fundamental differences between the two forms of primary hyperparathyroidism exist. Parathyroid glands removed from patients were frozen, fixed in optimal cutting temperature (OCT) matrix and cut into 8 μm slices and stained. Three groups of cells—adenomatous, hyperplastic, and normal—were selectively harvested from these slices with laser capture microdissection (LCM). These cells were then lysed with detergent, and the proteins were denatured with urea and assayed with a Ciphergen time of flight (TOF) matrix assisted laser desorption ionization mass spectrometer (MALDI-MS). The resulting proteomic profiles of the three groups were analyzed and compared via statistical tests and classification. As a proof of principle the raw data were used; an alternative method is the use of scaled spectra or calculated peak lists.

"COMBINED TRANSCRIPTOME, PROTEOME, AND METABOLOME ANALYSIS AT A CELL TYPE-SPECIFIC LEVEL"

Julia Kehr

Max-Planck-Institut für Molekulare Pflanzenphysiologie;
Am Mühlenberg 1; 14476 Golm/ Potsdam, Germany

Knowledge of the local distribution of transcripts, proteins, and metabolites between different cells and tissues is a presupposition for understanding multi-cellular organisms like higher plants. However, most experimental approaches in plant science as yet average several different cell types.

The use of microcapillaries is one well-established method to obtain the contents of individual cells from living plants. Caused by the small volumes that can be extracted, however, highly specialised strategies for downstream analyses have to be employed. Laser microdissection coupled to laser pressure catapulting (LMPC) has only recently entered plant science and allows the enrichment of specific cell types from histological sections in a completely touch-free manner. This technique makes tissue-specific samples accessible to standard analytical strategies. However, truly comprehensive molecular profiling is still a challenge. One major difficulty is the preparation of tissue sections that retain enough histological detail to reliably identify the cell type of interest while keeping all analytes intact, extractable and localised at their in vivo position. We have implemented methods that allow a comprehensive analysis of transcripts, proteins, and metabolites in parallel samples at a cell type-specific level.

"Multiplex Mirna Profiling from Single Embryonic Stem Cells and Early Embryos"

Ralph Vogelsang
Applied Biosystems, Frankfurter Str. 129B, D-64293 Darmstadt, Germany

MicroRNAs are short (17-25 nucleotides), non-coding RNAs that play critical roles in gene regulation and cellular differentiation during development. Recently developed miRNA microarray techniques have contributed greatly to miRNA research but require too large an RNA sample to be used in many crucial studies, such as developmental studies involving primordial tissue samples from embryos or laser-captured samples from developing tissues. Here we report a real-time PCR 220-plex miRNA expression profiling method that is sensitive and accurate enough to profile miRNA from samples as small as a single cell. MicroRNA expression profiles have been measured for single mouse embryonic stem cells and for early mouse embryos at the zygote, two-cell, and four-cell stages. This method should also be useful in cancer studies where only limited amounts of material are available from tissue biopsies or archived material.

"Intraneuronal Crystallin Beta B-2 (Crybb2) Triggers Elongation of Axons"

Thomas Liedtke, Uwe Schröer,
Jens Christian Schwamborn and Solon Thanos
Experimental Ophthalmology, Domagkstrasse 15, 48149 Münster, Germany

Mature neurons of the central nervous system (CNS) fail to regenerate their axons within their natural environment after injury and are induced to degenerate. Adult retinal ganglion cells (RGCs) can regenerate their axons both in vivo and in vitro and serve as a model to examine mechanisms of brain repair. Using regenerating retina tissue in culture and proteomics, we have discovered that the supernatants of cultured retinas contain isoforms of crystallins, with crystallin beta-b2 (crybb2) being clearly upregulated in the regenerating retina when compared with controls. Immunohistochemistry revealed the expression of crybb within the retina, including in filopodial protrusions and axons of RGCs. Cloning and overexpression of crybb2 in RGCs and hippocampal neurons increased axonogenesis, which in turn could be blocked with antibodies against beta crystallin. Conditioned medium from crybb2-transfected cell cultures also supported the growth of axons. Finally, real-time imaging of the uptake of GFP-tagged-crybb2 fusion protein in single cells showed that this protein becomes internalized and redistributed within the cell including the axons. These data are the first to show that axonal regeneration is related to crybb2 movement. The results suggest that neuronal crystallins constitute a novel class of neurite-promoting factors that likely operate through an autocrine mechanism, and that they could be used in neurodegenerative diseases.

"Proteins in Action: Monitored by TR (Time-Resolved) FTIR Spectroscopy"

Klaus Gerwert

Lehrstuhl für Biophysik, Ruhr-Universität Bochum,
Gebäude ND 04 Nord; 44780 Bochum, Germany

In the Postgenom era proteins are coming into the focus in the life sciences. Proteins are the nanomachines that perform the work in living organisms or are the receptors and mediators for external signals. By NMR and x-ray the three dimensional structural architecture of proteins are determined. In order to elucidate the function, time-resolved methods have to be applied.

FTIR difference spectroscopy can be used to monitor the reactions within proteins at the atomic level with ns time-resolution up to days [1]. In combination with site directed mutagenesis or isotopically labelling the IR bands can be clear cut assigned to specific amino acids or ligands. This provides in combination with structural models also spatial resolution.

Based on fast scan studies on bacteriorhodopsin the key catalytic residues, asp 85 and asp 96 an their protonation kinetics are identified and summarized in a first detailed proton pump model [2]. Their structural arrangement as resolved in succeeding x-ray experiments by several groups supports this proposal. The X-ray structural model at 1.55 A resolves in addition the oxygens of internal water molecules. Based on succeeding step scan FTIR measurements the interplay between these water molecules, a strong hydrogen bonded water, a dangling water and a protonated water complex is elucidated in detail. It results in a controlled Grothuis proton transfer from the central proton binding site to the protein surface. [3,4]. A similar mechanism might apply in the photosynthetic reaction center [5] and the cytochrome oxydase [6]. The step scan approach is also successfully applied to the photoactive yellow protein [7].

The difference technique requires fast triggering of the protein reaction, which is easy to accomplish for the chromoproteins as described before. Progress for the investigation of non chromophoric proteins is aquired by developing a micro mixing cell for FTIR studies, allowing mixing times in the sub ms time range. This cell is used to investigate protein folding reactions [8].

Alternatively, photolabile caged compounds can be applied. Using caged GTP the GTPase mechanism of the protooncogen Ras is investigated [9, 10]. Also its protein protein interaction with the GAP protein could be studied time-resolved [11,12]. This provides a detailed insight into the catalytic mechanism by which GAP activates the GTPase by five orders of magnitude. The activation by GAP proteins is a central process in the signal transduction. In oncogenic ras this activation process is inhibited and involved in uncontrolled cell growth. The study proves that the approach can be extended to protein-protein interactions.

REFERENCES

[1] Kötting, C., Gerwert, K., *Chem. Phys. Chem,* 6, 881-888 (2005)
[2] Gerwert, K., Souvignier, G. and Hess, B., *Proc. Natl. Acad. Sci. USA,* 87, 9774-9778, (1990)
[3] Garczarek, F., Brown, L.S., Lanyi, J.K., Gerwert, K., *Proc. Natl. Acad. Sci. USA,* Vol. 102, No. 10, 3633-3638 (2005)
[4] Garczarek, F., Gerwert, K., *Nature,* 439, 109-112 (2006)
[5] Remy, A., Gerwert, K., *Nature Struct. Biol.* 10, 637-644 (2003)
[6] Lübben, M., Gerwert, K., *FEBS Letters* 397, 303-307 (1996)
[7] Brudler, R., Rammelsberg, R., Whoo, T. T., Getzoff, E. D., Gerwert, K., *Nature Structural Biology,* 8, 265–270, (2001)
[8] Kauffmann, E., Darnton, N. C., Austin, R. H., Gerwert, K., *Proc. Natl. Acad. Sci., USA* 98, 6646–6649, (2001)
[9] Cepus,V., Goody, R.S., Gerwert, K., *Biochemistry,* 1998, 37, 10263-10271, (1998)
[10] Allin, C., Gerwert, K., *Biochemistry,* 40, 3037–3046, (2001)
[11] Allin, C., Ahmadian, M. R., Wittinghofer, A., Gerwert, K., *Proc. Natl. Acad. Sci., USA* 98, 7754–7759, (2001)
[12] Hermes S, Stachnik JM, Onidas D, Remy A, Hofmann E, Gerwert K, *Biochemistry,* in press (2006)

SINGLE CELL DETECTION BY TIME OF FLIGHT SECONDARY ION MASS SPECTROMETRY"

Daniel Breitenstein, E. Tallarek, M. Fartmann, R. Kersting and B. Hagenhoff

TASCON GmbH, Materials Characterisation,
Gievenbecker Weg 15, 48149 Münster, Germany

Time-of-Flight Secondary Ion Mass Spectrometry (TOF-SIMS) has proved to be a powerful tool in organic surface analysis. Meanwhile also the analysis of single cells has become within reach. Driven by instrumental advancement now two-dimensional and three-dimensional images on the distribution of organic chemical components within a cell preparation are possible. This contribution reviews the state of the art in single cell detection by TOF-SIMS.

As for all mass spectrometric techniques, for TOF-SIMS an ionization process has to be take place prior to separation and detection of the analyte. In TOF-SIMS this process occurs upon impact of primary ions (e.g. Bi1+) of a highly focussed ion beam. This impact leads, among others, to the desorption of secondary ions from the surface of the investigated sample. These secondary ions can be mass separated and detected by a Time-of-Flight analyzer. For single cell detection two TOF-SIMS application modes are relevant: The imaging mode and the so called 3D microarea analysis mode.

TOF-SIMS imaging is based on the two-dimensional mass spectrometric detection of chemical components. As the focussed primary ion beam can be rastered over the surface of the sample, the intensity of all detected analytes in each rastered pixel can be displayed along a colorscale. This procedure leads to so called mass resolved images mapping each detected signal. TOF-SIMS imaging is to date applied to tissue slides as well as cell culture samples allowing a lateral resolution ca. 0.3 μm. Sample destruction upon ion impact can nowadays be limited using cluster ions as primary ions (e.g. Bi3+, C60+).

A prolonged primary ion bombardment of the sample leads to sputter erosion of the surface. Mass analysis of the sputtered material allows to identify the vertical and horizontal composition of the sample. This so called 3D microarea analysis mode of TOF-SIMS allows a three-dimensional mapping of the detected analytes. The vertical resolution for cell material was shown to be better than 100 nm. Nevertheless, a vertical resolution of 30 nm might be achievable, as this was already shown for organic polymer films.

"MULTIPLEX RT-PCR OF DNA MODIFICATION GENES IN SINGLE BLASTOMERES OF MOUSE PREIMPLANTATION EMBRYOS"

Andreas May, Ulrich Zechner and Thomas Haaf
Institut für Humangenetik, Klinikum der Johannes Gutenberg-Universität; Langenbeckstr. 1, D-55131 Mainz, Germany

Over the last years, RT-PCR has become the preferred method for validating results obtained from microarray analyses and other techniques that evaluate gene expression changes on a global scale. RT-PCR is the most sensitive technique for mRNA detection and quantification currently available. On the other hand, there is an increasing requirement of techniques for expression analysis at the single cell level. We are specifically interested to study the expression of known and candidate genes for nuclear reprogramming in individual cells of preimplantation embryos. In a pilot study, we are using the AmpliGrid technology (Advalytix, Brunnthal) for comparing expression profiles in mouse blastomeres.

The AmpliGrid is a glass chip with a surface structure that allows for the specific positioning of one single microliter on each of the 48 discrete reaction sites (enabeling to run 48 1 μl reactions in parallel). Single cells are placed under optical control onto these reaction sites followed by direct amplification (RT-PCR).

In a first set of experiments we have identified appropriate control genes, which are expressed at more or less constant levels in early embryonic cells, and compared expression of a dozen study genes that are involved in DNA modification and epigenetic reprogramming of the two germ line genomes after fertilization in individual cells of mouse morula-stage embryos. Preliminary results suggest that the transcriptome encoding the reprogramming machinery and, by extrapolation, nuclear reprogramming differs between blastomeres.

"Non-Contact Microdissection for Pure DNA, RNA, Proteins and Live Cells" (Bridging the Gap between Microscopy and Molecular Analysis)

Renate Burgemeister

P.A.L.M. Microlaser Technologies AG, Am Neuland 9+12,
82347 Bernried, Germany

Modern molecular biomedical research relies on the capability of pure sample preparation. Amongst various options to achieve homogeneous samples, only laser microdissection and micromanipulation offers high-resolution control of sample composition by selecting or rejecting individual cells.

Tissue preparation and extraction protocols allow the utilization of microsamples for qualitative and quantitative molecular and proteomic analyses like, e.g., PCR and RT-PCR amplification, microarray analysis, and MALDI/SELDI spectrometry. Laser Microdissection and Pressure Catapulting (LMPC) results in an eminent increase in the specificity of downstream analyses.

The PALM MicroBeam combines laser technology with high quality robotic tools for precise microdissection of specimens, whilst the patented pressure catapulting feature allows for non-contact collection with no impairment to the recovery of DNA, RNA or protcins. Thc integration of image analysis platforms to the LMPC technology fully automates screening, identification and finally subsequent high-throughput sample handling.

We will show our results of single cell LMPC and subsequent measurement of expression profiles of individual cells. Using LMPC and real-time PCR in combination with a new mRNA extraction kit we analyzed the expression of murine porphobilinogen deaminase (PBGD) gene from single microdissected frozen tissue cells, and of human PBGD gene from single live cells from cell culture, respectively.

Furthermore, in forensics the isolation of single cells is also an important issue. Collection of spermatozoa and epithelial cells from smear preparations is feasible in a fast and convenient way.

The technology of LMPC enables a big step forward to the contamination-free analysis of small samples. It opens a wide field of interesting applications in cell and developmental biology or pharmacology where homogeneous cell clusters or defined clones are needed.

"MOLECULAR PROFILING OF HERBAL AND ANIMAL TISSUE BY DIRECT INFRARED LASER DESORPTION IONIZATION MASS SPECTROMETRY"

Klaus Dreisewerd
Institute of Medical Physics and Biophysics, University of Münster,
Robert-Koch-Str. 31, D-48149 Muenster, Germany

We have developed infrared laser desorption ionization mass spectrometry (IR-LDI-MS) to generate molecular profiles directly from native tissue or from natural products with high mass accuracy. The method requires little sample preparation besides for an eventual dissection of the areas of interest and drying of particular water-rich samples. The spatial resolution of the analysis is to first order defined by the focal laser spot size and the ablation depth of material removal of a few micrometer per pulse. Moreover, the amount of ablated material can be controlled by the laser pulse energy and focal spot size, and the laser thus employed as a precision surgical tool. Various types of small molecules are visualized in herbal tissue. Among these are carbohydrates, phospholipids, triglycerides, waxes, and flavonoids. The potential of the method for a "3-dimensional" molecular analysis is furthermore demonstrated. Substantially different molecular profiles are for example detected from the membrane surrounding and the inside of seeds.

"FROM MASS IMAGING TO CLASS IMAGING – NEW HORIZONS IN MALDI MOLECULAR IMAGING"

Sören-Oliver Deininger, Arnd Ingendoh and Martin Schürenberg
Bruker Daltonik GmbH, Fahrenheitstr. 4,
28359 Bremen, Germany

MALDI Imaging is a powerful new tool for the proteomic characterization of tissue sections. The conventional approach towards MALDI imaging is to acquire spectra with spatial resolution over the tissue and assemble images based on the intensities of selected mass signals. This approach is usually sufficient to locate histological features (such as the tumor area) on the tissue. This approach lacks the possibility to classify unknown tissue, because single peptide/protein markers are not sufficient for classifications and the possibilities to perform statistical analyses are limited. Here we present a fundamentally different approach to MALDI imaging: The mass spectra are used to generate statistical models to allow a classification of the spectra based on similarity. The parameter used for generating the molecular image is then the similarity of the mass spectra with the reference model. This allows the unambiguous classification of tissue by mass spectrometry, and also allows the correlation of whole mass spectra with clinical data. We estimate this approach as very relevant for the diagnostic and prognostic use of MALDI imaging of tissue samples.

"ADVANCES IN TISSUE IMAGING USING A MALDI QUADRUPOLE OA-TOF MASS SPECTROMETER"

Ronan O´Malley, Marten Snel, Daniel Kenny, Emmanuelle Claude, Richard Tyldesley-Worster and James Langridge

Waters MS Technology Centre, Manchester M22 5PP, UK

Imaging the spatial distribution of molecules in tissue using MALDI mass spectrometers is a rapidly developing technique. The acquisition of accurate mass data in this type of experiment can be hampered in axial MALDI Tof systems. It has long been recognised that even small changes in sample position and laser energy in the source region of this type of mass spectrometer affect mass measurement accuracy and mass spectral resolution. Here, we show how the use of an orthogonal Tof MALDI mass spectrometer circumvents these problems by decoupling the MALDI source from the mass analyser

Imaging data are acquired on a MALDI Q-Tof Premier (Waters, Manchester) mass spectrometer. The tissue sections are mounted on a target plate and moved in a raster pattern relative to the laser. To reduce interference from the biological matrix and enhance specificity the instrument is operated in MS/MS mode, a quadrupole is used for precise precursor ion selection. The sensitivity of specific ions can be further enhanced by synchronising the high voltage push of the Tof mass analyser with the arrival of ions of appropriate m/z in the acceleration region.

MALDI imaging information was obtained from thin sections of rat brain from animals doped with the well studied D_2/D_3 dopamine receptor antagonist Raclopride. Data obtained on the spatial distribution of Raclopride and endogenous adenosine monophosphate are presented. Challenges in sample preparation are discussed.

"IMAGING OF PROTEINS AND SMALL MOLECULES USING THE 4800 MALDI TOF/TOF ANALYZER"

Andrew James, Matthew Willetts, Matthias Glueckmann and Volker Kruft

Applied Biosystems, Frankfurter Str. 129B, D-64293 Darmstadt, Germany

Introduction

The emerging technique of MALDI mass spectrometry imaging (MALDI MSI) offers the potential of profiling biomolecules and small molecules and their metabolites directly on tissue samples. The technology involves mounting thinly sliced tissue sections onto a MALDI sample plate followed by the application of a MALDI matrix solution. The matrix facilitates the desorption and ionization of labile analytes. A two dimensional array of mass spectra is obtained, combined to an ion intensity map allowing to locate analytes of interest on the tissue.

The 4800 MALDI TOF/TOF analyzer is an ideal instrument for tissue imaging due to its high precision of sample stage positioning of about 1 um and the approx. laser beam diameter on sample of about 50 um and its high speed of acquisition.

Experimental

Target tissue were harvested from rats, snap frozen and sliced using a cryostat. Tissue surfaces were coated using a-cyano-4-hydroxycinnamic acid solution. All MALDI MSI measurements were carried out using the 4800 MALDI TOF/TOF Analyzer in MS or MSMS mode. Data were acquired using the MSI acquisition tool available at http://www.maldi-msi.org. Data evaluation was performed using BioMAP Software available from the same webpage. Some sections were kindly provided by Novartis, Basel, Switzerland.

Results

The 4800 MALDI TOF/TOF Analyzer combined with BioMAP software represents a powerful platform for the imaging of peptides, proteins and small molecules.

Results obtained for imaging from rat brain sections are presented. Imaging results in MS reflector and linear mode data is presented. The MSMS capability of the 4800 instrument allowed identification of protein features of interest from rat brain sections.

In further MSMS experiments on small molecules the highly specific transition of a selected precursor mass to a fragment ion mass is used to obtain specific ion intensity maps of the analyte of interest. Data on rat sections of a complete rat dosed with a pharmaceutical compound is presented.

POSTER

Andreas Bauwens[1], Martina Bielaszewska[1], Helge Karch[1], Henning Schweppe[1], Alexander Friedrich[1], Rudolf Reichelt[2], Björn Kemper[3], Patrik Langenhanenberg[3], Gert von Bally[3], Jana Hüve[4], Martin Kahms[4], Reiner Peters[4], Michael Mormann[2], Jasna Peter-Katalinić[2] and Johannes Müthing[2]

[1] Institute for Hygiene, [2]Institute for Medical Physics and Biophysics,
[3] Laboratory of Biophysics,
[4] Institute for Medical Physics and Biophysics and CeNTech
University of Münster, D-48149 Germany

"Application of Nanotechnologies for Analyzing Virulence Factors of Pathogenic Bacteria: Shiga Toxin 1-Caused Damage of Human Brain Microvascular Endothelial Cells"

Shiga toxin-producing *Escherichia coli* (STEC) strains, especially of serotype O157:H7, are responsible for life-threatening zoonotic food- or water-borne illness consisting of diarrhea, hemorrhagic colitis, and hemolytic uremic syndrome (HUS). Serious systemic complications of STEC infection such as HUS, and the ability of these organisms to cause large outbreaks make these bacteria one of the most important emerging pathogens. HUS is histopathologically characterized by microvascular lesions in the renal glomeruli, the gastrointestinal tract, and other organs, e.g., pancreas, lungs, and brain, which result from the injury of microvascular endothelial cells by Shiga toxins (Stx), the major virulence factors of STEC. It is now evident that, when death occurs during HUS, it is most commonly associated with the damage to the brain. Specific therapeutic or preventive strategies are presently not available.

The integrity of the blood-brain barrier is critical for normal brain function. Human brain microvascular endothelial cells (HBMECs) form a unique, tightly interconnected, cellular monolayer. This highly selective impermeable barrier strictly controls the exchange of material between the blood circulation and the brain. Several studies suggest that Stx1 is involved in manifestation of neurological dysfunction, which is associated with endothelial cell injury in microcapillaries of the central nervous system. The processes by which Stx1 causes these pathological changes are not well understood.

Stx1 is a member of the so-called AB_5-class of bacterial toxins, consisting of a single A subunit (~33 kDa) and a pentamer of identical B subunits (~7.7 kDa each). The catalytic A subunit has RNA N-glycosidase activity and inhibits eukaryotic protein synthesis. The pentameric B subunit is responsible for binding to the functional cell-surface receptor glycosphingolipid (GSL) globotriaosylceramide (Gb3Cer/CD77). Gb3Cer (Galα1-4Galβ1-4Glcβ1-1Cer) consists of a lipophilic ceramide and a hydrophilic trisaccharide moiety. This oligosaccharide with the unusual terminal Galα1-4Gal-sequence represents the binding structure for Stx1.

Due to the clinical importance of this life-threatening toxin, a cooperation was established connecting biological and biophysical nanotechnologies for the *in vitro* and *in vivo* receptor-based functional analysis of Stx1-mediated interaction with HBMECs.

FT-ICR Mass Spectrometry

In comparison to other mass analyzers FT-ICR MS provides unique specifications with respect to resolving power, mass accuracy, high sensitivity, and dynamic range in molecular analysis. According to these properties FT-ICR MS, combined with immunochemical detection of GSLs, represents a perfect tool for analysis of GSLs with structural heterogeneities in the oligosaccharide and ceramide moieties. Here, we report about the structural characterization of several Gb3Cer species obtained from HBMECs. For this purpose, HBMECs were propagated on microcarriers in bench-scale bioreactors. The isolated GSLs were separated on high-performance thin-layer chromatography plates and the Stx1-receptors were detected by an ELISA analogous procedure directly on the HPTLC plate. The

Stx1-positive GSLs were extracted from the silica gel and submitted to FT-ICR MS analysis. Several chemically distinct Gb3Cer species with variations in their fatty acid chain lengths from C16 to C24 fatty acids were fully structurally characterized and identified as the functional Stx1-receptors.

4Pi Fluorescence Microscopy

The subcellular localization of the Stx1 receptor Gb3Cer has been analyzed using confocal and 4Pi microscopy in HBMECs. 4Pi microscopy enables three-dimensional imaging with an axial resolution down to ~110 nm by coherently adding the spherical wavefronts of two opposing lenses and utilization of two-photon excitation. Therefore a resolution along the optical axis is obtained which is 3-7 times better than in classical confocal microscopy. Cells were grown on quartz cover slips and processed for immunofluorescence by fixation, permeabilization, and successive incubation with first and secondary antibodies. Subsequent confocal imaging revealed the subcellular distribution pattern of the Stx1 receptor. Gb3Cer was found to be highly clustered in the plasma membrane, but a significant fraction was also located inside the cell probably associated with the endomembrane system. Furthermore, these intracellular receptor association sites have been structurally characterized in more detail using 4Pi microscopy.

High-Resolution Scanning Electron Microscopy

Scanning electron microscopy (SEM) of HBMECs exhibited the typical "cobble-stone pattern" of endothelial cells with strict contact inhibition. Upon incubation of confluent grown microcarriers with Stx1, the monolayers of HBMECs were found to be partly damaged. The most obvious toxin-mediated changes of the cellular shape were indicated by the loss of the monolayer integrity due to increase of spaces between the cells and a considerable loss of cell surface microvilli. This clearly reflects a moderate toxic effect of Stx1 towards HBMECs, which have been suggested for a long time being unsusceptible to the exposure of Stx1. Thus, Stx1-mediated mild deformation of HBMECs upon treatment with Gb3Cer-binding Stx1 could be clearly demonstrated.

Modular Digital Holographic Microscopy

An inverse modular digital holographic microscope with included incubator for stabilized temperature was applied for quantitative phase contrast imaging of Stx1-treated living single HBMECs. Long term *in vivo* investigations of dynamic changes of the cells' shape and thickness were carried out by time lapse measurements. 16 h after the addition of the toxin, first reactions of the cells such as cell rounding were detected that effect a local increase of the cell thickness up to 50 percent. Simultaneously, a significant decrease of intracellularly induced fluctuations in the holographic phase contrast images was observed. The begin of the final cell collapse was detected after 41 h by decrease of the quantitative phase contrast due to corrosion that finished after 50 h. These findings are in accordance with the SEM results, showing that digital holographic microscopy allows minimal invasive analysis of dynamic cellular changes caused by Stx1 towards HBMECs.

Conclusion

The integrated nanotechnology-based approach reveals an appropriate strategy for analyzing morphology changes of Stx1-targeted cells, initiated and based on the biological principle of carbohydrate-protein viz. Gb3Cer-Stx1-interaction. The multidisciplinarity will likely provide new insights into the virulence factors of pathogenic bacteria. Furthermore, clinical proofs of Stx receptor analogues for preventing systemic toxemic complications of STEC infections are currently in progress. Experimental animal models of HUS that would reproduce the whole spectrum of the pathological changes seen in humans and would be thus useful for testing potential therapeutic approaches have not been developed. We believe that our attempt to analyze the interaction of Stx1 with the host cell using nanotechnology techniques will contribute to the optimizing of therapeutic strategies to defeat the life-threatening infections caused not only by STEC but also by other toxin-producing pathogenic bacteria.

"NATIVE UV LASER INDUCED FLUORESCENCE DETECTION FOR SINGLE CELL ANALYSIS IN MICROFLUIDIC DEVICES"

Dominik Greif, Wibke Hellmich,
Dario Anselmetti and Alexandra Ros
Experimental Biophysics and Applied Nanosciences; University of Bielefeld;
Universitaetsstrasse 25; 33615 Bielefeld, Germany

For proteome research, effective, sensitive and quantitative separation and detection techniques are essential and thus a subject of intense investigations. Microfluidic devices have the potential to fulfil these requirements, which is impressively demonstrated by the transfer of proteom relevant separation techniques to the microfluidic format such as protein electrophoresis, isoelectric focusing and two-dimensional protein separation techniques [1, 2]. However, averaging effects from cell-cycle dependent states, the different and inhomogeneous cellular response to external stimuli, or the introduction of genomic and proteomic variability during cell proliferation, are difficult to address upon studying cell ensembles. Novel and efficient single cell analysis tools are therefore mandatory and have the potential to provide considerable insights for proteom research as well as systems nanobiology. However, low abundant proteins in a single cell require a high sensitivity with a detection limit in the range less than 100 nM (assuming a copy number of 105 proteins in a cell with a diameter of 10 μm). We chose UV-laser induced fluorescence (LIF) as detection method on chip because in capillary format, such as capillary electrophoresis, label-free LIF detectors for proteins based on the UV fluorescence of the aromatic amino acids have been reported with ultimate sensitivity in detection.

Here, we extend our previous studies on single cell analysis in poly(dimethylsiloxane) (PDMS) microfluidic devices with native label-free UV-LIF detection [3] in order to improve separation efficiency and detection sensitivity. With the use of poly(oxyethylene) based coatings we were able to improve the separation efficiency, resulting in an increase in plate height to 46 μm compared to previous studies [3]. Furthermore, we investigated the reduction of the background fluorescence by the adaptation of the confocal volume to the microfluidic channel dimensions as well as with bulk modification of PDMS by the use of carbon black particles. With the use of optimised parameters we enhanced the detection limit to 25 nM, thereby reaching the relevant concentration ranges necessary for the label-free detection of low abundant proteins in single cells [4]. Taking these results into account in combination with a more recently proposed method of our group for single cell analysis combining navigation and steering of single cells with optical tweezers, on-chip cell lysis and electrophoretic separation of proteins with subsequent LIF detection [3, 5], we were able to perform single cell electropherograms on chip combined with UV-LIF detection. We will demonstrate the first electropherogram from an individual spodoptera frugiperda (Sf9) cell with native label-free UV-LIF detection in a microfluidic chip in comparison to an electropherogram of four Sf9 cells [4]. Detailed chip design and operations are described in references [3, 5]. Our future work is dedicated toward label-free protein fingerprinting from single cells on chip.

REFERENCES

[1.] N. Lion, et al., *Electrophoresis* 24 (2003) 3533.

[2.] P.A. Auroux, D. Iossifidis, D.R. Reyes, A. Manz, *Anal. 376 Chem.* 74 (2002) 2637.

[3.] W. Hellmich, C. Pelargus, K. Leffhalm, A. Ros, D. Anselmetti, *Electrophoresis* 26 (2005) 3689.

[4.] W. Hellmich, D. Greif, C. Pelargus, D. Anselmetti, A. Ros, *J. Chromatogr.* A, in press.

[5.] A. Ros, W. Hellmich, J. Regtmeier, T.T. Duong, D. Anselmetti, *Electrophoresis* 27 (2006) 2651.

"SINGLE-CELL MULTI-ANALYSIS (FISH, IHC, SINGLE-CELL CGH) OF EARLY FLAT UROTHELIAL LESIONS"

Stella Koufou, S. Langer, M. Speicher and R. Knüchel

Universitätsklinikum RWTH; Institut für Pathologie;
Pauwelsstr. 30; 52074 Aachen, Germany

Aims: Our aim was to investigate first genetic events in proliferating cells of early flat urothelial lesions (hyperplasia and dysplasia). Methodically multi-colour fluorescence in situ hybridisation (FISH) and Ki-67 immunohistochemistry were the basis for consequent analysis of single cells by comparative genomic hybridisation (CGH).

Methods: Double staining of fluorescence in situ hybridisation (Urovysion, Vysis/Abbott) and Ki-67 immunohistochemistry was carried out on frozen tissue sections from 15 patients with precancerous lesions (hyperplasia and dysplasia) of the bladder. In addition single-cell suspensions made of frozen tissue sections were used. Ki-67 positive and negative cells were Laser-microdissected (PALM) and analysed in addition with single-cell comparative genomic hybridisation (CGH).

Results: Combination of these three methods (FISH, IHC and CGH) at single-cell level was successful. Using this multi-method analysis tool, our results are: In early flat urothelial lesions (hyperplasia and dysplasia) genetic aberrations for the Urovysion probe set are also present in proliferating cells. For example, in dysplasia proliferating cells have more aneusomies (mono- and trisomies) for centromere probe 3 than Ki-67 negative cells. Proliferating basal cells show a normal diploid FISH signal and no loss of the chromosomal 9p21 (p16) locus. CGH data confirmed the loss of the chromosomal region 9p21 in proliferating cells of precancerous lesions as the most common event, however, other significant aberrations, e.g. amplification of the Y-chromosome, occurred.

Conclusion: The linkage of these three methods (FISH, IHC and CGH) opens up new vistas in single-cell analysis. The facility to combine these three methods (FISH, IHC, CGH) at single-cell level is apt to show that genetic aberrations detected in early bladder lesions or normal urothelium are biologically relevant since found in proliferating cells. This work has been supported by the German Science Foundation (DFG, grant no: Kn263/9-2)

"Immune Modulators Expressed in Laser Microdissected Keratinocytes During the Early Phase of Experimental Leishmaniasis and the Potential Role of I-TAC for Th2-Cell-Differentiation"

K. Roebrock[1], J.M. Ehrchen[1], D. Foell[1], H.J. Schuberth[2] and J. Roth[1]

[1] Institute of Experimental Dermatology, Westfalian Wilhelms-University Muenster, Germany

[2] Institute for Immunology, University of Veterinary Medicine Hannover, Germany

Studies on Leishmania major (L. major) have been largely responsible for the characterisation of the Th1/Th2 paradigm associated with susceptibility or resistance of different inbred mice strains to intracellular infection. Little is known about early events in the skin preceding final T-cell priming in lymphnodes. In order to further clarify the role keratinocytes might play in the process of T-cell-differentiation, approximately 1000 keratinocytes were isolated from infected and non-infected skin of susceptible Balb/c and resistant C57BL/6 mice by laser microbeam microdissection and laser pressure catapulting. cDNA was preamplified by means of random based PCR amplification, prior to gene expression analysis by real time PCR. In addition, we investigated the gene expression kinetics in whole skin samples from these mice for selected genes during the first three days after L. major infection. For the first time it was shown, that keratinocytes are involved in the production of immune-modulators rapidly after infection with L. major. The expression level of the majority of genes was clearly higher in keratinocytes of resistant C57BL/6 mice than in susceptible Balb/c mice. An exception to this was the expression of interferon inducible T-cell-alpha chemoattractant (I-TAC), which was significantly stronger induced in Balb/c compared to C57BL/6 mice. This is the first report of the expression of I-TAC in experimental leishmaniasis. I-TAC expression in the epidermis was confirmed by immunofluorescence and in situ hybridization. The biological relevance of this differential I-TAC expression in Balb/c and C57BL/6 mice was underlined by a negative influence of I-TAC administration on the course of infection in C57BL/6 mice.

Differential expression of immune-modulators in keratinocytes of susceptible and resistant mice after infection with L. major indicates that early induction of epidermal gene expression has an impact on the resulting specific immune response and hence on resistance and susceptibility in experimental leishmaniasis.

"Neprilysin and Angiotensin-Converting Enzyme: Peripheral Control of Proopiomelanocortin Peptide Bioactivity by Endothelial Cell-Derived Peptidases"

homas E. Scholzen[1], Thomas A. Luger and Simone König[2]

[1] Ludwig-Boltzmann Institute and Department of Dermatology,
[2] Integrated Functional Genomics, Interdisciplinary Center for Clinical Research,
Medical Faculty, University of Münster, Germany

Dermal microvascular endothelial cells (EC) are an established source and target of the immunomodulatory proopiomelanocortin (POMC) peptides adrenocorticotropin (ACTH) and α-melanocyte-stimulating hormone (α-MSH). Despite a well-characterized POMC peptide generation in intracellular secretory vesicles, much less is known on their extracellular cleavage by the neuropeptide specific zinc metalloproteases neprilysin (neutral endopeptidase; NEP) and angiotensin-converting enzyme (ACE), which may regulate the local POMC peptide bioavailability and modulate activation of ACTH/α-MSH-specific melanocortin 1 receptors (MC1). EC membranes prepared from ACEhigh/NEPlow expressing primary dermal EC (HDMEC) and the ACElow/NEPhigh expressing EC cell line HMEC-1 degraded ACTH1-39 over time resulting in temporary increased α-MSH immunoreactivity. MALDI-TOF MS peptide mapping and ESI-MS sequencing identified several stable fragments generated from ACTH1-39, ACTH1-24 and α-MSH by EC membranes or recombinant NEP and ACE. Whereas some fragments could be clearly assigned to a cell-specific NEP or ACE activity, other unexpected degradation products require additional enzyme activity. Selected peptides, such as α-MSH2-12 generated from ACTH or α-MSH by both EC membranes and recombinant NEP were equipotent in comparison to the parental peptides in inducing intracellular cAMP in MC1 receptor transfected HMEC-1 or HEK293 cells. Likewise, pharmacologic NEP inhibition promoted cAMP induction by ACTH and α-MSH. In a murine model of allergic contact dermatitis, α-MSH2-12 retained a similar anti-inflammatory activity in comparison to α-MSH or ACTH. Thus, NEP and ACE significantly contribute to the EC processing of stress hormones (ACTH) and anti-inflammatory peptides (α-MSH), which modulates MC1 activation, but does not completely inactivate the peptide ligand. Since NEP and ACE are regulated by inflammatory mediators and UV light, this may be important for ACTH/MSH-modulated skin inflammation.

"Hepatic Targeting Using Transplantation of Single Cells"

Andree Zibert, J. Haberland and H.H.-J. Schmidt
Transplantationshepatologie; University of Münster;
Domagkstr. 3A; D-48129 Münster, Germany

Minimally invasive procedures for targeting liver disease using cell based approaches are currently introduced into the clinic. Most protocols involve the transplantation of non-modified cells, e. g. hepatocytes, which are, however, rarely available for clinical applications. Adult stem cells can be highly expanded *in vitro* and when differentiated into hepatocyte-like cells may have enormous potential for cell based therapy.

We hypothesize that both isolated hepatocytes and adult stem cells bioengineered *in vitro* with appropriate bioactive factors by e. g. gene therapy methods are superior targets for efficient cell based therapy of liver. We will monitor the differentiation of stem cells after expression of bioactive molecules *in vitro* as well as *in vivo* using established animal models of liver disease (LEC rat, CCL_4 rat). Animal models mimic the natural course of human liver fibrosis, a major issue in the clinical setting.

As a proof of principle we will translate the known antifibrotic effects of the factor adiponectin (244 amino acids) to cell based therapy of liver fibrosis in various animal models. Monitoring the status of cell differentiation and gene expression will be followed in long-term experiments by established state-of-the-art methodology. Characterization of the differentiation status (proteomics) and gene expression (gene array technology, qPCR) of stem cells will allow a better understanding of the underlying mechanisms of cell therapy. *In vivo* characterization of stem cell differentiation and repopulation of liver will be carried out by intravital microscopy, immunohistology, FISH and RT-qPCR. Therapeutic effects will be determined by clinical and biochemical parameters.

The current status of this project will be presented. Our model is highly relevant to other alternative strategies in organ transplantation.

In: Biomacromolecular Mass Spectrometry Yearbook. Volume 1 ISBN: 978-1-60876-167-8
Ed: Simone Koenig

Chapter 15

THE USE OF TIME-OF-FLIGHT SECONDARY ION MASS SPECTROMETRY IN BIOCHEMISTRY*

Daniel Breitenstein*
Heisenbergstr. 15
48155 Muenster, Germany

ABSTRACT

Time-of-Flight Secondary Ion Mass Spectrometry (ToF-SIMS) is an analytical tool for surface analysis. Applying ToF-SIMS the impact of primary ions leads, among others, to the desorption of secondary ions that are characteristic for the composition of the sample's surface. Recent progress in instrumental development now facilitates a considerable use of this technique for the analysis of biological samples.

Several aspects make ToF-SIMS a beneficial tool for bioanalysis: It just probes the uppermost three monolayers of the sample, offers imaging capabilities with a lateral resolution as good as 300 nm and, recently, a 3D imaging option for biological samples was established. The latter option allows a vertical resolution of 30 nm.

This paper reviews the state-of-the-art technology and results on spectrometry, imaging, depth profiling and 3D analysis on biological samples such as cells, tissue sections and Langmuir-Blodgett films in order to give an overview on the chances and limits of this technique.

ABBREVIATIONS

EI-source	Electron Impact Source
LMIG	Liquid Metal Ion Gun
NRK-Cells	Normal Rat Kidney Cells
ToF-Analyzer	Time-of-Flight Analyser

---

* A version of this chapter was also published in *Biomacromolecular Mass Spectrometry Volume 1, Number 4*, published by Nova Science Publications, Inc. It was submitted for appropriate modifications in an effort to encourage wider dissemination of research.

* Phone: +49-251-1622-400; Fax: +49-251-1622-401 ; E-mail : daniel.breitenstein@tascon-gmbh.de

ToF-SIMS Time-of-Flight Secondary Ion Mass Spectrometry

INTRODUCTION

The tremendous development of Time-of-Flight Secondary Ion Mass Spectrometry (ToF-SIMS) in the field of biological applications within the last years becomes visible when comparing the comprehensive review by Pacholski and Winograd published in 1999 [1] with other more recent reviews [2, 3] and the description here. Pacholski and Winograd had to point out just eight years ago that in so called "static" ToF-SIMS experiments „images are usually acquired with only a few counts/pixel". They emphasized that "most of the examples [at that time] involved the detection of elemental species (…) rather than molecular species (…), and this trend will surely change in the future".

Aim of this review is to outline on the one hand technical developments that made this prognosis become correct and that has made ToF-SIMS a useful method in the field of biochemistry. On the other hand examples for biochemical approaches of the four ToF-SIMS application modes are given to indicate the potentials and limitations of this technique. As most existing literature on ToF-SIMS is written with a physicist background and as the great Dutch author Harry Mulish expects a exponential decay of listeners (readers) upon mentioning of mathematical formulas [4] this article attempts to limit these aspects to a minimal value.

Due to limited space necessarily a number of aspects cannot be addressed: One of it is the historical development of the ToF-SIMS technology. This aspect is described in detail by Benninghoven [5]. Furthermore, as the macromolecular mass spectrometry is the main field of interest of this journal, the elemental distribution of substances (e.g. within cells) is not in the focus of this article. Therefore, ToF-SIMS studies focusing on this aspect such as "dynamic" SIMS studies (SIMS-microscopy; recently reviewed by Guerquin-Kern [6]) are excluded from this review.

This review is devided into 8 chapters: Following this introduction (chapter 1-3) the basic principle of ToF-SIMS is explained (chapter 4). Some more detailed information on the technological set-up of ToF-SIMS machines are available in the fifth chapter. Examples on biochemical application of the four ToF-SIMS operational modes are given in chapter 6. The final chapter describes the current challenges in ToF-SIMS instrumentation and should allow the reader a realistic view on the current capabilities of the technique.

BASIC PRINCIPLE

Like all mass spectrometric techniques ToF-SIMS requires the formation, mass-separation and detection of analyte ions. In ToF-SIMS the formation of secondary ions is a result of the impact of primary ions onto the surface of the sample whereas the separation and detection is performed by a Time-of-Flight (ToF) analyzer.

Generation of Secondary Ions

The fundamental event in a ToF-SIMS experiment is the sputtering and ionization of the surface components after impact of primary ions. Whereas the desorption of neutral particles can be described fairly well on basis of the results of practical experiments as well as molecular dynamics calculations [7], the effects leading to the formation of charged species are not yet understood.

Collision Cascade (Monoatomic Bombardment)

It is consensually believed that elastic collisions of the primary ion with surface atoms displace these surface atoms (primary recoils) and that this movement induces a movement of further surface atoms (secondary recoils) [8]. In case the momentum of the atoms and molecules recoiled by this collision cascade is partly directed towards the surface normal only those particles are able to leave the surface the energy of which can overcome the surface binding energy. Therefore, 95 % of the particles desorbed by the impact of a monomolecular primary ion are originating from the 1-3 outermost monolayers of the sample [9].

Close to the track of the impacting projectile penetrating into the bulk of the sample a massive fragmentation of molecules takes place resulting in destruction of organic material and emission of mainly atoms. The farer the distance to the point of impact the more intact organic molecules survive and some of these molecules may be emitted – mainly as neutrals and elctrons ($\geq$ 99%) but partly as positively or negatively charged secondary ions [10]. Most of the secondary ions obtained in ToF-SIMS experiments are singly charged [11]. More details on the collision cascade under monoatomic bombardment are provided by Urbassek [12].

Considering the massive fragmentation of the surface near molecules under monoatomic bombardment, for an artefact-free measurement it is mandatory that a collection of data from already damaged surface areas is avoided. This can statistically be ensured by keeping the primary ion dose density below a value guaranteeing that only 1 % of the surface structure is influenced. The value of this so called "static" SIMS limit varies among the secondary ion species but is typically ranged at a primary ion dose density of 10^{11}-10^{13} ions/cm^2. Exceeding this static limit one reaches "dynamic" SIMS conditions including operational modes which are based on the removal of surface material by sputtering [13, 14]. However, using monoatomic bombardment most organic secondary ions are highly fragmented in such an approach [15, 16]. Therefore, "dynamic SIMS" experiments allow predominantly the detection of elements and their distribution within the sample [6].

Effect of Cluster Primary Ions (Polyatomic Bombardment)

Cluster ions like SF_5^+ [17, 18], or the just recently available C_{60}^+ [19-21], Au_3^+ [22] or Bi_3^+ [23] consist of multiple atoms. Upon impact on the sample these clusters break apart [24, 25] leading to a multiple, simultaneous bombardment of a small spot. One of the observations possible under these conditions is that the secondary ion yield, i.e. the number of secondary ions generated per impacting primary ion, is drastically increased compared to monoatomic bombardment [26, 27]. The increase is non-linear with respect to the number of projectile particles: E.g. changing the primary ion from Au_1^+ to Au_3^+ on an organic polymer additive sample (Irganox 1010) does not lead to the threefold but to an almost 45-fold increase of the secondary ion yield for the intact molecule (quasi-molecular-ion) on thick organic substrates [28].

The key observation for the use of cluster primary ions in molecular ToF-SIMS analysis is that the yield enhancement is not accompanied by a likewise increase in surface damage [25, 28-32]. The traditional distinction between "static" and "dynamic" ToF-SIMS conditions does therefore not play such a critical role for polyatomic projectiles and thick organic layers: Meanwhile a number of groups reported the detection of molecular secondary ions even upon depth profiling.

The non-linear effects are assumed to originate on the one hand from the surface nearer deposition of the collision energy, as cluster ions have a lower kinetic energy per constituent than atomic primary ions have [25, 33]. On the other hand it is additionally and/or alternatively explained by the overlap of the collision cascades of each individual constituent of the polyatomic projectile [7]. More details on the effect of the use of cluster ions on ToF-SIMS results can be found in the substantial review by Wucher [26]. Nevertheless, the physics and chemistry of cluster ion bombardment are far from being well understood and are currently widely discussed in the SIMS community.

TECHNOLOGICAL ASPECTS

Secondary Ion Separation and Detection: Time-of-Flight Analyzer

As the surface material is limited and as upon an impact of primary ions only a limited number of secondary ions are generated, an efficient (and parallel) detection of most generated secondary ions is of high interest. For this purpose the Time-of-Flight (ToF) analyzer is superior to most other mass spectrometric analyzer systems [34]. A linear ToF-analyzer consists of an accelerating section and a drift section. In the accelerating section all generated secondary ions of a given polarity are accelerated to the same nominal kinetic energy. The velocities of these secondary ions depend on their mass to charge ratio. The accelerated ions are transferred into the drift section. The drift section allows a field free drift of the secondary ions. As the velocity of the secondary ions differs they are mass-separated on their way to the detector: faster – i.e. lighter – secondary ions arrive earlier at the detector than slower - i.e. heavier - do.

As the secondary ions have an initial velocity distribution due to the strength and the direction of the recoils of the collision cascade, this separation process is hampered. To compensate for this effect most ToF-SIMS instruments are nowadays equipped with an energy focusing unit. For this purpose electrostatic fields can be used that act as an ion mirrow (so called reflectron, Figure 1). As high energy ions of one ion species penetrate deeper into this electrostatic field, their flight path is longer than that of ions of lower start energy. Under optimal alignment of the reflectron all secondary ions of one mass to charge ratio hit the detector after the same flight time. A second option for energy focusing used in ToF-SIMS is the electrostatic sector analyzer (Figure 1). Within the radius of the electrostatic sector analyzer the drift path of the ions of higher energy is longer than the one for lower energy [34].

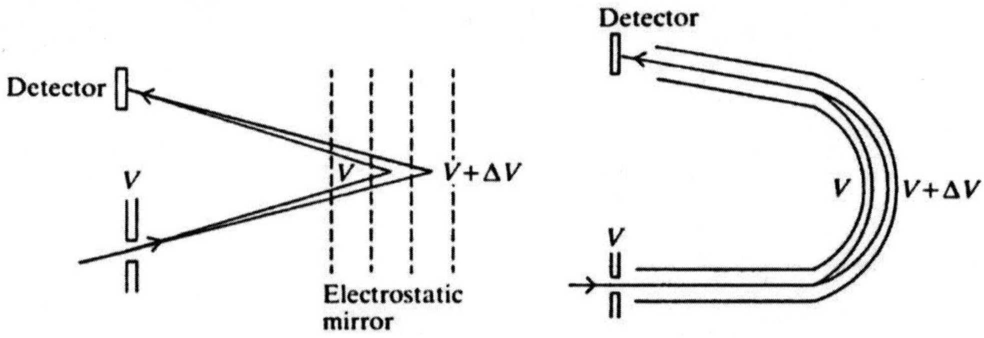

Figure 1. Schematic design of a reflectron (electrostatic mirrow, right) and a sector field analyser (left).

For ion detection multichanelplate systems are used: Each impact of incoming ions causes an electrical signal which is registered.

Technology of Primary Ion Formation: Primary Ion Sources

To enable the sputtering and ionization process on the surface, primary ions have to be generated, accelerated and focussed onto the target surface. The two systems of ion guns mainly used for this purpose are liquid metal ion guns and electron impact sources. The description of surface ionization sources and duoplasmatron sources go beyond the scope of this review and can be found elsewhere [35].

Liquid Metal Ion Guns (LMIG)

The primary ion sources offering the highest spatial resolution is the liquid metal ion gun (LMIG). Figure 2 gives a schematic view on a LMIG: It consists of a needle and a reservoir coated with the metal used for primary ion generation. The metal can be melted by a heating filament. From the tip of the needle the primary ions are generated by an extraction voltage resulting in an ion energy of 5 to 25 kV. Similar to the ionization process in Electrospray Mass Specrometry, the liquid metal forms a taylor cone at the tip of the needle and the primary ions can be continuously extracted towards the sample.

The advantage of the LMIG against the electron impact source lies in the well defined point of primary ion extraction which has a size down to 10 nm in diameter [35]. This feature allows a focussing of the primary ion beam onto the target. Focussing is optimized by an ion optical lens system inside the primary ion column.

The primary ion species used depends on several critical parameters: First, the material has to be ionized easily. Moreover, as the primary ions have to be created from a liquid phase, a low melting point is of high interest from a technological point of view. Furthermore, as a consequence of the Time-of-Flight principle within the ion gun, the element used as primary ion should be either monoisotopic or easily available in one isotopic species. In this perspective also the price of the monoisotopic material is of interest. Finally, it has been shown that a high mass of and a polyatomic nature of the primary ion is of high interest for improved results on organic samples.

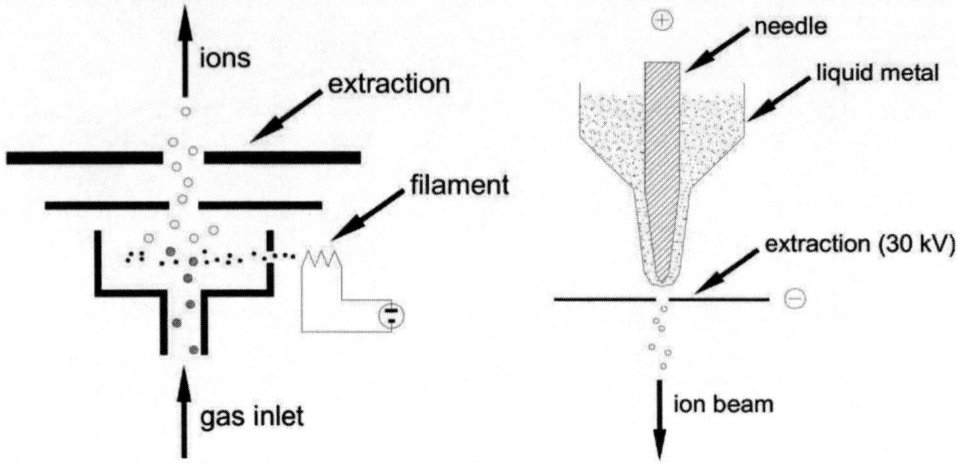

Figure 2. Schematic design of an electron impact source (EI source, left) and a liquid metal ion gun (LMIG, right).

In this perspective, the „classical“ primary ion used is Gallium (^{69}Ga, two Ga isotopes, melting point (mp): 302.9 K). Whereas Indium (^{115}In, two In isotopes, mp: 429,8 K) can be seen as a developing step, more recently Gold (^{197}Au, monoisotopic, used as an eutectic mixture with Germanium; mp ~400 K [22]; pure Gold, mp: 1337,3 K) and Bismuth (^{209}Bi, monoisotopic, mp: 544.4 K) were introduced. The introduction of Gold and Bismuth primary ion sources was the breakthrough for a number of organic or biological applications: As Gold and Bismuth sources already profit from the high mass of the monoatomic primary ion, the main improvement compared to the other primary ion elements is the formation of polyatomic cluster ions ($M^{x+}{}_{y}$, whith x= 1, 2 and y=1, 3, 5, 7). In a recent study to highlight the performance of the different primary ion species available was compared [23].

Electron Impact Sources

The electron impact source requires the continuous insertion of gas or vapor into the ionization chamber. There the gas (or vapor) molecules (or atoms) are ionized by low energetic electrons orbiting inside the chamber (Figure 2). Electrons are produced by a filament cathode. Ionized gas or vapor particles are captured by an electrical extraction field and are accelerated through an ion optical system onto the target.

Classically noble gases, oxygen or SF_5^+ were used in gas sources. Especially the oxygen source became one of the dominant sources for depth profiling of inorganic samples [36]. In the field of biological application the recent development of a C_{60} source [21] is of great interest: Applying the ionization process to sublimed buckminsterfullerenes opens the possibility to generate a high mass (720 Da) and multi-atomic ion species.

While the polyatomic nature of the C_{60}^+, C_{60}^{++} and C_{60}^{+++} ions is reported to be advantageous for biochemical applications, the ionization process only allows a focusing of the primary ion beam down to a diameter of appr. 1 µm. Thus, the imaging capabilities of the C_{60}-source are to date inferior to the LMIG-sources [23].

Separation and Pulsing of Primary Ions

The continuous extraction of primary ions from the primary ion sources would lead to a continuous bombardment of the samples surface and would therefore lead to the continuous generation of secondary particles. In such a scenario the mass separation principle of the ToF-analyzer can not easily be applied. Thus, for ToF-SIMS analysis the surface bombardment with primary ions often is pulsed and the obtainable mass resolution depends – beside others - on the length of the primary ion pulse. Pulsing is performed by blanking the primary ion beam by deflecting and undeflecting it out of and into the optical axis of the machine.

In most primary ion sources a broad range of primary ions are generated. E.g. a gold-source produces Ge_x^{y+} and Au_x^{y+} ions and buckminsterfullerene primary ions can be found as C_{60}^{y+} with x and y being a natural number between 1 and 5 and 1 and 3, respectively. All these ions are accelerated by the extraction field with the same energy but do have a different velocity due to their different mass to charge ratios. As the differences in velocity will elongate the primary ion pulse a mass separation has to take place.

In order to achieve a high mass resolution pulses should be as short as possible. On commercially available ToF-SIMS machines pulse lengths of 1 ns can be obtained allowing a mass resolution ($m/\Delta m$) of >10.000 [37]. However, the shortest pulse length are achieved either of cost of focus quality or measuring time.

Focusing of Primary Ions

Focusing of the ion beam is performed by ion optical lens systems. As the set-ups of the lense systems vary between the manufacturers on the one hand and depend on the analytical goal of the analysis on the other hand, just a brief general overview is given here. More details can be found elsewhere [35].

The main components of the ion optical system are the apertures and the electrostatic lenses. Electrostatic lenses allow to focus the primary ion beam. Noteworthy, like for optical lenses, aberrations should be avoided upon use of the lens system. In this review, only two types of aberrations are discussed:

The most important aberration in ion optical systems is the sperical aberation, meaning that ions with a movement parallel with but far from the axis of the beam are focussed to a different spot than those moving on the axis of the beam. To avoid this type of aberation, apertures are introduced into the primary ion beam. The circular pupil allows only those primary ions to pass through that are in line with the beam axis. Thus, apertures allow to cut out those particles of the primary ion that are not in the exact axis of the beam. Noteworthy, cutting out these primary ions increases the focus, but decreases the amount of primary ions hitting the sample – and thus the measuring time needed.

The second important aberration to deal with is chromatic aberration: Ions with different velocities are brought to a different focus. As reducing the pulse length can lead to an increase in energy spread (e.g. when maintaining the beam current; "bunching") it is difficult to optimize focus quality and mass resolution in one set-up.

Technology of Charge Compensation

The impact of charged primary ions on the surface and the desorption of charged particles from the surface leads to a charging of the bombarded area. Such local charging e.g. affects the extraction of secondary ions: Depending on the samples conductivity it is either completly inhibited or artefacts derived from non-uniform extraction field distribution might be visible. To avoid such charging, in modern ToF-SIMS instruments charge compensation by flooding the sample with low energetic electrons derived from a electrical filament is performed. Noteworthy, this feature can lead to a destruction of organic bonds. Thus, a minimization of charge compensation is to aspire [38, 39].

Technology of Sputtering

In ToF-SIMS, depth profiling is performed by alternating an analysis (imaging or spectrometry) procedure with a sputter erosion of the sample surface. Within the sputtering process the surface is eroded under massive bombardment by primary ions. For this erosion either the primary ion beam utilized for analysis can be used or, in the more convenient case, the instrument is equipped with a second ion beam system performing this task (dual beam approach [36]). There are two advantages of dual beam depth profiling compared to single beam depth profiling:

On the one hand the number of primary ions in an analysis mode is not adequate for a quick sputtering of the surface. Thus, in the single beam set-up a second setting allowing an enhanced bombardment of the surface has to be used. Subsequent switching from one setting to another is time consuming. In contrast, in the dual beam mode (in the ideal case) the time needed for analysing the secondary ions can already be used for sputtering (interlaced mode).

On the other hand in the dual beam approach the sputter ion and its energy can be chosen independently from the analysis ion. Thus, for sputtering low energy ions can be used in order to obtain a high depth resolution whereas high energy ions which can be well focussed are used to analyse the centre of the crater. Refering to the advantages and disadvantages of the different ion sources e.g. C_{60}^{y+} ions can be used for sputtering and Bi_3^+ can be used for highly focussed analysis of the sample [40].

Operational Modes

Overview

As shown in table 1 ToF-SIMS analysis can be performed in four different operational modes. One can mainly differentiate between those operational modes where (almost) no sample erosion takes place and operational modes where sample erosion is performed.

Furthermore, one can distinguish between measurements where the lateral distribution of analytes are probed or not.

Table 1. Operational modes in ToF-SIMS

	with low erosion (sputtering)	with erosion (sputtering)
without lateral resolution	Spectrometry (Figure 3)	depth profiling (Figure 5)
with lateral resolution	imaging (Figure 4)	3D microarea analysis (Figure 6)

The use of the four application modes is demonstrated on one cell sample (Figure 3-6). It is a Normal Rat Kidney (NRK) cell culture sample that was chemically fixed with glutaraldehyde.

The most simple operational mode is the spectrometry mode (low sample erosion, no lateral probing). The result of this measurement is displayed in a spectrum, as it is well known from other mass spectrometric techniques (Figure 3). From this spectrum information on the chemical composition of the surface can be gained.

More complex is the imaging mode (low sample erosion, lateral probing). The primary ion beam is in most cases highly focused and is rastered pixelwise in x and y direction over the sample surface leading to spectral information on every pixel. Analysing e.g. a square of 256x256 pixels leads to a dataset of 65536 individual spectra. The raw data taken from this kind of experiments can be represented in various ways, e.g. one spectrum of each individual pixel or of a region of interest could be presented. Nevertheless, most frequently this data is represented in mass resolved images (Figure 4): The intensity of one secondary ion species in each of the pixels is represented according to a colorscale leading to a two-dimensional image of the surface. Noteworthy, ToF-SIMS allows the simultanous generation of a mass resolved image for any detected secondary ion.

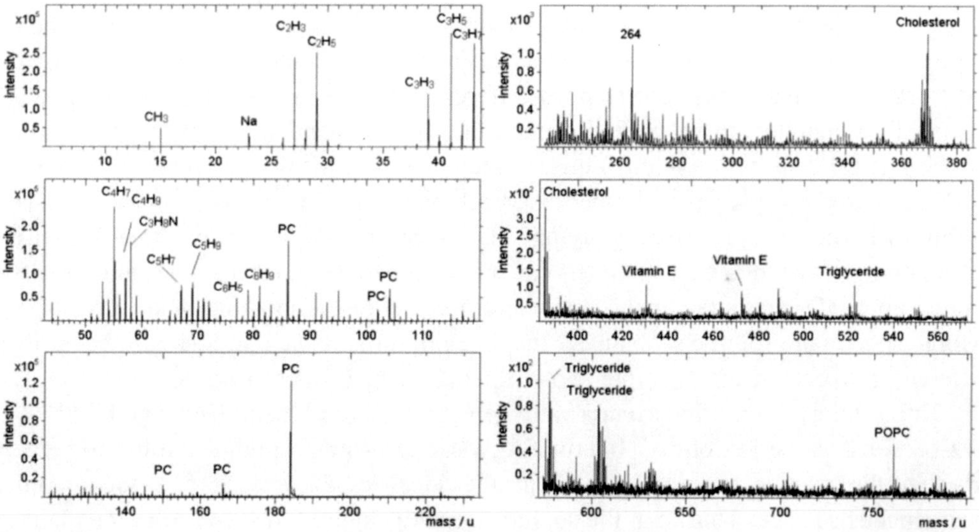

Figure 3. ToF-SIMS spectrum of positively charged secondary ions detected on a NRK cell layer (Abbreviations: PC: phosphatidylcholine-fragment, POPC: palmitoyl-oleyl-phosphatidylcholine).

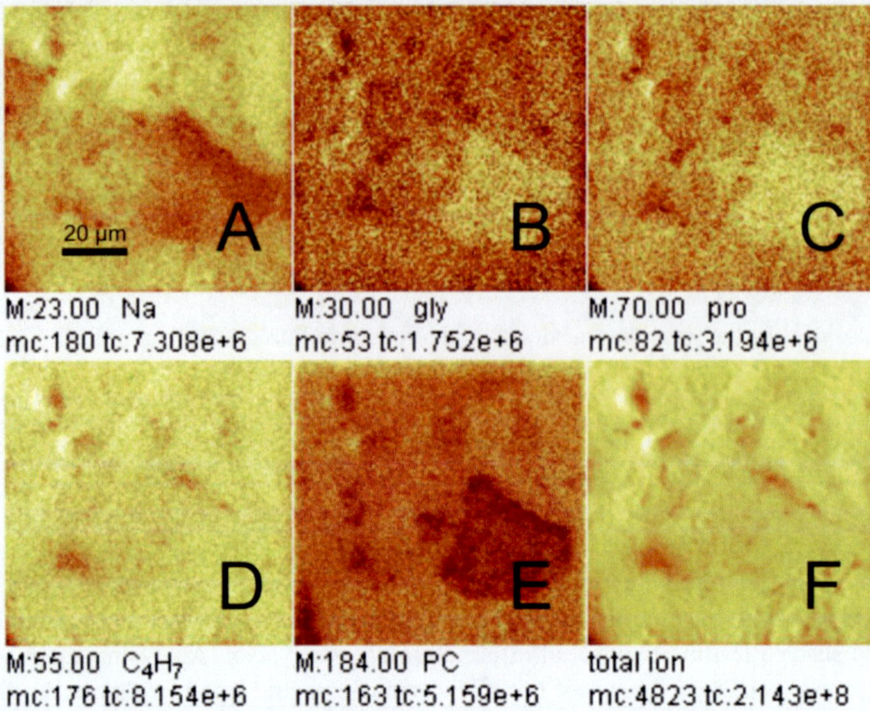

Figure 4. Mass resolved images of NRK cell layer. The intensity distribution of the following secondary ions is shown: sodium (A), glycine (B), proline (C), C_4H_7 (D), the phosphatidylcholine-headgroup (E) and the sum of all secondary ions detected (F). Abbreviations: mc: maximum counts (the number of secondary ions detected in the brightest pixel); tc: total counts (the number of secondary ions detected in all pixels of the respective mass resolved image).

Performing a depth profile the spectrometric data acquisition is alternately interrupted by a sputter process eroding the surface. Thus, the sample is destroyed upon this process: the erosion process leads to the formation of a sputter crater, the analysis area chosen to be in the center of this crater. For every sputter and analysis event the complete spectrum is stored in a raw data stream. Thus, mass spectra from different depth of the sample can be gained. The intensity of an analyte can be plotted against the time demanded for sputtering (Figure 5). Thus, the vertical distribution of any detected secondary ion within the sample can be probed.

3D microarea analysis is performed by combining the imaging mode with a subsequent sputtering of the sample. By applying 3D microarea analysis a three dimensional representation of the detected analytes can be obtained (Figure 6). The amount of data generated upon 3D microarea analysis is enormous: As in the imaging mode already 65536 individual spectra are routinely obtained, this imaging process can additionally be performed in different z-layers. In the example shown in Figure 6, images were performed in 120 z-levels. Thus, almost 7.9 million spectra were generated. To represent the data, virtual xy-, xz- and yz-sections can be performed. Interpreting these data-representations, it has to be kept in mind, that the sputter process erodes (in the ideal case) uniformly a topographically inhomogeneous surface. Thus, e.g. the sodium signal in figure 8 derived from the plain glass substrate does not show up in an even z-layer.

In the following sections the operational modes will be discussed in more detail. The discussion will focus on their use for biochemical applications.

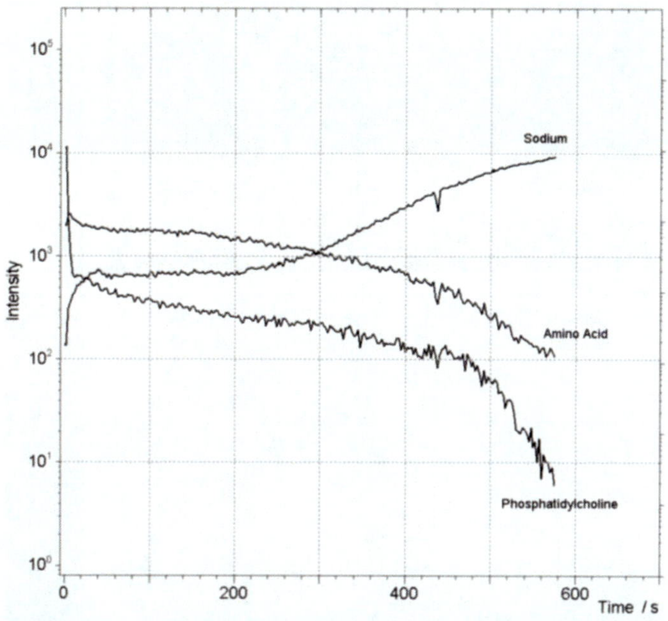

Figure 5. Depth profiling of a NRK cell layer: The intensity of characteristic secondary ions (sodium, amino acid and phosphatidylcholine derived signal) is plotted against the sputter time.

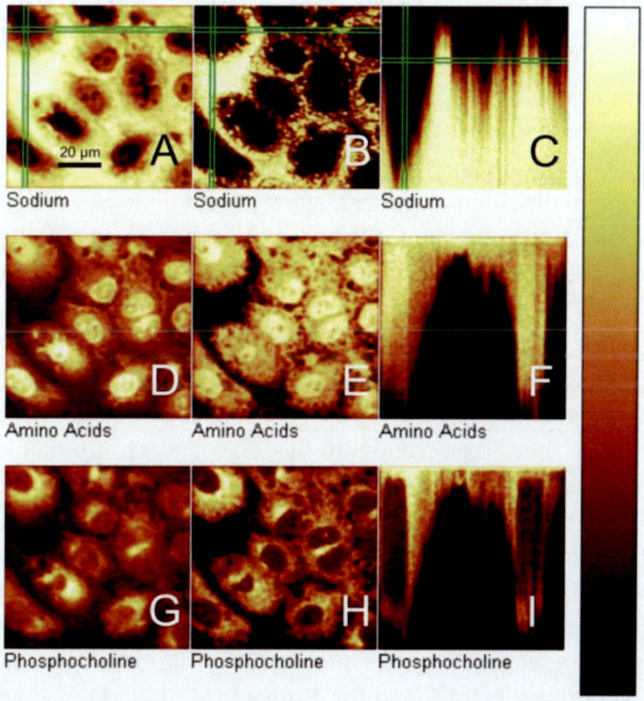

Figure 6. 3D microarea analysis of a NRK-cell sample. The amino acid (D, E, F) and phosphatidylcholine (G, H, I) derived signals as well as the sodium signal (A, B, C) of the glass substrate are shown. The sum of all z-layers (A, D, G), a xy-section (B, E, H) and a yz-section (C, F, I) of these signals is shown according to the green lines in the sodium image.

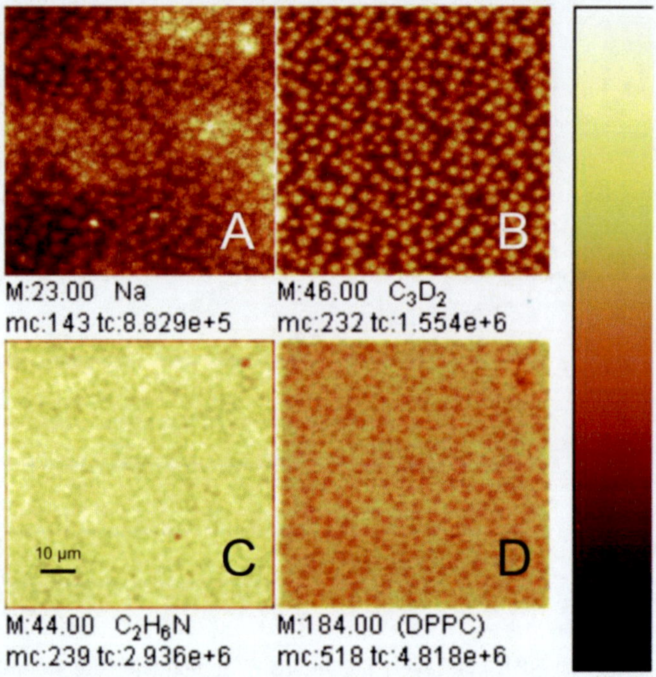

Figure 7. Mass resolved images of a Langmuir-Blodgett transferred lipid model system (DPPC/d62DPPG/SP-B). The sodium signal (A) correlates with the d62 derived signal (B), while the distribution of characteristic protein (C) and the DPPC (D) signals are equal.

Spectrometry

In the spectrometry mode the uppermost one to three monolayers are probed. No information on the lateral distribution of the analytes is obtained. As can be seen for the NRK cell example in Figure 3, a number of substances can be detected in a mass range below 1000 Da: E.g. small unspecific hydrocarbons as well as amino acid and lipid fragments. Additionally, lipids like triglycerides, phospholipids, tocopherol (vitamin E) and cholesterol can be detected as quasimolecular ions.

The analysis of *lipids* is one of the key techniques possible with ToF-SIMS. Whereas biopolymers usually fragment and most oftenly statistical evaluation procedures have to be used for identification, lipids can routinely be detected as quasimolecular ions. More information on lipid detection can be found in the imaging section (see chapter 0).

Within spectra of *proteins* usually the fragment ions of the individual amino acids (NH_2CHR^+, with R = amino acid residue) dominate [41, 42]. Lists of the composition and mass to charge ratio of typical amino acid derived secondary ions can be found e.g. in a study by Michel et al. [43]. Nevertheless, small proteins and peptides up to a mass of app. 10000 Da can be detected upon adequate sample preparation. E.g. the detection of angiotensin (M=1045 Da) and its derivatives [42] and the B-chain (M=3496 Da) as well as the complete molecule (M=5778 Da) of insulin [44] is possible. Additionally, the detection of porcine renin (M=1759 Da) and bovine ubiquitin (M=8563 Da) was shown [45]. Beside these single component systems, protein mixtures of multicomponent adsorbed protein films were analysed by Wagner et al. [46]. A very interesting approach is the detection and identification of tryptic peptides of a protein by ToF-SIMS [47-49].

A quite comprehensive overview on the ToF-SIMS fragmentation patterns of *nucleobases, nucleosides* and *nucleotides* is given by May et al. [50]. They were also able to distinguish different *DNA-oligomers* by statistical evaluation procedures (see capter 0). DNA-oligomers were also addressed by other studies of this [51] and other [45, 52] groups. Combined DNA and *peptide nucleic acids (PNA)* were analysed by Arlinghaus et al. [53].

Analysing a number of different *monosaccharides* leads to the detection of quasimolecular ions as well as typical fragmentation patterns [54]. In other studies spectra of *fluorophors* [55-57] or *pharmaceuticals* [23, 58, 59] are presented.

In principle any biochemically relevant substance that is vacuum sustainable can be analysed. Nevertheless, whereas organic molecules lower than 2000 Da generally can be detected as quasimolecular ion, bigger molecules are generally fragmented.

Imaging

Mass resolved images are most often interpreted to represent at least semi-quantitatively the distribution of the analyte on the surface. Nevertheless, it has to be kept in mind that the so called matrix effects may lead to misinterpretations (In SIMS the number of desorbed secondary ions is not only a function of the surface concentration but also of the respective chemical environment). The three main fields of interest for ToF-SIMS imaging on biochemical samples are to date the analysis of (Langmuir-Blodgett) monolayer films, of tissue sections and of cell culture samples.

Langmuir-Blodgett Films/Self Assembled Monolayers

As one of the first biological samples for molecular imaging by ToF-SIMS Langmuir Blodgett films were analyzed [1]. To generate a Lagmuir-Blodgett film lipid mixtures are spread onto the water or buffer surface of a film balance and transferred onto a substrate [60]. To allow an easier ionization and to avoid a charging of the surface usually gold-coated glass [61, 62] or mica [63, 64] substrates are used for ToF-SIMS applications. Alternatively SiO_2/Si-substrates can be used [65].

ToF-SIMS seems to be an appropriate tool on the one hand to detect phase separation processes, i.e. two dimensional condensation processes, in lipid model systems. On the other hand demixing processes can be observed. One example for the observation of a demixing process can be seen in Figure 7: Here, a ternary system of a lipid (DPPC) a deuterated lipid (d62DPPG), and a protein (SP-B) is transferred at high surface pressure. As can be seen, the fragments of the deuterated lipid are distributed at different areas as the fragments of the two undeuterated components.

One early publication observing phase separation as well as demixing processes was published by Leufgen et al. [56]. They detected the distribution of a native lipid (DPPC) and one of its fluorescent dye analogs. Thus, the distribution of the dye could be probed by two different techniques - by Fluorescent Light Microscopy as well as by ToF-SIMS. As can be seen in this and other studies fluorescence microscopy [61, 66, 67] and also atomic force microscopy [66, 68] can be used to validize ToF-SIMS results.

A number of publications deal with artificial model systems of pulmonary surfactant [61, 62, 66] – a lipid/protein mixture preventing alveolar collapse in mammalian lung. These studies aim to detect a protein induced lipid demixing, necessary for lipid enrichment processes in the mammalian body. They are accompanied by experiments, where the complete extract of lipids and hydrophobic proteins from bovine lungs is analyzed [67].

The use of deuterated lipids in Langmuir Blodgett studies as also shown in Figure 1 is a relatively new aspect [63, 66]. Using this type of systems, similar lipids can be distinguished by their isotopic label. Necessarily care has to be taken that the mixing behavior does not change upon isotopic labeling [66].

In recent years the raft-theory increased the interest in protein/lipid model systems. Therefore, recent studies have dealed with lipid and lipid/protein model systems expected to be relevant for raft formation [69]. Nevertheless, it has to be pointed out that the expected size of the raft (down to 25 nm [70]) is yet smaller than the lateral resolution of ToF-SIMS (down to 100 nm, [23]).

Noteworthy, artifacts were observed using ToF-SIMS for semiquantitative analysis of Langmuir-Blodgett films. Sostarecz et al. reported higher yields of DPPC specific fragments in an expanded film than in a condensed film [65]. This does not correlate with the expected molecule density because the area per molecule in a condensed film is lower than in a expanded film. Biesinger et al. reported an inversion of yield contrast upon exceeding the static limit on Langmuir-Blodgett films [64]. As both studies were performed with monoatomic bombardment of the sample further studies under polyatomic bombardment would be of interest.

In addition to Langmuir-Blodgett films self assembled mono- and multilayers layers offer another opportunity to benefit from the surface sensitivity of ToF-SIMS [71-75]. Most often they are used as model systems for the determination of ToF-SIMS specific features.

Tissue Sections

A number of publications deal with the analysis of tissue sections. The advantage of this preparation is the direct availability of the area of interest at the surface. The accurate approach of the sample is crucial for an adequate analysis. Mostly cryo sectioning followed by freeze drying is used for this purpose. Noteworthy, avoiding the contamination of the sample by lubricants or other unwanted substances is difficult but crucial for a successful analysis. From a technological point of view, the recent development of a combined cryosectioning and ToF-SIMS instrument is interesting [76] as the dehydration step can be avoided.

Various observations on tissue sections could be made by ToF-SIMS: The distribution of triglycerides, phospholipids, cholesterol and fatty acids in mouse brain [77-79] and freshwater snail brain [80] sections was analysed. In other studies a part of a rat brain, the cerebellar cortex, was probed by ToF-SIMS imaging. It was possible to detect the distribution of cholesterol and galactosylceramides in different cell types of this tissue section [37, 81]. In an earlier work the same group had been able to visualise the distribution e.g. of phosphatiylcholine [82] and with the help of metal enhanced SIMS the distribution of cholesterol [83] in rat kidney. In a leg section of model mice with Duchenne muscular dystrophy information on the cellular metabolism could be obtained by others by identifying the distribution of various lipids within the sample [84].

A recent study (and review article about biological tissue imaging by ToF-SIMS) to highlight is the comparison of ToF-SIMS and MALDI-MS results on one single rat brain section sample [3]. This study shows the complementary features of these two techniques: Whereas first a number of (small sized) lipids could be probed by ToF-SIMS – noteworthy without further sample treatment and "non"-destructive – afterwards MALDI-MS was able to probe (bigger sized) peptides on the same sample - after application of a matrix and under ablation of the surface material.

Cell Preparations

As for tissue sections the adequate preparation is crucial for cell preparations. A number of approaches are reported:

Freeze drying of cells was performed and the cell surface was analyzed [85, 86] showing e.g. the distribution of cholesterol on cell samples. Nevertheless, such approaches may lead to artifacts due to contamination derived from the cell culture medium. The same aspects have to be considered using the cell preserving chemical fixative glutaraldehyde as shown in Figure 4.

Fartmann et al. analyzed freeze fractured and freeze dried osteoblasts cultured on silicon obtaining hydrocarbon signals [87]. The freeze fracturing was performed by using a sandwich method: Polystyrene beads were added to the buffer solution before adding a second silicon wafer on top of the sample. After rapid cool down in nitrogen cooled propane, the fracturing was performed by disruption of the two silicon wafer pieces [57, 88, 89]. A similar sample preparation on Tetrahymena thermophila cultures was performed by Ostrowski et al. [89]. Nevertheless, whereas the quality of the sample preparation is to highlight and the quality of the reaults was state-of-the-art at the publication date the interpretation on the distribution of lipids on a cell surface at the comparatively low signal-to-noise ratio shown in this study might be a bit offensive.

Monroe et al. were successfully able to detect the distribution of vitamin E on a single neuron [90], not giving information on their sample preparation procedure.

Sjövall and coworkers used an imprinting method for preparation of leukocyte samples: following a rinsing and freeze drying procedure, imprints of cells were obtained by attaining pressure to a silver foil covering the sample. Thus, it was possible to detect phosphocholine and cholesterol derived signals [91].

Finally, chemical fixation procedures allow a ToF-SIMS view into the cell: Incubating cells with methanol/acetic acid flushes away the lipidic membrane components [92]. First, introduced into dynamic SIMS by Levi-Setti [16, 93], recently Rommel et al. were able to detect a xenobiotic fluorophor inside a cell as well by confocal laser scanning microscopy as by (static) ToF-SIMS [94].

Other Imaging Applications

A number of studies deal with model systems or real world samples from other fields than mentioned before.

The analysis of pulp [95] and paper [96] is of interest for the respective industries. Also wood containing artwork were analyzed [97]. Biochips [98, 99] and micropatterned biomolecules containing surfaces [100] were as well analyzed as protein adsorption on surfaces relevant for medical applications [101].

DEPTH PROFILING

In ToF-SIMS the vertical probing of the sample without obtaining information on the lateral distribution of the analyte is called depth profiling. Thus, this approach is useful for well defined layered samples. For most biological samples this application mode is not suitable. To get an idea on the type of data usually obtained by such an approach, in Figure 5 a depth profile of a NRK cell sample is shown: upon increasing depth, i.e. ongoing sputter process, the intensity of signals derived from organic cell compounds decreases, while the sodium signal of the glass substrate increases.

Although not ideally suited for biochemical analysis, a limited numbers of publications dealing with depth profiling are available. Most of them focus on the detection of ToF-SIMS specific values. E.g. Sostarecz et al. [102] analyzed multiple Langmuir Blodgett film of barium arachidate by C_{60} depth profiling, finding the sputter yield to be 100-fold bigger than that of Gallium. Cheng and Winograd depth profiled a peptide doped trehalose film with a C_{60} source that was spin coated on silicon [103, 104]. They were able to detect intact peptide ions up to a mass of 500 Da after sputtering. Although the intensity of the peptide secondary ions is decreasing upon continuous C_{60} bombardment these results suggest the possibility to 3D depth profile biological samples.

3D Microarea Analysis

In a recent publication the first example for a three dimensional imaging of a biological sample was given [40]. Although scientific output of this feasibility study was limited - it was possible to detect the phosphatidylcholine and amino acid distribution in Normal Rat Kidney (NRK) cells – from a technological point of view the results are very promising. The samples – similar to those presented in Figure 6 – were chemically fixed with glutardialdehyde to prevent a cell damaging under high vacuum conditions. Thus, in a central compartment of the cell - corresponding to the nucleus - amino acids were detected with high intensities whereas phosphatidylcholine fragments were detected in the surrounding of this structure. The approach shown there might open the opportunity to detect xenobiotic substances within the cell as well as to identify the distribution of biochemical components in a number of cell compartments.

Improvement of Data Quality and Evaluation Strategies

Besides further instrumental developments currently two options of improvement for data acquisition and data evaluation are under debate: On the one hand, the enormous amount of data hampers an easy manual evaluation of the data. Therefore, statistical methods are applied to either generate or test hypothesizes. On the other hand, the ionization probability – at least under monoatomic bombardment - is relatively low. An enhancement of ionization would be beneficial for ToF-SIMS analysis.

Enhancement of Ionization

Only up to 1 % of the ejected particles upon primary ion impact is charged and thus available for mass spectrometric analysis. The increase of the ionization probability is therefore a useful aim to obtain better results.

One option for improvement is the laser-postionization of the ejected neutral particle in Laser Secondary Neutral Mass Spectrometry (Laser-SNMS) – a technique that is not in the focus of this article. As there is no recent review article available on this technique, the reader is referred to book sections [105, 106] and a recent research article [107] for further information.

Using standard ToF-SIMS instrumentation an enhancement of ionization can be performed using matrices known from MALDI (ME-SIMS) [44, 59, 80, 86, 108]. Another possibility to enhance the ion yield is the use of metal cation adduct on the secondary ions. Especially the use of thin gold or silver coatings or the use of such substrates enhances the ion yield (Meta-SIMS) [37, 58, 85, 86, 109-112]. Whereas the use of metals leads to a cationization of the molecule ($[\text{Molecule+Metal}]^+$), the effects of MALDI matrices is multifacetted [113]. Nevertheless, all of these effects can be directly or indirectly allocated to an increased sputter desorption of surface material.

Most of the studies were performed under monoatomic primary ion bombardment. Therefore, the question if the enhancement is still detectable under polyatomic bombardment is currently under investigation. The state of the art in matrix- and metal-enhanced SIMS has recently be described by Delcorte [113].

Reading this sections the reader has to keep in mind that all three approaches discussed will not be generally applicable and limited to special conditions and samples.

Statistical Evaluation

The extraordinary amount of data collected in each ToF-SIMS experiment requires elaborous evaluation routines. While still the manual evaluation routine is most frequently used, the use of statistical methods is becoming more and more relevant. It would certainly go beyond the scope of this review to describe the different available statistical approaches in detail. An impression on different statistical approaches for ToF-SIMS data evaluation available can be achieved in reading the introductive articles by Graham et al. [114], Tyler [115], Milillo and Gardella [116] as well as Smentkowski et al. [117]. To get an idea which problems can be addressed by these statistical tools some examples are given here:

Kulp et al. were able to differentiate cell lysates by PCA [118]. Berman and al. were able to distinguish a number of monosaccharides [54]. May et al. performed PCA on DNA-oligomers [50] and Thompson et al. were able to discern bacilli by statistical methods [119].

Another interesting approach is the determination of conformational changes in proteins: As in native proteins the hydrophobic amino acids are more commonly found in the interior of the protein than in denatured proteins, the intensity of these amino acids on different samples of the same protein give indication on the conformational state of the protein [120]. This initial observation can meanwhile nicely be addressed by PCA [43].

The statistical tools are useful in the hand of an expert to validate or generate hypothesizes. Nevertheless, it has to be pointed out that in the hand of a non-expert they may lead to misinterpretation of the data.

Specifications of ToF-SIMS

Taking ToF-SIMS into account for addressing an analytical problem a scientist should be aware of the physical and practical limits of ToF-SIMS. Although some of them were already mentioned before, a condensed overview might (have) save(d) time for the "scan and skim" reader. Gratuitous to point out that all of the mentioned limitations are matter of recent scientific and engineering research and development.

The sample has to sustain the vacuum conditions of the instrument. The sample size is limited by the instrument but may be several cm in length and breadth. As area used for analysis routinously up to 500x500 μm^2 can be measured – on flat samples up to 9x9cm^2 can be scanned.

The mass range of ToF-SIMS as of all Time-of-flight techniques is principally unlimited. Nevertheless, the ionization process limits the practical achievable mass range to about 10.000 Da [10]. The strength of ToF-SIMS is in the range of 0-2000 Da.

The detection limit is expected to be in the femtomol to attomol per μm^2 range. E.g. $2x10^{-20}$ mol on an area of 100x100 nm was reported [23]. Some authors even claim to be able to detect an amount of substance in the zeptomol range (4 attomol on a ca. 2 mm diameter spot) [121]. Additionally, e.g. for proteins a amount of 0.1 ng/cm^2 on the surface could be detected [122]. From these results it may be assumed that the picomol/μm^2 range can be routinously be obtained.

The surface sensitivity is limited to the three uppermost monolayers [9]. This extreme surface sensitivity is a big advantage of ToF-SIMS, but especially for biochemical samples this aspect necessitates a contamination free sample preparation.

Quantification is difficult with ToF-SIMS. In general, semiquanitative information upon comparison of two similar samples or different positions on one sample can be gained.

The mass resolution, defined - at least for two singly charged species - as the ratio of a mass divided by the mass difference of the next neighboring peak possible to distinguish at 50% half width maximum ($R=m/\Delta m$), obtainable by ToF-SIMS is in the range of 10000. I.e. that a signal at mass 100.00 Da can be distinguished from a signal at mass 100.01 Da. It is rarely mentioned that mass resolution, lateral resolution and measuring time are part of the so called "magic triangle" in ToF-SIMS: For technical reasons only two of these three parameters can be optimized to an acceptable value per measurement.

The lateral resolution routinely obtained in imaging mode is 350 nm [79]. Nevertheless, meanwhile unpublished studies show a better resolution. Ultimately, the size of the collision cascade limits the lateral resolution in ToF-SIMS to app. 30 nm [11].

The vertical resolution in depth profiling and 3D microarea analysis was found to be 30 nm in organic samples [123]. As for inorganic samples a resolution of 1 nm can already be obtained [36], the ultimate border of vertical resolution in organic samples might not yet have been reached.

With a view on these specifications and limitations of ToF-SIMS a recent article was titled "Why don't biologists use SIMS?"[124]. The answer certainly is: "Some do, but most of them just start!".

ACKNOWLEDGEMENT

The author thanks Dr. Reinhard Kersting and Dr. Birgit Hagenhoff (both from Tascon GmbH, Muenster) for their proof reading. The samples shown in this review were prepared by PD Dr. Joachim Wegener, Christina Rommel and Michael Seifert (all from the University of Muenster). Writing this review has been financially supported by the European Union Framework Program 6 under grant number FP6-513698 (Toxdrop-Project).

REFERENCES

[1] Pacholski, M.L. and Winograd, N., Imaging with Mass spectrometry. *Chem. Rev.*, 1999. 99: p. 2977-3005.

[2] Belu, A.M., Graham, D.J., and Castner, D.G., Time-of-flight secondary ion mass spectrometry: techniques and applications for the characterization of biomaterial surfaces. *Biomaterials*, 2003. 24(21): p. 3635-3653.

[3] Brunelle, A., Touboul, D., and Laprevote, O., Biological tissue imaging with time-of-flight secondary ion mass spectrometry. J. Mass Spectrom., 2005. 40: p. 985-999.

[4] Mulisch, H., *The Procedure*. 2002: Penguin Books.

[5] Benninghoven, A., The History of Static SIMS: A Personal Perspective, in ToF-SIMS. *Surface Analysis by Mass Spectrometry*, J.C. Vickerman, Editor. 2001, IM Publications: Huddersfield. p. 41-74.

[6] Guerquin-Kern, J.-L., Wu, T.-D., Quintana, C., and Croisy, A., Progress in analytical imaging of the cell by dynamic secondary ion mass spectrometry (SIMS microscopy). *Biochim. Biophys. Acta*, 2005. 1724: p. 228-238.

[7] Garrison, B.J., Atoms, clusters and photons: Energetic probes for mass spectrometry. *Appl. Surf. Sci.*, 2006. 252: p. 6409–6412.

[8] Hagenhoff, B. and Rading, D., Ion Beam Techniques: Surface Mass Spectrometry., in *Handbook of surface and Interface Analysis.*, J.C. Riviére and S. Myhra, Editors. 1998, Marcel Dekker, Inc: New York, Basel, Hong Kong. p. 209-346.

[9] Vickerman, J.C., *TOF-SIMS: An overview, in ToF-SIMS - Surface analysis by mass spectrometry*, J.C. Vickerman and D. Briggs, Editors. 2001, IMPublications: Manchester/Chichester.

[10] Hagenhoff, B., High Resolution Surface Analysis by TOF-SIMS. *Mikrochim. Acta*, 2000. 132: p. 259-271.

[11] Benninghoven, A., Rüdenauer, F.G., and Werner, H.W., Secondary Ion Mass Spectrometry. *Basic concepts, instrumental aspects, applications and trends*. 1987, John Wiley and Sons: New York. p. 761-949.

[12] Urbassek, H.M., Status of Cascade Theory, in *ToF-SIMS. Suface Analysis by Mass Spectrometry.*, J.C. Vickerman, Editor. 2001, IM Publications: Huddersfield. p. 139-160.

[13] Chandra, S., 3D subcellular imaging in cryogenically prepared single cells. *Appl. Surf. Sci.*, 2004(231-232): p. 467-469.

[14] Galle, P., Levi-Setti, R., Lamperti, A., Bourahla, K., and Escaig, F., Subcellular localization of aluminium and indium in the rat kidney. *Appl. Surf. Sci.*, 2004. 231-232: p. 475-478.

[15] Chandra, S., Smith, D.R., and Morrison, G.H., Subcellular imaging by dynamic SIMS ion microscopy. *Anal. Chem*, 2000. 72(3): p. 104A-114A.

[16] Levi-Setti, R., Gavrilov, K.L., Strissel, P.L., and Strick, R., Ion microprobe imaging of 44Ca-labeled mammalian chromosomes. *Appl. Surf. Sci.*, 2004. 231-232: p. 479-484.

[17] Gillen, G. and Roberson, S.V., Preliminary evaluation of an SF-(5)+ polyatomic primary ion beam for analysis of organic thin films by secondary ion mass spectrometry. *Rapid Communication in Mass Spectrometry*, 1998. 12: p. 1303-1312.

[18] Kötter, F. and Benninghoven, A., Secondary ion emission from polymer surfaces under Ar+, Xe+ and SF-(5)+ ion bombardment. *Appl. Surf. Sci.*, 1998. 133: p. 47-57.

[19] Weibel, D.E., Lockyer, N., and Vickerman, J.C., C60 cluster ion bombardment of organic surfaces. *Appl. Surf. Sci.*, 2004. 231-232: p. 146-152.

[20] Wong, S.C.C., Hill, R., Blenkinsopp, P., Lockyer, N.P., Weibel, D., and Vickerman, J.C., Development of a C60+ ion gun for static SIMS and chemical imaging. *Appl. Surf. Sci.*, 2003. 203-204: p. 219-222.

[21] Weibel, D., Wong, S., Lockyer, N., Blenkinsopp, P., Hill, R., and Vickerman, J.C., A C60 primary ion beam system for time of flight secondary ion mass spectrometry: its development and secondary ion yield characteristics. *Anal. Chem*, 2003. 75(7): p. 1754-64.

[22] Davies, N., Weibel, D., Blenkinsopp, P., Lockyer, N., Hill, R., and Vickerman, J.C., Development and experimental application of a gold liquid metal ion source. *Appl. Surf. Sci.*, 2003. 203-204: p. 223-227.

[23] Kollmer, F., Cluster primary ion bombardment of organic materials. *Appl. Surf. Sci.*, 2004. 231-232: p. 153-158.

[24] Russo, M.F., Wojciechowski, I.A., and Garrison, B.J., Sputtering of amorphous ice induced by C60 and Au3 clusters. *Appl. Surf. Sci.*, 2006. 252: p. 6423-6425.

[25] Winograd, N., The Magic of Cluster SIMS. *Analytical Chemistry*, 2005: p. 143A-149A.

[26] Wucher, A., Molecular Secondary ion formation under cluster bombardment: A fundamental review. *Appl. Surf. Sci.*, 2006. 252: p. 6482-6489.

[27] Benguerba, M., Brunelle, A., Della-Negra, S., Depauw, J., Joret, H., Le Beyec, Y., Blain, M.G., Schweikert, E.A., Ben Assayag, G., and Sudraud, P., Impact of Slow Gold

Cluster on Various Solids: Non-linear Effects in Secondary Emission. Nucl. Instrum. *Methods Phys. Bes.* B, 1991. 62: p. 8-22.

[28] Kersting, R., Hagenhoff, B., Kollmer, F., Moellers, R., and Niehues, E., Influence of Primary Ion Bombardment Conditions on the Emission of Molecular Secondary Ions. *Appl. Surf. Sci.*, 2004. 231/232: p. 261-264.

[29] Castner, D., View from the edge. *Nature*, 2003. 422: p. 129-130.

[30] Kötter, F. and Benninghoven, A., Secondary ion emission from polymer surfaces under Ar+, Xe+ and SF5+ ion bombardment. *Appl. Surf. Sci.*, 1998. 133: p. 47-57.

[31] Stapel, D., Brox, O., and Benninghoven, A., Secondary ion emission from arachidic acid LB-layers under Ar+, Xe+, Ga+ and SF-(5)+ primary ion bombardment. *Appl. Surf. Sci.*, 1999. 140: p. 156-167.

[32] Stapel, D. and Benninghoven, A., 2001. *Appl. Surf. Sci.*, Application of atomic and molecular primary ions for TOF-SIMS analysis of additive containing polymer surfaces. 174: p. 261-270.

[33] Appelhans, A.D. and Delmore, J.E., Comparison of polyatomic and atomic primary beams for secondary ion mass spectrometry of organics. *Anal. Chem*, 1989. 61: p. 1087-1093.

[34] Schueler, B.W., Time-of-Flight Mass Analysers, in *ToF-SIMS. Surface Analysis by Mass Spectroemtry*, J.C. Vickerman, Editor. 2001, IM Publications: Huddersfield. p. 75-94.

[35] Hill, R., Primary Ion Systems, in *ToF-SIMS. Surface Mass Spectrometry*, J.C. Vickerman, Editor. 2001, IM Publications: Huddersfield. p. 95-112.

[36] Niehuis, E. and Grehl, T., Dual Beam Depth Profiling., in *TOF-SIMS: Surface Analysis by Mass Spectrometry.*, J.C. Vickerman, Briggs, D., Editor. 2001, IM Publications: Charlton. p. 753-778.

[37] Nygren, H., Borner, K., Hagenhoff, B., Malmberg, P., and Mansson, J.E., Localization of cholesterol, phosphocholine and galactosylceramide in rat cerebellar cortex with imaging TOF-SIMS equipped with a bismuth cluster ion source. *Biochim. Biophys. Acta,* 2005. 1737(2-3): p. 102-10.

[38] Hagenhoff, B., Van Leyen, D., Niehuis, E., and Benninghoven, A., Time-of-Flight Secondary Ion Mass-Spectrometry of Insulators with Pulsed Charge Compensation By Low-Energy Electrons. *Journal Of Vacuum Science And Technology A-Vacuum Surfaces And Films*, 1989. 10(5): p. 3056-3064.

[39] Gilmore, I.S. and Seah, M.P., Electron flood gun damage in the analysis of polymers and organics in time-of-flight SIMS. *Appl. Surf. Sci.*, 2002. 187: p. 89-100.

[40] Breitenstein, D., Rommel, C.E., Moellers, R., Wegener, J., and Hagenhoff, B., The Chemical Composition of Animal Cells and their Intracellular Compartments Reconstructed from 3D Mass Spectrometry. *Angew. Chem. Int.* Ed., 2006. submitted for publication.

[41] Mantus, D.S., Ratner, B.D., Carlson, B.A., and Moulder, J.F., Static secondary ion mass spectrometry of adsorbed proteins. *Anal. Chem.*, 1993. 65: p. 1431-1438.

[42] McArthur, S.L., Vendettuoli, M.C., Ratner, B.D., and Castner, D., Methods for Generating Protein Molecular Ions in ToF-SIMS. *Langmuir*, 2004. 20: p. 3704-3709.

[43] Michel, R., Pasche, S., Textor, M., and Castner, D.G., Influence of PEG Architecture on Protein Adsorption and Conformation. *Langmuir*, 2005. 21(26): p. pp 12327 - 12332; (Research Article) DOI: 10.1021/la051726h.

[44] Wittmaack, K., Szymczak, W., Hoheisel, G., and Tuszynski, W., Time-of-Flight Secondary Ion Mass Spectrometry of Matrix-Diluted Oligo- and Polypeptides Bombarded With Slow and Fast Projectiles: Positive and Negative Matrix and Analyte Ion Yields, Background Signals, and Sample Aging. *J. Am. Soc. Mass Spectrom.*, 2000. 11: p. 553–563.

[45] Wu, K.J. and Odom, R.W., *Biological Material Analysis by Matrix-Enhanced SIMS, in Secondary Ion Mass Spectrometry* XI., G. Gillen, R. Lareau, J. Bennett, and F.A. Stevie, Editors. 1998, John Wiley and Sons: Chichester. p. 525-528.

[46] Wagner, M.S., Horbett, T.A., and Castner, D.G., Characterizing multicomponent adsorbed protein films using electron spectroscopy for chemical analysis, time-of-flight secondary ion mass spectrometry, and radiolabeling: capabilities and limitations. *Biomaterials*, 2003. 24(11): p. 1897-1908.

[47] Jabs, H.U., Assmann, G., Greifendorf, D., and Benninghoven, A., High performance liquid chromatography and time-of-flight secondary ion mass spectrometry: a new dimension in structural analysis of apolipoproteins. *J. Lipid Res.,* 1986. 27(6): p. 613-21.

[48] von Eckardstein, A., Walter, M., Holz, H., Benninghoven, A., and Assmann, G., Site-specific methionine sulfoxide formation is the structural basis of chromatographic heterogeneity of apolipoproteins A-I, C-II, and C-III. *J. Lipid Res.*, 1991. 32(9): p. 1465-76.

[49] von Eckardstein, A., Funke, H., Henke, A., Altland, K., Benninghoven, A., and Assmann, G., Apolipoprotein A-I variants. Naturally occurring substitutions of proline residues affect plasma concentration of apolipoprotein A-I. *J. Clin. Invest.,* 1989. 84(6): p. 1722-30.

[50] May, C.J., Canavan, H.E., and Castner, D.G., Quantitative X-ray photoelectron spectroscopy and time-of-flight secondary ion mass spectrometry characterization of the components in DNA. *Anal. Chem.*, 2004. 76(4): p. 1114-22.

[51] Lee, C.-Y., Canavan, H.E., Gamble, L.J., and Castner, D.G., Evidence of Impurities in Thiolated Single-Stranded DNA Oligomers and Their Effect on DNA Self-Assembly on Gold. *Langmuir*, 2005. 21: p. 5134-5141.

[52] Cheran, L.-E., Vukovich, D., and Thompson, M., Imaging TOF-SIMS analysis of oligonucleotide microarrays. *Analyst,* 2003. 128: p. 126–129.

[53] Arlinghaus, H.F., Ostrop, M., Friedrichs, O., Feldner, J., Gunst, U., and Lipinsky, D., DNA sequencing with ToF-SIMS. *Surf. Interface Anal.*, 2002. 33(35-39).

[54] Berman, E.S., Kulp, K.S., Knize, M.G., Wu, L., Nelson, E.J., Nelson, D.O., and Wu, K.J., Distinguishing monosaccharide stereo- and structural isomers with TOF-SIMS and multivariate statistical analysis. *Anal. Chem.*, 2006. 78(18): p. 6497-503.

[55] Roddy, T.P., Cannon, D.M., Jr., Meserole, C.A., Winograd, N., and Ewing, A.G., Imaging of freeze-fractured cells with in situ fluorescence and time-of-flight secondary ion mass spectrometry. *Anal. Chem.*, 2002. 74(16): p. 4011-9.

[56] Leufgen, K.M., Rulle, H., Benninghoven, A., Sieber, M., and Galla, H.J., Imaging Time-of-Flight Secondary Ion Mass Spectrometry Allows Visualization and Analysis of Coexisting Phases in Langmuir-Blodgett Films. *Langmuir*, 1996. 12: p. 1708-1711.

[57] Roddy, T.P., Cannon, D.M., Jr., Ostrowski, S.G., Winograd, N., and Ewing, A.G., Identification of cellular sections with imaging mass spectrometry following freeze fracture. *Anal. Chem.*, 2002. 74(16): p. 4020-6.

[58] Adriaensen, L., Vangaever, F., and Gijbels, R., Metal-Assisted Secondary Ion Mass Spectrometry: Influence of Ag and Au Deposition on Molecular Ion Yields. *Anal. Chem.*, 2004. 76(22): p. 6777-6785.

[59] Adriaensen, L., Vangaever, F., Lenaerts, J., and Gijbels, R., Matrix-enhanced secondary ion mass spectrometry: the influence of MALDI matrices on molecular ion yields of thin organic films. *Rapid Communications in Mass Spectrometry*, 2005. 19: p. 1017–1024.

[60] Ulman, A., *Ultrathin Organic Films*. 1994, San Diego: Academic Press, Inc.

[61] Bourdos, N., Kollmer, F., Benninghoven, A., Ross, M., Sieber, M., and Galla, H.J., Analysis of lung surfactant model systems with time-of-flight secondary ion mass spectrometry. *Biophys. J.*, 2000. 79(1): p. 357-69.

[62] Bourdos, N., Kollmer, F., Benninghoven, A., Sieber, M., and Galla, H.J., Imaging of Domain Structures in a One-Component Lipid Monolayer by Time-of-Flight Secondary Ion Mass Spectrometry. *Langmuir*, 2000. 16: p. 1481-1484.

[63] Biesinger, M.C., Miller, D.J., Harbottle, R.R., Possmayer, F., McIntyre, N.S., and Petersen, N.O., Imaging Lipid Distributions in Model Monolayers by ToF-SIMS with selectively deuterated components and principal component anlysis. *Appl. Surf. Sci.*, 2006. 252: p. 6957-6065.

[64] Biesinger, M.C., Paepegaey, P.Y., McIntyre, N.S., Harbottle, R.R., and Petersen, N.O., Principal component analysis of TOF-SIMS images of organic monolayers. *Anal. Chem.*, 2002. 74(22): p. 5711-6.

[65] Sostarecz, A.G., Cannon, D.M., Jr., McQuaw, C.M., Sun, S., Ewing, A.G., and Winograd, N., Influence of molecular environment on the analysis of phospholipids by time-of-flight secondary ion mass spectrometry. *Langmuir*, 2004. 20(12): p. 4926-32.

[66] Breitenstein, D., Batenburg, J.J., Hagenhoff, B., and Galla, H.J., Lipid specificity of surfactant protein B studied by time-of-flight secondary ion mass spectrometry. *Biophys. J.*, 2006. 91(4): p. 1347-56.

[67] Harbottle, R.R., Nag, K., McIntyre, N.S., Possmayer, F., and Petersen, N.O., Molecular Organization Revealed by Time-of-Flight Secondary Ion Mass Spectrometry of a Clinically Used Extracted Pulmonary Surfactant. *Langmuir*, 2003. 19: p. 3698-3704.

[68] Sostarecz, A.G., McQuaw, C.M., Ewing, A.G., and Winograd, N., Phosphatidylethanolamine-induced cholesterol domains chemically identified with mass spectrometric imaging. *J. Am. Chem. Soc.*, 2004. 126(43): p. 13882-3.

[69] McQuaw, C.M., Sostarecz, A.G., Zheng, L., Ewing, A.G., and Winograd, N., Investigating lipid interactions and the process of raft formation in cellular membranes using ToF-SIMS. *Appl. Surf. Sci.*, 2006. 252: p. 6716-6718.

[70] Munro, S., Lipid rafts: elusive or illusive? Cell, 2003. 115(4): p. 377-88.

[71] Wolf, K.V., Cole, D.A., and Bernasek, S.L., High-resolution TOF-SIMS study of varying chain length self-assembled monolayer surfaces. *Anal. Chem.*, 2002. 74(19): p. 5009-16.

[72] Wong, S.C.C., Lockyer, N.P., and Vickerman, J.C., Mechanisms of secondary ion emission from self-assembled monolayers and multilayers. *Surface and Interface Analysis*, 2005. 37: p. 721-730.

[73] Francis, J.T., Nie, H.Y., McIntyre, N.S., and Briggs, D., ToF-SIMS investigation of octadecylphosphonic acid monolayers on a mica substrate. *Langmuir*, 2006. 22: p. 9244-50.

[74] Schröder, M., Sohn, S., Lipinsky, D., and Arlinghaus, H.F., Influence of primary ion species on the secondary cluster ion emission process from SAMs of hexadecanethiol on gold. *Appl. Surf. Sci.* 252(19): p. 6566-6569.

[75] Rading, D., Kersting, R., and Benninghoven, A., Secondary Ion Emission from Molecular Overlayers: Thiols on Gold. *Proceedings of SIMS XI conference.*, 1997(11): p. 455-458.

[76] Möller, J., Beumer, A., Lipinsky, D., and Arlinghaus, H.F., Introduction of a cryosectioning-ToF-SIMS instrument for analysis of non-dehydrated biological samples. *Appl. Surf. Sci.*, 2006. 252(19): p. 6709-6711.

[77] Touboul, D., Halgand, F., Brunelle, A., Kersting, R., Tallarek, E., Hagenhoff, B., and Laprevote, O., Tissue molecular ion imaging by gold cluster ion bombardment. *Anal. Chem.*, 2004. 76(6): p. 1550-9.

[78] Sjovall, P., Lausmaa, J., and Johansson, B., Mass spectrometric imaging of lipids in brain tissue. *Anal. Chem.*, 2004. 76(15): p. 4271-8.

[79] Touboul, D., Kollmer, F., Niehues, E., Brunelle, A., and Laprevote, O., Improvement of Biological Time-of-Flight Secondary Ion Mass Spectrometry Imaging with Bismuth Cluster Ion Source. *J. Am. Soc. Mass Spectrom.*, 2005. 16: p. 1608-1618.

[80] McDonell, L.A., Sander, R.P., Altelaar, A.F.M., Mize, T.H., Luxembourg, S.L., Verhaert, P.D.E.M., van Minnen, J., and Heeren, R.M.A., Subcellular imaging mass spectrometry of brain tissue. *J. Mass Spec.*, 2005. 1005(40): p. 160-168.

[81] Borner, K., Nygren, H., Hagenhoff, B., Malmberg, P., Tallarek, E., and Mansson, J.E., Distribution of cholesterol and galactosylceramide in rat cerebellar white matter. *Biochim. Biophys. Acta,* 2006. 1761(3): p. 335-44.

[82] Nygren, H., Borner, K., Malmberg, P., Tallarek, E., and Hagenhoff, B., Imaging TOF-SIMS of rat kidney prepared by high-pressure freezing. *Microsc. Res. Tech.*, 2005. 68(6): p. 329-34.

[83] Nygren, H., Malmberg, P., Kriegeskotte, C., and Arlinghaus, H.F., Bioimaging TOF-SIMS: localization of cholesterol in rat kidney sections. *FEBS Lett*, 2004. 566(1-3): p. 291-3.

[84] Touboul, D., Brunelle, A., Halgand, F., De La Porte, S., and Laprevote, O., Lipid imaging by gold cluster time-of-flight secondary ion mass spectrometry: application to Duchenne muscular dystrophy. *J. Lipid Res.*, 2005. 46(7): p. 1388-95.

[85] Nygren, H. and Malmberg, P., Silver deposition on freeze-dried cells allows subcellular localization of cholesterol with imaging TOF-SIMS. *J. Microsc.,* 2004. 215(Pt 2): p. 156-61.

[86] Altelaar, A.F., Klinkert, I., Jalink, K., de Lange, R.P., Adan, R.A., Heeren, R.M., and Piersma, S.R., Gold-enhanced biomolecular surface imaging of cells and tissue by SIMS and MALDI mass spectrometry. *Anal. Chem.*, 2006. 78(3): p. 734-42.

[87] Fartmann, M., Dambach, S., Kriegeskotte, C., Wiesmann, H.P., Wittig, A., Sauerwein, W., Lipinsky, D., and Arlinghaus, H.F., Characterization of cell cultures with ToF-SIMS and laser-SNMS. *Surf. Interface Anal.*, 2002. 34: p. 63-66.

[88] Wittig, A., Wiemann, M., Fartmann, M., Kriegeskotte, C., Arlinghaus, H.F., Zierold, K., and Sauerwein, W., Preparation of cells cultured on silicon wafers for mass spectrometry analysis. *Microsc. Res. Tech.*, 2005. 66(5): p. 248-58.

[89] Ostrowski, S.G., Van Bell, C.T., Winograd, N., and Ewing, A.G., Mass spectrometric imaging of highly curved membranes during Tetrahymena mating. *Science*, 2004. 305(5680): p. 71-73.

[90] Monroe, E.B., Jurchen, J.C., Lee, J., Rubakhin, S.S., and Sweedler, J.V., Vitamin E Imaging and Localization in the Neuronal Membrane. *J. Am. Chem. Soc.,* 2005. 127(35): p. 12152-12153.

[91] Sjovall, P., Lausmaa, J., Nygren, H., Carlsson, L., and Malmberg, P., Imaging of membrane lipids in single cells by imprint-imaging time-of-flight secondary ion mass spectrometry. *Anal. Chem.*, 2003. 75(14): p. 3429-34.

[92] Hoetelmans, R.W., Prins, F.A., Cornelese-ten Velde, I., van der Meer, J., van de Velde, C.J., and van Dierendonck, J.H., Effects of acetone, methanol, or paraformaldehyde on cellular structure, visualized by reflection contrast microscopy and transmission and scanning electron microscopy. *Appl. Immunohistochem. Mol. Morphol.*, 2001. 9(4): p. 346-51.

[93] Levi-Setti, R., Gavrilow, K.L., and Neilly, M.E., Cations in Mammalian Cells and Chromosomes: Sample Preparation Protocolls Affect Elemental Abundances by SIMS. *Appl. Surf. Sci.*, 2005. 252: p. 6765-6769.

[94] Rommel, C.E., Breitenstein, D., Hagenhoff, B., and Wegener, J., ToF-SIMS Analysis of Chemically Fixed Cell Samples. *Journal of Mass Spectrometry*, 2006. in preparation.

[95] Fardim, P., Gustafsson, J., von Schoultz, S., Peltonen, J., and Holmbom, B., Extractives on fiber surfaces investigated by XPS, ToF-SIMS and AFM. Colloids and Surfaces A: *Physicochem. Eng. Aspects,* 2005. 255: p. 91-103.

[96] Fardim, P. and Holmbom, B., ToF-SIMS imaging: a valuable chemical microscopy technique for paper and paper coatings. *Appl. Surf. Sci.*, 2005. 249: p. 393–407.

[97] Mazel, V., Richardin, P., Touboul, D., Brunelle, A., Walter, P., and Laprevote, O., Chemical imaging techniques for the analysis of complex mixtures:New application to the characterization of ritual matters on African wooden statuettes. *Analytica Chimica Acta,* 2006. 570(1): p. 34-40.

[98] Aoyagi, S. and Kudo, M., Effective monitoring of protein reaction on glass plate surfaces by TOF-SIMS. *Biosens Bioelectron,* 2005. 20(8): p. 1626-30.

[99] Hellweg, S., Jacob, A., Hoheisel, J.D., Grehl, T., and Arlinghaus, H.F., Mass spectrometric characterization of DNA microarrays as a function of primary ion species. *Appl. Surf. Sci.,* 2006. 252: p. 6742–6745.

[100] Belu, A.M., Yang, Z., Aslami, R., and Chilkoti, A., Enhanced TOF-SIMS imaging of a micropatterned protein by stable isotope protein labeling. *Anal. Chem.,* 2001. 73(2): p. 143-50.

[101] Aoyagi, S., Kudo, M., Hayama, M., Hasegawa, U., Sakai, K., Tozu, M., and Hoshi, T., TOF-SIMS Imaging of Protein Adsorption on Dialysis Membrane by means of Information Entropy. *Surf. Sci. Nanotech.,* 2003. 1: p. 67-71.

[102] Sostarecz, A.G., McQuaw, C.M., Wucher, A., and Winograd, N., Depth profiling of Langmuir-Blodgett films with a buckminsterfullerene probe. *Anal. Chem.,* 2004. 76(22): p. 6651-8.

[103] Cheng, J. and Winograd, N., Depth profiling of peptide films with TOF-SIMS and a C60 Probe. *Anal. Chem.,* 2005. 77(3651-3659).

[104] Cheng, J., Wucher, A., and Winograd, N., Molecular Depth Profiling with Cluster Ion Beams. *J. Phys. Chem.* B, 2006. 110: p. 8329-8336.

[105] Arlinghaus, H.F., Laser-SNMS, in *Surface and Thin Film Analysis. Principles, Instrumentation, Applications*, B.H.a.J. H., Editor. 2002, Wiley-VCH: Weinheim. p. 132-139.
[106] Wucher, A., Laser Post-Ionisation: Fundamentals, in *ToF-SIMS - Surface analysis by mass spectrometry*, J.C. Vickerman, Editor. 2001, IMPublications: Manchester/Chichester. p. 347-374.
[107] Arlinghaus, H.F., Fartmann, M., Kriegeskotte, C., Dambach, S., Wittig, A., Sauerwein, W., and Lipinsky, D., Subcellular imaging of cell cultures and tissue for boron localization with laser-SNMS. *Surf. Interface Anal.*, 2004. 36: p. 698-701.
[108] Altelaar, A.F., van Minnen, J., Jimenez, C.R., Heeren, R.M., and Piersma, S.R., Direct molecular imaging of Lymnaea stagnalis nervous tissue at subcellular spatial resolution by mass spectrometry. *Anal. Chem.*, 2005. 77(3): p. 735-41.
[109] Nygren, H., Johansson, B.R., and Malmberg, P., Bioimaging TOF-SIMS of tissues by gold ion bombardment of a silver-coated thin section. *Microsc. Res. Tech.,* 2004. 65(6): p. 282-6.
[110] Delcorte, A., Medard, N., and Bertrand, P., Organic secondary ion mass spectrometry: sensitivity enhancement by gold deposition. *Anal. Chem.*, 2002. 74(19): p. 4955-68.
[111] Delcorte, A., Bour, J., Aubriet, F., Muller, J.F., and Bertrand, P., Sample metallization for performance improvement in desorption/ionization of kilodalton molecules: quantitative evaluation, imaging secondary ion MS, and laser ablation. *Anal. Chem.*, 2003. 75(24): p. 6875-85.
[112] Kim, Y.P., Oh, E., Hong, M.Y., Lee, D., Han, M.K., Shon, H.K., Moon, D.W., Kim, H.S., and Lee, T.G., Gold nanoparticle-enhanced secondary ion mass spectrometry imaging of peptides on self-assembled monolayers. *Anal. Chem.*, 2006. 78(6): p. 1913-20.
[113] Delcorte, A., Matrix-enhanced secondary ion mass spectrometry: The Alchemist's solution? *Appl. Surf. Sci.*, 2006. 252(19): p. 6582-6587.
[114] Graham, D.J., Wagner, M.S., and Castner, D.G., Information from complexity: Challenges of TOF-SIMS data interpretation. *Appl. Surf. Sci.* 252(19): p. 6860-6868.
[115] Tyler, B.J., Multivariate statistical image processing for molecular specific imaging in organic and bio-systems. *Appl. Surf. Sci.,* 2006. 252(19): p. 6875-6882.
[116] Milillo, T.M. and Gardella, J.A., Spatial statistics and interpolation methods for TOF SIMS imaging. *Appl. Surf. Sci.*, 2006. 252(19): p. 6883-6890.
[117] Smentkowski, V.S., Keenan, M.R., Ohlhausen, J.A., and Kotula, P.G., Multivariate statistical analysis of concatenated time-of-flight secondary ion mass spectrometry spectral images. Complete description of the sample with one analysis. *Anal. Chem.,* 2005. 77(5): p. 1530-6.
[118] Kulp, K.S., Berman, E.S., Knize, M.G., Shattuck, D.L., Nelson, E.J., Wu, L., Montgomery, J.L., Felton, J.S., and Wu, K.J., Chemical and biological differentiation of three human breast cancer cell types using time-of-flight secondary ion mass spectrometry. *Anal. Chem.,* 2006. 78(11): p. 3651-8.
[119] Thompson, C.E., Ellis, J., Fletcher, J.S., Goodacre, R., Henderson, A., Lockyer, N.P., and Vickerman, J.C., ToF-SIMS studies of Bacillus using multivariate analysis with possible identification and taxonomic applications. *Appl. Surf. Sci.*, 2006. 252(19): p. 6719-6722.

[120] Tidwell, C.D., Castner, D.G., Golledge, S.L., Ratner, B.D., Meyer, K., Hagenhoff, B., and Benninghoven, A., Static time-of-flight secondary ion mass spectrometry and x-ray photoelectron spectroscopy characterization of adsorbed albumin and fibronectin films. *Surf. Interface Anal.,* 2001. 31(724-733).

[121] Quong, J.N., Knize, M.G., Kulp, K.S., and Wu, K.J., Molecule-specific imaging analysis of carcinogens in breast cancer cells using time-of-flight secondary ion mass spectrometry. *Appl. Surf. Sci.,* 2004. 231-232: p. 424-427.

[122] Wagner, M.S., McArthur, S.L., Shen, M., Horbett, T.A., and Castner, D.G., Limits of detection for time of flight secondary ion mass spectrometry (ToF-SIMS) and X-ray photoelectron spectroscopy (XPS): detection of low amounts of adsorbed protein. *J. Biomater. Sci. Polym.* Ed., 2002. 13(4): p. 407-28.

[123] Wagner, M.S., Molecular Depth Profiling of Multilayer Polymer films Using Time-of-Flight Secondary Ion Mass Spectrometry. *Anal. Chem.,* 2005. 77: p. 911-922.

[124] Heeren, R.M.A., McDonnell, L.A., Amstalden, E., Luxembourg, S.L., Altelaar, A.F.M., and Piersma, S.R., Why don't biologists use SIMS?: A critical evaluation of imaging MS. *Appl. Surf. Sci.* 252(19): p. 6827-6835.

In: Biomacromolecular Mass Spectrometry Yearbook. Volume 1 ISBN: 978-1-60876-167-8
Ed: Simone Koenig

Chapter 16

HYPOTHETICAL NEW CLASS OF CODING RNAS IN INSECT GENOMES*

Jan M. Buellesbach
University of Cologne, Departement of Genetics,
Zuelpicher Str. 47, 50674 Cologne,
Germany

ABSTRACT

The recent discovery of a polycistronic mRNA in the red flour beetle *Tribolium castaneum* qualifying as a gap gene albeit encompassing only four very short open reading frames offers a great potential for mass spectrometric studies specifically targeting at small peptides. Addressing the question if more putative candidate peptides are indeed directly translated into their functional form, those studies might greatly help in investigating this postulated new class of coding RNAs, determining whether they are a rather rare and special occurrence, or much more common than previously assumed and just as yet hidden from the routinely applied gene prediction methods due to their unexpected properties.

The transcription factors characterized as gap segmentation genes play an early and important role in the hierachical genetic cascade during the embryo development of several insect species (Peel et al., 2005). Their main functions constitute the subdivision of the early embryo into broad domains, the cross-regulation with each other during this segmentation process, and the involvement in the initiation of the next step in the cascade, the segment polarity genes (Pankratz and Jackle, 1990). Recently, Savard et al. were able to assign all attributes of a gap gene to a polycistronic transcript in the red flour beetle *Tribolium castaneum*, despite its surprising and unusual structure (Savard et al., 2006). The discovered gene, dubbed "*mille pates*" (*mlpt*), constitutes a transcript encompassing four comparably short open reading frames. (Figure 1 A) Depletion of this gene results in stubby larvae

* A version of this chapter was also published in *Biomacromolecular Mass Spectrometry Volume 1, Number 4*, published by Nova Science Publications, Inc. It was submitted for appropriate modifications in an effort to encourage wider dissemination of research.

missing their terminal abdominal segments, as well as having the identity of their seven anterior abdominal segments transformed into thoracical, initiating the growth of up to seven additional pairs of legs, also stubbier than their wild type counterparts. Initially identified during an EST expression screen in *Tribolium* embryos (Savard, 2004), it was first assumed that *mlpt* belongs to the class of mRNA-like non-coding RNAs (Tupy et al., 2005). But further investigation revealed that the peptides display a high degree of conservation in certain motifs over a wide variety of different insect species (Savard et al., 2006). Especially *mlpt* orthologs identified in several members of the genus *Drosophila* bear striking resemblance in the organization as well as in the structure (Figure 1 B-E).

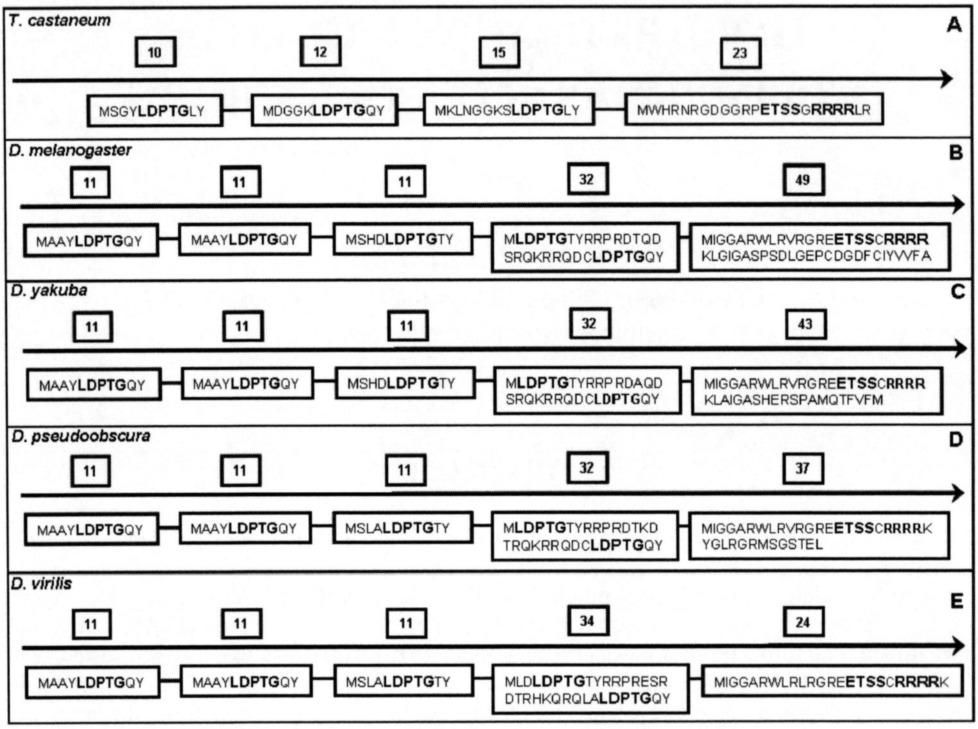

Figure 1. Schematic representation of the peptide organization from the *mlpt*-gene in *Tribolium castaneum* (A) and its four orthologs discovered in different *Drosophila* species (B-E). Conserved sequence motifs are indicated in bold face. The boxes above the peptides indicate the number of amino acids.

Additionally, most of the peptides possess a Kozak consensus sequence (Kozak, 1987) in the vicinity of their start codon. These findings render the possibility that all short open reading frames are translated into functional peptides much more likely then the initial classification as non-coding RNA, which might have been premature.

Concerning the still cryptic function of these peptides, several speculations have yet to be verified experimentally. It has been argued that they might act as signaling peptides via tagging specific target molecules (Tour and McGinnis, 2006). The transfer across cell-membranes is thought to be another likely mechanism, especially considering the C-terminal peptide with four repeated Arginine residues in its conserved core (rightmost peptides in *Tribolium* and the four *Drosophila* species, respectively, Figure 1), since Arginine-rich peptides are a well characterized class among cell-penetrating proteins (Melikov and Chernomordik, 2005). Furthermore, there is a clear role of *mlpt* exceeding that of a gap gene, since the appendage formation takes place at a much later stage in development, and larvae depleted of that gene develop stubby legs as well. However, although short peptides are generally known to be very well capable of exerting various functions, most prominently displayed in a number of short neuropeptides and hormones in higher vertebrates, they have not been described to be directly translated from such comparably short open reading frames as yet. Rather, these peptides are always processed into their active form from larger precursors, previously transcribed and translated from longer open reading frames (Autelitano et al., 2006). Moreover, the genomic organization of the discovered gene seems to be an even more puzzling aspect. Common in prokaryotes as operons, polycistronic transcripts were thought to be a rare exception in eukaryotes till now. Though existent as Pre-mRNAs, they are almost exclusively further processed to mRNAs with single open reading frames before their actual translation (Blumenthal, 2004). However, some rare dicistronic transcripts have been reported to occur in *Drosophila* (Andrews et al., 1996) and in general, the potential to deal with polycistronic transcripts can be expected from eukaryotic organisms as well, demonstrated by the correct translation of polycistronic viral mRNAs (Ryabova et al., 2002).

This raises the interesting question whether genes with such an unusual structure and organization are a much more general and common class of coding RNAs which just have been overseen in eukaryotes so far. Taking into account that in all discovered *mlpt* orthologs the first two open reading frames seem to be repetitions with a high degree of similarity, specifically directed computational searches in numerous other insect genomes for potential short open reading frame repeats displaying a similar genetic structure already yielded promising candidate lists (Buellesbach and Wuenschiers, unpublished data). One of the problems encountered here is the degree of certainty whether these candidates are actually expressed or not. The screening of EST-databases might be one possible solution to assign validity to the potential candidates, although the key issue would remain, namely whether they are also translated into functional peptides. Antibody staining proved to be a quite difficult detection method for such short peptides (Aranda, pers. comm.). Therefore, specific mass spectrometric studies seem to be a much more appropriate and solid approach for an unambiguous detection. Peptide mass fingerprinting already yielded reliable results in *Drosophila* embryos (Kurono et al., 2006) and appears to be a suitable analytical method in this context. Additionally, new light on the possible function of these peptides might be shed by their exact cellular localization as well. Tissue samples from Tribolium embryos in various stages are quite easy to obtain, and since the expression of at least *mlpt* is well known and described (Savard et al., 2006), a screening of the corresponding tissue regions via high-throughput mass spectrometric methods (e.g. capillary HPLC-MS) seems particularly suitable. This would hint at the possible domains of interaction, and thus provide one large further step to unravel the possible functions of these peptides.

ACKNOWLEDGMENTS

The author would like to thank Roebbe Wuenschiers and Diethard Tautz for introducing me to the subject and Manuel Aranda for valuable discussions concerning the specific experimental setup.

REFERENCES

Andrews, J, Smith, M, Merakovsky, J, Coulson, M, Hannan, F, and Kelly, L E. 1996. The stoned locus of Drosophila melanogaster produces a dicistronic transcript and encodes two distinct polypeptides. *Genetics*, 143(4), 1699–1711.

Autelitano, Dominic J, Rajic, Antonio, Smith, A Ian, Berndt, Michael C, Ilag, Leodevico L, and Vadas, Mathew. 2006. The cryptome: a subset of the proteome, comprising cryptic peptides with distinct bioactivities. *Drug Discov.. Today*, 11(7-8), 306–314.

Blumenthal, Thomas. 2004. Operons in eukaryotes. *Brief Funct. Genomic. Proteomic*, 3(3), 199–211.

Kozak, M. 1987. An analysis of 5'-noncoding sequences from 699 vertebrate messenger RNAs. *Nucleic. Acids Res.*, 15(20), 8125–8148. Comparative Study.

Kurono, Sadamu, Kurono, Tamie, Komori, Naoka, Niwayama, Satomi, and Matsumoto, Hiroyuki. 2006. Quantitative proteome analysis using D-labeled N-ethylmaleimide and 13C-labeled iodoacetanilide by matrix-assisted laser desorption/ionization time-of-flight mass spectrometry. *Bioorg. Med. Chem.*, 14(24), 8197–8209.

Melikov, K, and Chernomordik, L V. 2005. Arginine-rich cell penetrating peptides: from endosomal uptake to nuclear delivery. *Cell Mol. Life Sci.*, 62(23), 2739–2749.

Pankratz, M J, and Jackle, H. 1990. Making stripes in the Drosophila embryo. *Trends Genet*, 6(9), 287–292.

Peel, Andrew D, Chipman, Ariel D, and Akam, Michael. 2005. Arthropod segmentation: beyond the Drosophila paradigm. *Nat. Rev. Genet.*, 6(12), 905–916.

L.A. Ryabova, M.M. Pooggin and T. Hohn, Viral strategies of translation initiation: ribosomal shunt and reinitiation, *Prog. Nucleic Acid Res. Mol. Biol.* 72 (2002), 1–39.

Savard, J. 2004. *A genomic approach to the study of Tribolium castaneum genetics, development and evolution*. PhD thesis, University of Cologne, Cologne, Germany.

Savard, Joel, Marques-Souza, Henrique, Aranda, Manuel, and Tautz, Diethard. 2006. A segmentation gene in tribolium produces a polycistronic mRNA that codes for multiple conserved peptides. *Cell*, 126(3), 559–569.

Tour, Ella, and McGinnis, William. 2006. Gap peptides: A new way to control embryonic patterning? *Cell*, 126(3), 448–449. Comment.

Tupy, Jonathan L, Bailey, Adina M, Dailey, Gina, Evans-Holm, Martha, Siebel, Christian W, Misra, Sima, Celniker, Susan E, and Rubin, Gerald M. 2005. Identification of putative noncoding polyadenylated transcripts in Drosophila melanogaster. *Proc. Natl. Acad. Sci. USA*, 102(15), 5495–5500.

In: Biomacromolecular Mass Spectrometry Yearbook. Volume 1 ISBN: 978-1-60876-167-8
Ed: Simone Koenig

Chapter 17

COUPLING SURFACE PLASMON RESONANCE WITH MASS SPECTROMETRY – TOWARDS NOVEL PROTEIN ARRAYS*

Rafal Marszalek*

Integrated Functional Genomics, Interdisciplinary Center for Clinical Research, Medical Faculty, University of Muenster, Roentgenstrasse 21, 48149 Muenster, Germany, and
Department of Neurobiochemistry,
Faculty of Chemistry, Jagiellonian University,
Ingardena Str. 3, 30-060 Krakow, Poland

ABSTRACT

Nowadays surface plasmon resonance and mass spectrometry are fully developed technologies for biomolecule characterization. Considerable efforts have been made in recent years to combine these techniques, since SPR and MS provide complementary data – quantitative and qualitative, respectively. An extremely important aspect of SPR experiments is the correct choice of chip surface and surface modification. These factors need to be carefully considered especially when planning the SPR-MS analyses. There are two basic experimental setups for SPR-MS experiments: one of these uses the microrecovery technique, whilst the other is an on-chip-MS based approach. For high-throughput arrays only the second setup has significant value. Both SPR and MS are technologies with developed array solutions. Therefore these techniques may be coupled to create the SPR-MS biomacromolecule array platform.

Keywords: *biomolecular interaction analysis; MALDI-MS; protein arrays; SPR.*

---

* A version of this chapter was also published in *Biomacromolecular Mass Spectrometry Volume 1, Number 4*, published by Nova Science Publications, Inc. It was submitted for appropriate modifications in an effort to encourage wider dissemination of research.

* E-mail: r.marszalek07@ic.ac.uk

ABBREVIATIONS

BIA	biomolecular interaction analysis
EDC	N-(3-dimethylaminopropyl)-N'-ethylcarbodiimide
NHS	N-hydroxysuccinimide
SAM	self-assembled monolayer
SPRi	surface plasmon resonance imaging
TIR	total internal reflection

For the past decade the field of proteomics has been at the forefront of the biological sciences and has led to the development of new methodologies for protein identification and quantification using multi-technology approaches. Frequently mass spectrometry (MS) is combined with other techniques, such as 2D gel electrophoresis (2-DE) [1] or liquid chromatography (LC) [2]. Such approaches allow for proteins to be separated and characterized in terms of their physical or chemical properties and mass. Functions of many proteins, however, depend on their interactions with other biomacromolecules. It is therefore useful, if not crucial for understanding biochemical processes, to qualitatively and quantitatively describe these interactions.

One of the techniques that can provide quantitative data on biomolecule interactions is an analysis system using surface plasmon resonance (SPR) detection. This chip-based technology is commonly used in combination with mass spectrometry. In recent years some efforts have been made to enable high-throughput analysis using this multiplexed approach, leading to the development of the first SPR-MS array [3].

SURFACE PLASMON RESONANCE

SPR is the phenomenon that occurs when a beam of plane-polarized light is reflected under conditions of total internal reflection (TIR) from a surface covered with thin metal film. SPR measurements take the advantage of the creation of the evanescent field. Although under TIR conditions the whole incident light is reflected, only a component of this light – evanescent wave – penetrates the surface [4, 5].

When the light is reflected at a certain angle, the energy of the photons corresponds to the difference between the energy levels of the free electrons from the metal layer – the so called plasmons – and some part of it can be absorbed [6]. The intensity of the reflected light measured with the SPR instruments then reaches a minimum. The angle of reflection is called the surface plasmon resonance angle (Φ_{SPR}) [7].

The surface plasmon resonance angle is very sensitive to any changes in the refractive index of the medium near the surface [7]. When some form of analyte is bound to the surface, it causes a shift of the Φ_{SPR}. This change can be quantified since the angle shift corresponds to changes in the amount of the bound analyte. Therefore SPR is used as a method for quantitative measurements of biomolecule interactions, especially their thermodynamics [8], kinetics and affinity [9]. It may be also used as a qualitative method to control processes occurring at the surface.

SPR is a label-free technique and this enables the number of preparation steps to be reduced, the label-related interferences to be avoided and handling of the sample to be simplified [10]. The drawback of this method is the fact that the measured quantity is not compound-specific, i.e. that the change of refractive index might be caused by the analyte as well as by the contaminant or even the running buffer. This problem is circumvented by carrying out double-reference measurements.

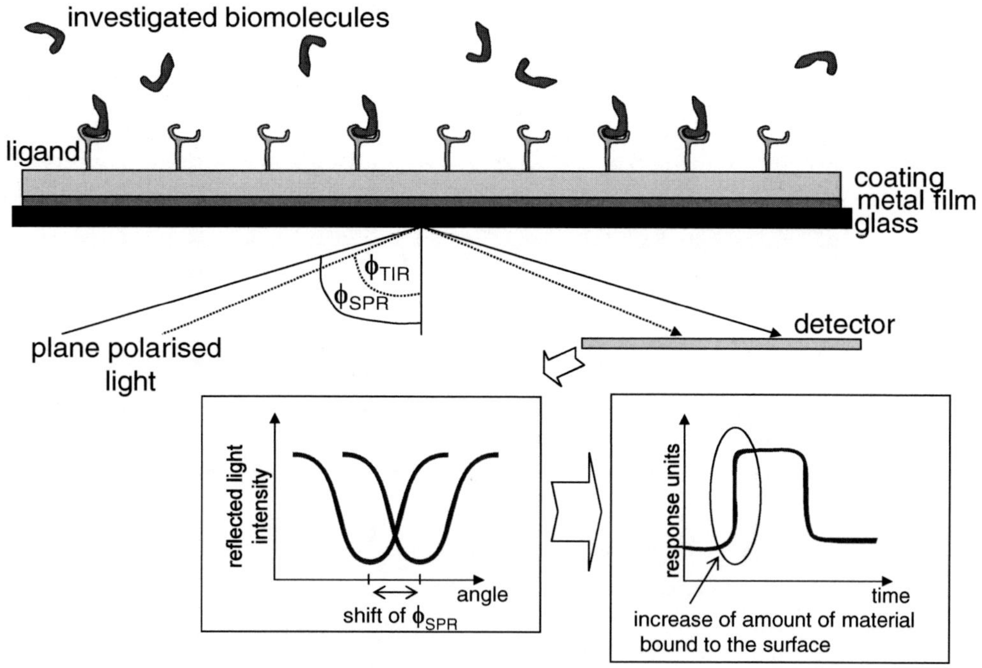

Figure 1. The principle of the surface plasmon resonance. The SPR angle is measured at the minimum in the intensity of reflected light. The shift of the SPR angle corresponds to the changes of the amount of molecules bound to the chip surface. The chip shown in the picture is covered with a coating – which might be for example modified dextran.

SPR BIOSENSORS

In the conventional surface plasmon resonance biomolecular interaction (SPR-BIA) experiment molecules of one substance (ligand) are immobilized on the surface which is subsequently treated with the analyte solution. The interaction between the ligand and analyte is monitored with SPR detection.

In the early applications of SPR for biomolecule analysis, the ligand was adsorbed directly onto a gold film [4, 11]. However, this approach has some disadvantages. The ligand layer is – because of its heterogeneity – not well defined. Additionally, since SPR is a label-free technique, it can be as sensitive to non-specific binding as to the interaction being studied. Thus every substance in the medium can contribute to the signal [12]. Finally, many proteins in contact with the metal film lose their biological activity [13].

To overcome these difficulties, the metal film may be covered with a self-assembled monolayer (SAM). SAMs may be later used to couple a polymer coat – usually carboxymethylated dextran – to the surface. The dextran may be also coupled directly to the chip. This type of coating shows much better determined properties compared to the plain chip surface [10]. The other advantage of this approach is the possibility of covalent immobilization of the ligand. The most popular immobilization method for basic SPR experiments is activation of dextran carboxymethyl groups with a mixture of N-(3-dimethylaminopropyl)-N'-ethylcarbodiimide (EDC) with N-hydroxysuccinimide (NHS) prior to the reaction with primary amines.

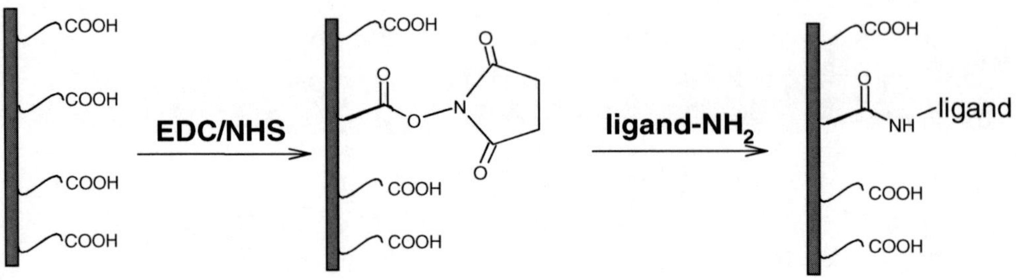

Figure 2. The principle of EDC/NHS amine coupling. The carboxymethyl groups are activated with EDC/NHS mixture. Then the amine groups of the ligand are coupled to the activated surface. In the streptavidin-biotin immobilization method streptavidin is first coupled to the surface with EDC/NHS chemistry.

EDC/NHS chemistry is suitable for most proteins and peptides (except some which are very acidic), but **is, however, nonspecific** and cannot be used against a chosen amine group in the ligand. Furthermore, the immobilized molecules have random space orientation that may become **a** significant problem [14]. Therefore, more specific immobilization techniques utilizing the affinity phenomenon have been developed, for example binding of the biotinylated analyte to the streptavidin modified surface or binding of His-tagged proteins to the Ni^{2+} chelating surfaces [15, 16].

SPR-MS

Whereas surface plasmon resonance can be used to gain information about the interaction itself and to quantify it if required, it cannot provide any structural information about the analyte. On the other hand, mass spectrometric measurements rarely can be quantified, but they give information about the structure of the analyte characterized by its unique molecular mass. Mass spectrometry can be used not only for identification of an analyte, but it can also help, for example, to characterize the post-translational modification of proteins [17, 18] or to distinguish between different isoforms [19, 20]. As well as for protein research it can be used for saccharide profiling [21, 22], lipid characterization [23, 24] or nucleic acid analysis [25]. Both techniques – SPR and MS – are often coupled as they provide complementary data.

STRATEGIES FOR SPR(BIA)-MS EXPERIMENT

There are two basic ways (Figure3.) in which SPR-MS experiment might be performed [26]. In the first approach the interaction experiment carried out at the chip surface is followed by the elution of the analyte and then undergoes MS analysis. A BIA instrument capable of being coupled on-line with MS has been introduced by Biacore company. This uses a technical solution known as sample microrecovery.

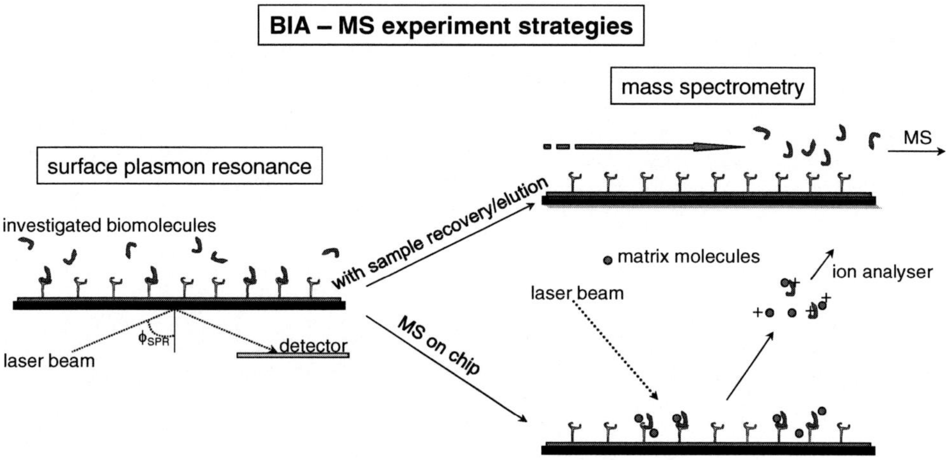

Figure 3. Strategies for SPR-biomolecular interaction – mass spectrometry experiment. Following the BIA experiment, the sample may be either eluted from the surface or the MS measurement might be performed directly on the chip.

In this microrecovery technology, following interaction the analyte is eluted with a small volume of eluent sandwiched by air bubbles ("sandwich elution") [27] from the chip and then analyzed with MS. The main advantage of this approach is the great variety of subsequent experiments that may be performed: the sample might be concentrated and then digested for peptide mapping with MALDI-MS [28] or used for ESI-MS/MS experiments [29]. It has, however, some disadvantages, e.g. longer analysis time, decreased sensitivity and detection limits. Additionally, there is also no practical possibility of high-throughput analyses.

In the second approach, after the interaction the matrix is applied directly on the biosensor chip and then MALDI-MS analysis is performed. This way of performing the SPR-MS experiment is simpler than that with sample elution, has better sensitivity and lower detection limits. It is also suitable for multiple spots analysis on a single chip [26].

APPLICATIONS

The BIA-MS combination is often used for screening experiments when the interaction partner is unknown, e.g. for inhibitors [30], binding partners [31] or low molecular weight ligands [32]. When the analyte is well known so called ligand fishing might be performed [28]. This approach is commonly used for quality controls. SPR-MS can be used for the quality control of food [33], therapeutical antibodies [34] or expression of recombinant proteins [35].

Technology for High-Throughput SPR-MS Arrays

BIA-MS methodology is an extremely useful tool for proteomic applications. However it originally had one great limitation: there were no solutions for making SPR-MS protein assays. The feasibility of such assays does not depend on MS, but rather on the possibilities of the biomolecular analysis instrumentation. In the early SPR instruments the number of analyses that could be performed on one chip was limited (e.g., four flow cells in the Biacore instrument) and no parallel measurements were possible.

Table 1. There is a great variety of experiments to be performed with the BIA-MS method. In most cases, however, it is possible only for the high affinity interactions

Reference	Strategy	Type of experiment	Type of interaction
Biacore Appl. Note 60[34]	with elution	quality check of monoclonal antibodies for therapeutical purposes	antibody – antigen
Borch et al.[30]	with elution	inhibitor screening	enzyme – inhibitor
Font et al.[31]	with elution	screening for DNA-binding partners	DNA – protein
Mattei et al.[36]	with elution	epitope mapping - post-translational modifications analysis	enzyme – substrate
Nedelkov et al.[33]	on-chip	quality control – analysis of toxins in food	antibody – toxin
Nedelkov et al.[37]	on-chip	design of multi-affinity surface for screening experiments	antibody – antigen
Nedelkov et al.[38]	on-chip	protein complexes investigation	antibody – protein
Soenksen et al.[32]	with elution	low molecular weight ligands investigation	enzyme – inhibitor
Zhukov et al.[28]	with elution	ligand fishing	receptor – ligand
Bouffartigues et al.[39]	with elution	screening for DNA-binding proteins; comparison of SPR and SPRi coupled to SELDI-MS	DNA-protein

The situation has, however, changed following the introduction of SPR imaging (SPRi). There are three modes in which the SPR experiment may be performed. In the first, monochromatic light is used and the intensity of reflected light at different angles is measured (angular mode). In the second mode, in which the light is polychromatic and the incident light angle is constant, the light intensity at the different wavelengths is measured (wavelength mode). In the last, most modern one – SPR imaging – the light beam is monochromatic and the angle is fixed near the SPR angle. In SPR imaging only changes in the light intensity are measured [40].

SPR imaging has one very distinct feature: it is able to detect from many sites on one chip at a time. SPRi technology has been so far mainly developed by Corn's group. Corn and his colleagues have used it for measurements of peptide and protein arrays [41, 42], investigation of proteins in cell lysate [43] and making antibody arrays against the low molecular weight markers [44]. SPRi, apart from being a breakthrough in biomolecular analysis technologies, is also an excellent solution for creation of the SPR-MS protein or peptide arrays.

First SPR-MS Arrays

Despite significant progress in instrumentation development, it is not possible for the high-throughput BIA-MS analyses to be performed with the "with elution" strategy. A few promising experiments have already been completed with the BIA(SPR) coupled with on-chip MALDI-MS. The first array-directed experiments performed by Nedelkov and Nelson [33, 37] were oriented to developing multi-affinity surfaces within one flow cell. The problem they recognized was how to increase the number of detected antigens, whilst retaining good sensitivity for mass spectrometric measurements at the same time. In their first experiment only two different compounds were detected [33]. Since then, however, they have managed to immobilize up to six types of antibodies, but with only five antigens detected afterwards with MS [37]. The main problem faced here in terms of high-throughput experiments was the incapability of the SPR instrument to measure large number of spots at a time. As already mentioned, this was later overcome with the introduction of SPR imaging.

As a next step towards SPR-MS arrays it has been shown by Nedelkov *et al.* [45] that the spatial resolution of mass spectrometry is sufficient for this type of experiment. They have been able to apply 440 spots on an area of ~4×4 mm and have measured the signal for each spot with MALDI-TOF mass spectrometer. The SPR resolution achieved in this experiment was still unsatisfactory. It should, however, be noted that the SPR imaging instrument was not used.

The diversity of currently available SPRi instruments [46] capable of imaging SPR arrays opens the way for the development of cheap and efficient SPR-MS protein arrays. Recently, the first platform for such experiments has been finally described by Nedelkov [3]. A small 10×10 antibody array has been used in proof-of-principle experiments and the robustness and reproducibility of arrays have been investigated. Although only standard protein solutions have been investigated, the results presented seem promising in terms of applicability of the method for real sample analyses.

Table 2. In recent years a few experiments demonstrating the feasibility of BIA-MS arrays have been successfully performed

Reference	Experiment
Nedelkov et al.[33]	Multi-toxin detection through multi-affinity flow cells (using the example of two staphylococcal toxins).
Nedelkov et al.[37]	Performance of multi-affinity surfaces (using the example of six immobilized antibodies per one flow cell).
Nedelkov et al.[45]	BIA-MS on the arrayed chips.
Nedelkov [3]	First high-throughput protein SPR-MS assay.

Concluding Remarks

In the last twenty years SPR has been experiencing a real renaissance as an analytical method for protein sciences. Since this revival has overlapped with the introduction of new mass spectrometric techniques for protein and peptide analysis, inevitably both methods have been employed in a coupled mode, either on-line or chip-based, thus profiting from the complementarity of the data provided by these techniques. The subsequent development of array solutions for both technologies finally led to creation of SPR-MS high-throughput arrays. The commercial instruments capable of performing array experiments are already available for SPR as well as for MS and, after the introduction of the first working SPR-MS array platform, it should be only a matter of time before other biopharmaceutical or analytical experiments are performed with this approach.

ACKNOWLEDGMENTS

I would like to thank Dr Simone König from University of Muenster for all the constructive comments on the manuscript. I would also like to thank Elizabeth Gardner from Imperial College London for **critical** proof-reading the text.

REFERENCES

[1] Ogorzalek Loo, R. R.; Mitchell, C.; Stevenson, T. I.; Martin, S. A.; Hines, W. M.; Juhasz, P.; Patterson, D. H.; Peltier, J. M.; Loo, J. A.; Andrews, P. C., Sensitivity and mass accuracy for proteins analyzed directly from polyacrylamide gels: Implications for proteome mapping. *Electrophoresis* 1997, 18, (3-4), 382-390.

[2] Kuester, B.; Mann, M., Identifying proteins and post-translational modifications by mass spectrometry. *Curr. Opin. Struct. Biol.* 1998, 8, 393-400.

[3] Nedelkov, D., Development of surface plasmon resonance mass spectrometry array platform. *Anal. Chem.* 2007, 79, (15), 5987-5990.

[4] Liedberg, B.; Nylander, C.; Lundstrom, I., Surface plasmon resonance for gas detection and biosensing. *Sensor Actuator.* 1983, 4, 299-304.

[5] Welford, K., Surface plasmon-polaritons and their uses. *Opt. Quant. Electron.* 1991, 23, 1-27.

[6] Alves, I. D.; Park, C. K.; Hruby, V. J., Plasmon resonance methods in GPCR signaling and other membrane events. *Curr. Protein Pept. Sci.* 2005, 6, (4), 293-312.

[7] Nylander, C., Chemical and biological sensors. *J. Phys. E: Sci. Instrum.* 1985, 18, 736-750.

[8] O'Shannessy, D. J.; Brigham-Burke, M.; Soneson, K. K.; Hensley, P.; Brooks, I., Determination of rate and equilibrium binding constants for macromolecular interactions using surface plasmon resonance: use of nonlinear least squares analysis methods. *Anal. Biochem.* 1993, 212, 457-468.

[9] Malmqvist, M., Biospecific interaction analysis using biosensor technology. *Nature* 1993, 361, (6408), 186-187.

[10] Yu, X.; Xu , D.; Cheng, Q., Label-free detection methods for protein microarrays. . *Proteomics* 2006, 6, 5493-5503.

[11] Daniels, P. B.; Deacon, J. K.; Eddowes, M. J.; Pedley, D. G., Surface plasmon resonance applied to immunosensing. *Sensor Actuator.* 1988, 15, 11-18.

[12] Cullen, D. C.; Brown, R. G. W.; Lowe, C. R., Detection of immune-complex formation via surface plasmon resonance on gold-coated diffraction gratings. *Biosensors* 1987/88, 3, 211-225.

[13] Lofas, S.; Johnsson, B., A novel hydrogel matrix on gold surfaces in surface plasmon resonance sensors for fast and efficient covalent immobilization of ligands. *J. Chem. Soc. Chem. Commun.* 1990, 1526-1528.

[14] O'Shannessy, D. J.; Brigham-Burke, M.; Peck, K., Immobilization chemistries suitable for use in the BIAcore surface plasmon resonance detector. *Anal. Biochem.* 1992, 205, 132-136.

[15] Nieba, L.; Nieba-Axmann, S. E.; Persson, A.; Hamalainen, M.; Edebratt, F.; Hansson, A.; Lidholm, J.; Magnusson, K.; Karlsson, A. F.; Pluckthun, A., BIACORE analysis of histidine-tagged proteins using a chelating NTA sensor chip. *Anal. Biochem.* 1997, 252, 217-228.

[16] Zhu, H.; Bilgin, M.; Bangham, R.; Hall, D.; Casamayor, A.; Bertone, P.; Lan, N.; Jansen, R.; Bidlingmaier, S.; Houfek, T.; Mitchell, T.; Miller, P.; Dean, R. A.; Gerstein, M.; Snyder, M., Global analysis of protein activities using proteom chips. *Science* 2001, 293, 2101-2105.

[17] Smith, J. B.; Thevenon-Emeric, G.; Smith, D. L.; Green, B., Elucidation of the primary structures of proteins by mass spectrometry. *Anal. Biochem.* 1991, 193, 118-124.

[18] Lennon, J. J.; Walsh, K. A., Locating and identifying posttranslational modifications by in-source decay during MALDI-TOF mass spectrometry. *Protein Sci.* 1999, 8, 2487-2493.

[19] Hardouin, J.; Hubert-Roux, M.; Delmas, A. F.; Lange, C., Identification of isoenzymes using matrix-assisted laser desorption/ionization time-of-flight mass spectrometry. *Rapid Commun. Mass Spectrom.* 2006, 20, 725-732.

[20] Helmke, S. M.; Yen, C.-Y.; Cios, K. J.; Nunley, K.; Bristow, M. R.; Duncan, M. W.; Perryman, M. B., Simultaneous quantification of human cardiac a- and b-myosin heavy chain proteins by MALDI-TOF mass spectrometry. *Anal. Chem.* 2004, 76, 1683-1689.

[21] Lattova, E.; Perreault, H., Profiling of N-linked oligosaccharides using phenylhydrazine derivatization and mass spectrometry. *J. Chromatogr. A* 2003, 1016, 71-87.

[22] Lattova, E.; Perreault, H., Labelling saccharides with phenylhydrazine for electrospray and matrix-assisted laser desorption-ionization mass spectrometry. *J. Chromatogr. B* 2003, 793, (1), 167-179.

[23] Jones, J. J.; Borgmann, S.; Wilkins, C. L., Characterising the phopholipid profiles in mammalian tissues by MALDI FTMS. *Anal. Chem.* 2006, 78, 3062-3071.

[24] Jones, J. J.; Stump, M. J.; Fleming, R. C.; Lay, J. O. j.; Wilkins, C. L., Strategies and data analysis techniques for lipid and phospholipid chemistry elucidation by intact cell MALDI-FTMS. *J. Am. Soc. Mass Spectrom.* 2004, 15, (11), 1665-1674.

[25] Griffin, T. J.; Smith, L. M., Single-nucleotide polymorphism analysis by MALDI-TOF mass spectrometry. *Trends Biotech.* 2000, 18, (2), 77-84.

[26] Nedelkov, D.; Nelson, R. W., Surface plasmon resonance mass spectrometry: recent progress and outlooks. . *Trends Biotech.* 2003, 21, (7), 301-305.

[27] Soenksen, C. P.; Nordhoff, E.; Jansson, O.; Malmqvist, M.; Roepstorff, P., Combining MALDI mass spectrometry and biomolecular interaction analysis using biomolecular interaction analysis instrument. *Anal. Chem.* 1998, 70, 2731-2736.

[28] Zhukov, A.; Schuerenberg, M.; Jansson, O.; Areskoug, D.; Buijs, J., Integration of surface plasmon resonance with mass spectrometry: automated ligand fishing and sample preparation for MALDI-MS using a Biacore 3000 biosensor. *J. Biomol. Tech.* 2004, 15, 112-119.

[29] Natsume, T.; Nakayama, H.; Isobe, T., BIA-MS-MS: biomolecular interaction analysis for functional proteomics. *Trends Biotech.* 2001, 19, (10), S28-33.

[30] Borch, J.; Roepstorff, P., Screening for enzyme inhibitors by surface plasmon resonance combined with mass spectrometry. *Anal. Chem.* 2004, 76, 5243-5248.

[31] Font, M.-P.; Cubizolles, M.; Dombret, H.; Cazes, L.; Brenac, V.; Sigaux, F.; Buckle, M., Repression of transcription at the human T-cell receptor Vb2.2 segment is mediated by a MAX/MAD/mSin3 complex acting as a scaffold for HDAC activity. *Biochem. Biophys. Res. Commun.* 2004, 325, 1021-1029.

[32] Soenksen, C. P.; Roepstorff, P., Capture and analysis of low molecular weight ligands by surface plasmon resonance combined with mass spectrometry. *Eur. J. Mass Spectr.* 2001, 7, 385-391.

[33] Nedelkov, D.; Rasooly, A.; Nelson, R. W., Multitoxin biosensor-mass spectrometry analysis: a new approach for rapid, real-time, sensitive analysis of staphylococcal toxins in food. *Int. J. Food Microb.* 2000, 60, 1-13.

[34] Biacore Application Note Nr 60. 2004.

[35] Natsume, T.; Taoka, M.; Manki, H.; Kume, S.; Isobe, T.; Mikoshiba, K., Rapid analysis of protein interactions: on-chip micropurification of recombinant protein expressed in Esherichia coli. *Proteomics* 2002, 2, 1247-1253.

[36] Mattei, B.; Cervone, F.; Roepstorff, P., The interaction between endopolygalacturonase from Fusarium moniliforme and PGIP from Phaseolus vulgaris studied by surface plasmon resonance and mass spectrometry. *Comp. Funct. Genom.* 2001, 2, 359-364.

[37] Nedelkov, D.; Nelson, R. W., Design and use of multi-affinity surfaces in biomolecular interaction analysis-mass spectrometry (BIA-MS): a step toward the design of SPR/MS arrays. *J. Mol. Recog.* 2003, 16, 15-19.

[38] Nedelkov, D.; Nelson, R. W.; Kiernan, U. A.; Niederkofler, E. E.; Tubbs, K. A., Detection of bound and free IGF-1 and IGF-2 in human plasma via biomolecular interaction analysis mass spectrometry. *FEBS Lett.* 2003, 536, 130-134.

[39] Bouffartigues, E.; Leh, H.; Anger-Leroy, M.; Rimsky, S.; Buckle, M., Rapid coupling of the surface plasmon resonance (SPR and SPRi) and ProteinChiptm based mass spectrometry for the identification of proteins in nucleoprotein interactions. *Nucleic acids research* 2007, 35, (6), e39/1-e39/10.

[40] Homola, J.; Vaisocherova, H.; Dostalek, J.; Piliarik, M., Multi-analyte surface plasmon resonance biosensing. *Methods* 2005, 37, 26-36.

[41] Wegner, G. J.; Lee, H. J.; Corn, R. M., Characterization and optimization of peptide arrays for the study of epitope-antibody interactions using surface plasmon resonance imaging. *Anal. Chem.* 2002, 74, 5161-5168.

[42] Wegner, G. J.; Wark, A. W.; Lee, H. J.; Codner, E.; Saeki, T.; Fang, S.; Corn, R. M., Real-time surface plasmon resonance imaging measurements for the multiplexed determination of protein adsorption/desorption kinetics and surface enzymatic reactions on peptide microarrays. *Anal. Chem.* 2004, 76, 5677-5684.

[43] Kyo, M.; Usui-Aoki, K.; Koga, H., Label-free detection of proteins in crude cell lysate with antibody arrays by a surface plasmon resonance imaging technique. *Anal. Chem.* 2005, 77, (22), 7115-7121.

[44] Lee, H. J.; Nedelkov, D.; Corn, R. M., SPR imaging measurements of antibody arrays for the multiplexed detection of low molecular weight protein biomarkers. *Anal. Chem.* 2006, 78, (18), 6504-6510.

[45] Nedelkov, D.; Tubbs, K. A.; Nelson, R. W., Surface plasmon resonance-enabled mass spectrometry analysis. *Electrophoresis* 2006, 27, 3671-3675.
[46] Mukhopadhyay, R., Surface plasmon resonance instruments diversify. *Anal. Chem.* 2005, 77, 313A-317A.

In: Biomacromolecular Mass Spectrometry Yearbook. Volume 1 ISBN: 978-1-60876-167-8
Ed: Simone Koenig

Chapter 18

ADDUCT IONS ATTRIBUTED TO COPPER OBSERVED AFTER MECHANICAL TREATMENT OF MALDI SAMPLE PLATES – A CAUTIONARY NOTE*

Simone König*
Integrated Functional Genomics,
Interdisciplinary Center for Clinical Research,
Medical Faculty, University of Münster, Germany

ABSTRACT

Adduct ions attributed to copper were observed at low abundance in peptide maps measured with MALDI mass spectrometry on some re-usable stainless steel targets. These ions were enhanced after mechanical treatment of the target using sandpaper. Although the copper source was not be satisfactorily elucidated the contaminant adduct ions seemed to be more present with target plates often cleaned. Such ions generated unwanted chemical noise at low intensity. In the examples shown here, peptide adduct ions of considerable abundance were observed. Their appearance can be cause for misinterpretation of the data in protein identification experiments and it should be prevented.

In our core facility, matrix-assisted laser desorption/ionization mass spectrometry (MALDI-MS) is routinely used for peptide mapping. To that end, either a Tof-Spec 2E or a MALDImicroMX instrument (both from Waters/Micromass, Manchester, UK) are employed that accommodate the same sample plates. These are designed for re-use and, according to the manufacturer, are made from 430 S17 stainless steel (the number gives the UK identification number; in the US the AISI number would be 430, whereas the German Werkstoffnummer is

---

* A version of this chapter was also published in *Biomacromolecular Mass Spectrometry Volume 1, Number 4*, published by Nova Science Publications, Inc. It was submitted for appropriate modifications in an effort to encourage wider dissemination of research.

* Correspondence to: Dr. Simone König, Integrated Functional Genomics, Röntgenstr. 21, 48149 Münster, Germany, ph: +49-(0)251-8357164. fax: +49-(0)251-8357255. e-mail: koenigs@uni-muenster.de

1.4016). Plates are regularly cleaned applying ultrasonic treatment with a sequence of different aqueous and organic solvents (acetic acid, methanol, isopropanol, dichloromethane, acetone) which depend on the measured samples. Initially, the bulk of the matrix and samples is removed with detergent (2 % Decon 90 recommended by the manufacturer) using a brush and the target is thoroughly rinsed with water and methanol. Although the treatment is rather soft, over time the target surface tends to loose its smoothness to some extent. This may be one reason for the observations described below which were enhanced when targets were treated with fine abrasive sandpaper. By standard, α-cyano-4-hydroxy cinnamic acid is used as a matrix at 2.5 mg/ml in 50/50 v/v ethanol/acetonitril containing 1 % aqueous 0.1 % trifluoroacetic acid (TFA). Protein digests are commonly presented in 80/20 v/v acetonitrile/water containing 0.1 % TFA. All reagents are typically of the highest analytical quality.

For some time, additional ions in peptide maps were observed which were characterized by an unusual isotopic pattern. Their relative abundance was generally below 2 % so that they were not a source of great concern. However, at one point their abundance went up to more than 20 % relative intensity for all spots on a 96-well target as is exemplified in Figure 1a.

The figure shows a peptide map of α-s1-casein as prepared from commercially available fresh milk employing 2D-gel electrophoresis. The 2D-gel protein spot was washed and digested according to well established proteomic procedures employed routinely in our laboratory. The two most abundant peptide peaks are labelled with their sequence (YLGYLEQLLR and HQGLPQEVLNENLLR). Depending on their overall abundance these protonated ion signals are accompanied by up to three adduct ion cluster sets exhibiting a mass difference of 62. The initial assumption that the adducts were derived from residual ethylene glycole polymer gas phase complexation was not supported, since the adduct ions exhibited a special isotope pattern where the M+3 isotope became more intense than the M+2 isotope (insert in Figure 1a). Both the ion patterns and the observed mass deficit suggested the addition of copper which has isotopes of 62.9296 and 64.9278 with natural abundances of 69.17 and 30.83 %, respectively. Tables 1 and 2 list the expected isotope distributions for both peptides assuming proton or copper addition, respectively, and the inset on the left in Figure 1a visualizes this isotope distribution for the smaller peptide YLGYLEQLLR. Thereby, the measured $[M+Cu]^+$ ion distribution resembles strongly the one shown for the larger peptide in the inset on the right. A direct comparison of the theoretical to the measured isotope distribution for the peptide at *m/z* 1759.94 is presented in Figure 1b. The agreement between the experimental and theoretical values confirms the hypothesis. Interestingly, the Zenobi group [2] has found that Cu(II) can be reduced under MALDI conditions and result in singly-charged Cu(I) ion. This was attributed to the fact that a MALDI sample target can be a source of photoelectrons which are scavenged by copper.

Although the 430 S17 stainless steel is not supposed to contain copper, treatment of the sample plate surface with fine sand paper dramatically enhanced the adduct ions. Although such procedures are not typically performed with sample plates, this nevertheless raised the question, if more common target cleaning processes, or damages made by the laser, may result in such effects. This cannot be entirely ruled out. The fact that even supposedly very low copper concentrations can result in such drastic effects also points to the necessity of avoiding any contamination by copper in the used solvents or in the preparation procedures.

Despite being an annoyance in our case, copper addition to peptides has been found to have its uses. In studies by the Russell group [3, 4], copper binding sites in peptides have been investigated in detail. It was found that the release of copper ions from copper target plates avoids suppression of ion signals that accompanies the addition of metal salts to the sample solution. In addition, a good correlation between the number of basic residues in peptides and the number of Cu^+ ions that add to a peptide were observed.

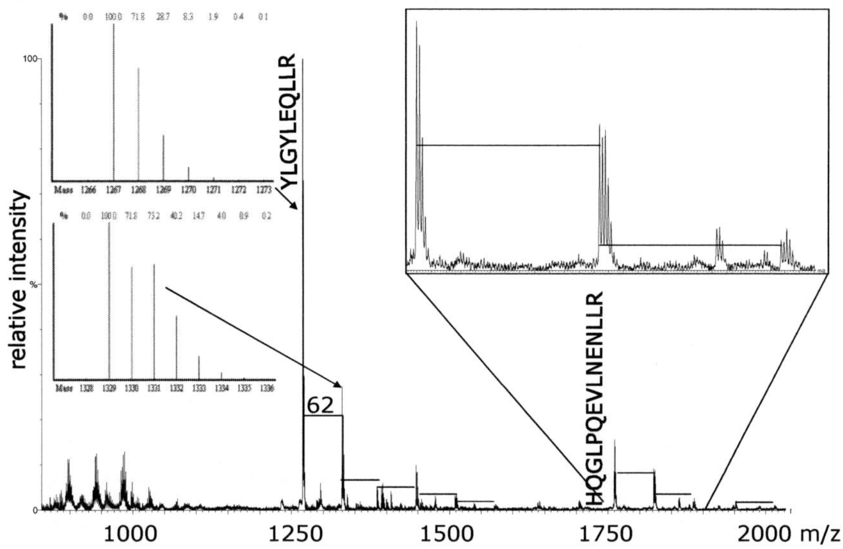

Figure 1a. Tryptic peptide **map of α**-s1-casein with the two peptides YLGYLEQLLR and HQGLPQEVLNENLLR. The inset on the right zooms in to the adducts of the latter to demonstrate the shift in the isotopic distribution. On the left, the expected isotopic distribution for peptide YLGYLEQLLR protonated (top) and containing one Cu (bottom) are shown (see Table 1) [1]. For the latter the isotope distribution corresponds to the one for the larger peptide as shown in the inset on the right.

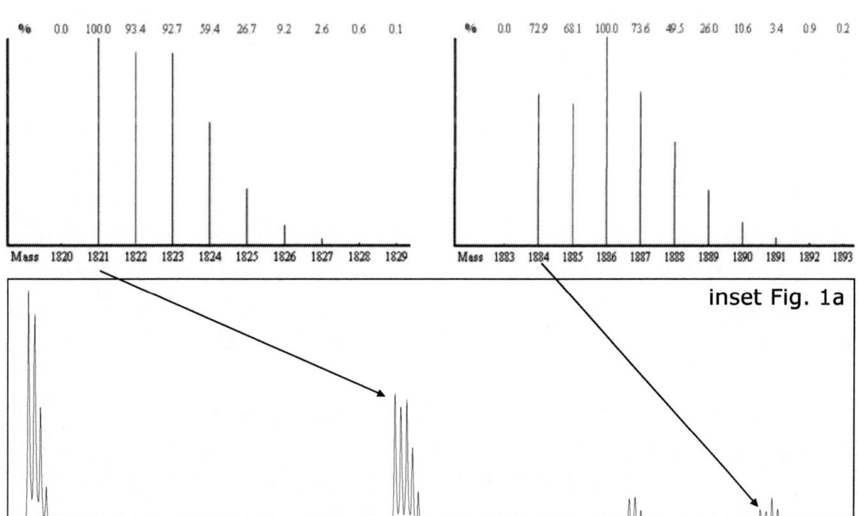

Figure 1b. Expected isotopic distribution for peptide HQGLPQEVLNENLLR containing one Cu (left) or two Cu (right) [1] in comparison to the original data as obtained from the inset in Figure 1a. See Table 2.

Table 1. Expected isotopic distribution of peptide YLGYLEQLLR ($C_{60}H_{95}N_{14}O_{16}$) [1] Molecular weight (MW), exact mass (EM)

$[M+H]^+$ MW 1268.4805 EM 1267.7050		$[M+Cu]^+$ MW 1331.0186 EM 1329.6268	
Mass	Abundance	Mass	Abundance
1267	100	1329	100
1268	71.8	1330	71.8
1269	28.7	1331	73.2
1270	8.3	1332	40.2
1271	1.9	1333	14.7
1272	0.4	1334	4.0
1273	0.1	1335	0.9
		1336	0.2

Table 2. Expected isotopic distribution of peptide HQGLPQEVLNENLLR ($C_{76}H_{127}N_{24}O_{24}$) [1]

$[M+H]^+$ MW 1760.9680 EM 1759.9455		$[M+Cu]^+$ MW 1823.5060 EM 1821.8673		$[M+2Cu]^+$ MW 1887.0520 EM 1884.7969	
Mass	Abundance	Mass	Abundance	Mass	Abundance
1759	100	1821	100	1884	72.9
1760	93.4	1822	93.4	1885	68.1
1761	48.1	1823	92.7	1886	100
1762	17.8	1824	59.4	1887	73.6
1763	5.2	1825	26.7	1888	49.5
1764	1.3	1826	9.2	1889	26.0
1765	0.3	1827	2.6	1890	10.6
1766	0.1	1828	0.6	1891	3.4
		1829	0.1	1892	0.9
				1893	0.2

Above, the presence of peptide adduct ions attributed to copper as assigned by their characteristic isotope pattern in MALDI-MS spectra, was discussed in conjunction with target plate treatment. The source of the copper remains ambiguous since both the target material and the chemicals involved cannot be entirely excluded. The note is intended as a word of caution.

ACKNOWLEDGEMENT

The author thanks Henry M. Fales (National Institutes of Health, Bethesda, MD, USA) for bringing copper into the discussion and Klaus Dreisewerd (Institute of Medical Physics and Biophysics, Münster) for the critical review of the manuscript. Technical assistance of Doreen Ackermann (IFG) is greatly appreciated.

REFERENCES

[1] J. Yan, Isotope Pattern Calculator v4.0;http://www.geocities.com/junhuayan/ pattern.htm

[2] M. Dashtiev, V. Frankevich, R. Zenobi, Signal enhancement in MALDI by doping with Cu(II) chloride. *Rapid Commun. Mass Spectrom.* 2005, 19, 289-291.

[3] S. J. Shields, B. K. Bluhm, D. H. Russell, Novel method for $[M+Cu]^+$ ion formation by matrix-assisted laser desorption ionization. *Intern. J. Mass Spectrom.* 1999, *182/183,* 185-195.

[4] B. K. Bluhm, S. J. Shields, C. A. Bayse, M. B. Hall, D. H. Russell, Determination of copper binding sites in peptides containing basic residues: A combined experimental and theoretical study. *Intern. J. Mass Spectrom.* 2001, 204, (1-3), 31-46.

In: Biomacromolecular Mass Spectrometry Yearbook. Volume 1 ISBN: 978-1-60876-167-8
Ed: Simone Koenig

Chapter 19

DOWNSCALING ANALYSIS: 4TH MÜNSTER CONFERENCE ON SINGLE CELL ANALYSIS; NOVEMBER 26-27, 2007*

Organizers: Rita Naskar, Ilka Romann, Björn Teichert and Simone König, Interdisciplinary Center for Clinical Research, Core Facility Integrated Functional Genomics, Münster, Germany

Chairs: Stephan Ludwig, Berenike Maier, Dietmar Fischer,Connie Jimenez and Simone König

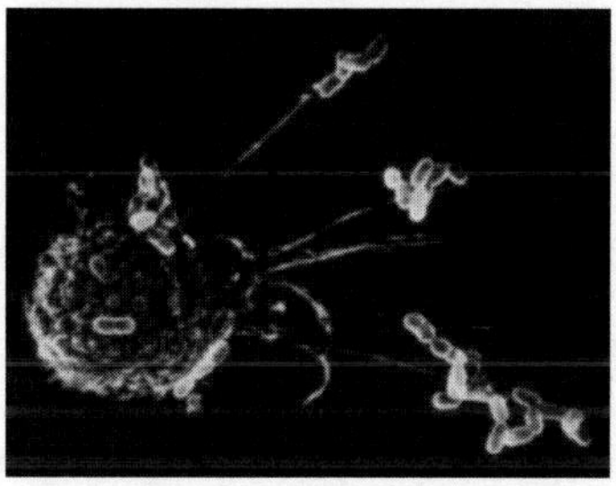

INTRODUCTION

Simone König*
Integrated Functional Genomics,
Röntgenstr. 21, 48149 Münster

---

* A version of this chapter was also published in *Biomacromolecular Mass Spectrometry Volume 1, Number 4*, published by Nova Science Publications, Inc. It was submitted for appropriate modifications in an effort to encourage wider dissemination of research.

* E-mail: koenigs@uni-muenster.de

The Münster Conference Series on Single Cell Analysis brings together scientists of very different fields – physicists, physicians, biologists, chemists – with the common goal of studying events on a cellular, sub-cellular, and molecular level. The impact of mass spectrometry in that area is increasing and, therefore, the conference is monitored in this journal.

The conference started with a keynote by Stephen D. Ginsberg of New York University on single cell gene expression analysis in human post-mortem brain tissues and animal models of neurodegeneration. Modern RNA amplification methods finally allow single cell RNA analysis so that relative gene expression level comparisons across cell types under different experimental conditions and disease states become feasible. In addition to amplification techniques, however, miniaturization is a critical issue in nanoanalytics. Andrea Robitzki (University of Leipzig) talked about novel cell based micro-sensors used as a (i) cardiomyocyte biosensor, (ii) mamma-carcinoma chip, and (iii) an embryonic cell microarray etc. coupled to a micro-laser manipulation platform and Petra Dittrich (ISAS Dortmund) discussed microfluidic platforms for the analysis of living cells.

Another highlight of the conference was the lecture given by Walter Schubert of Magdeburg University. He recently published the work of his group on the deciphering of the protein network in Nature and talked about the ambitious human TOPONOME project. More interesting tools for the researcher were introduced by Volker Dötsch (University of Frankfurt) discussing conformation and dynamics of proteins inside living cells by in-cell NMR spectroscopy and Jörg Enderlein (University of Tübingen) showing Two-focus Fluorescence Correlation Spectroscopy data on molecular diffusion.

Mass spectrometry is again represented well. Imaging continues to be developed and Connie Jimenez of the University of Amsterdam presented results on peptide profiling of single neurons using proteomic technology. Inductively coupled plasma mass spectrometry, on the other hand, served for the mapping of cancer biomarkers as illustrated by Jaume Seuma (University of Sheffield). A new exciting tool involving ion mobility mass spectrometry was introduced by Keith Compson of Waters Corporation. Ion mobility spectrometry adds another dimension of separation since isobaric ions can now be separated based on their differences in size, charge and shape.

The conference was again accompanied by an industry exhibition featuring the latest technology in sample preparation, microdissection and amplification. Participants came from as far as Africa and the United States which demonstrates the growing impact of the conference.

Speaker Abstracts

Single Cell Gene Expression Analysis in Human Post-Mortem Brain Tissues and Animal Models of Neurodegeneration

Stephen D. Ginsberg
Department of Psychiatry
New York University School of Medicine
550 First Avenue, New York, NY 10016, USA

Technical and experimental advances in microaspiration techniques, RNA amplification, quantitative real-time PCR (qPCR), and microarray analysis have led to an increase in the number of studies of single cell gene expression. In particular, the central nervous system (CNS) is an ideal structure to apply single cell gene expression paradigms. Unlike an organ that is comprised of one principal cell type, the brain contains a constellation of neuronal and non-neuronal populations of cells. A goal is to sample gene expression from similar cell types within a defined region without potential contamination by expression profiles of adjacent neuronal subpopulations and non-neuronal cells. The unprecedented resolution afforded by single cell RNA analysis in combination with RNA amplification methods and microarrays allows for relative gene expression level comparisons across cell types under different experimental conditions and disease states. The ability to analyze single cells is an important distinction from global and regional assessments of mRNA expression, and can be applied to optimally prepared tissues from animal models as well as post-mortem human brain tissues. This presentation illustrates the potential power of single cell gene expression studies within the CNS in relation to neurodegenerative disorders such as Alzheimer's disease (AD).

Dementia, or progressive memory loss and cognitive decline, describes a syndrome associated with a range of diseases that progressively impair brain functions and rob the afflicted of their ability to learn, reason, make judgments, communicate and carry out daily activities. Dementia knows no social, economic, ethnic or geographical boundaries. The full extent of its impact has only begun to be appreciated. The leading cause of dementia, AD, accounts for approximately 50-60% of all cases. Increasing age is the greatest risk factor for AD. Another clinical pathological entity, termed mild cognitive impairment (MCI,) describes a population whose memory or other aspects of cognition are not normal but who do not meet the clinical diagnosis of dementia. MCI has recently been recognized and may mark the earliest stages where a person shifts from a normal aging course towards dementia and ultimately AD. The likelihood of developing AD approximately doubles every five years after the age of 65. By age 85, the risk reaches nearly 50 percent. The duration of dementing illness may often vary from 3 to 20 years. It is imperative that we learn more about the causes of neurodegeneration that lead to AD in order to develop rational therapies that delay the onset of MCI, AD, and related dementing illnesses.

Biological substrates or mechanisms underlying the specific loss of certain types of neurons and synapses remain elusive. The overall goal of our research effort is to understand the molecular and cellular events that cause cell-type specific neuronal damage, or selective vulnerability in the brain. One experimental approach that we employ is to study a specific region of the brain called the hippocampus, a structure located in the forebrain that is involved intimately with learning and memory. Cells within the hippocampus are removed (termed microdissected) and cellular and molecular biological studies are performed to assess the cohort of genes within them that may lead to disease pathogenesis. The goal is to identify those messages or transcripts that change over time in relationship to the synaptic and cellular damage and discover a mechanism that explains why specific neurons degenerate, and whether plastic changes occur in the brain to compensate for these losses. By understanding these fundamental processes, we hope to develop a series of therapeutic strategies to prevent neurodegenerative diseases from turning into debilitating and life threatening illnesses.

A multidisciplinary approach of surgical, state-of-the-art molecular biology, immunohistochemical, and imaging techniques are utilized as part of the experimental design. Particular emphasis is placed upon analyzing single neurons *in vivo* as a means of understanding cellular events occurring locally at synaptic and somatodendritic sites. For example, post-mortem normal control, MCI, and AD tissues are combined with regional and single cell mRNA amplification techniques and microarray technology to assess several classes of transcripts simultaneously including glutamate receptors, dopamine receptors, synaptic proteins, cytoskeletal elements, neurotrophins, cell death genes, and transcription factors from individual neurons and their processes. These studies enable a "molecular fingerprint" of the hippocampus as well as specific neurons within the region. Furthermore, these studies are hypothesized to elucidate markers for early cell-specific synaptic and neurodegenerative changes that can be applied to other models of activity dependence and neurodegenerative disorders.

Frontiers in Nano- and Microstructures on Microarrays for Real Time and Online Bio-Monitoring of Cellular Physiology

Andrea A. Robitzki
Department of Molecular Biological-Biochemical Processing Technology
University of Leipzig, Centre for Biotechnology and Biomedicine
Deutscher Platz 5, 04103 Leipzig, Germany

The demand on innovative cell analysis and/or functional genomics/proteomics is the miniaturisation of devices and the optimisation of real time monitoring for a temporal and spatial high resolution of functional proteome. This report represents an overview and introduction of novel cell based micro-sensors used as a (i) cardiomyocyte biosensor, (ii) mamma-carcinoma chip, and (iii) an embryonic cell microarray etc. coupled to a micro-laser manipulation platform. The goal of the biosensor technology is the development of high-sensitive screening systems that allows the electronic detection of drugs, tissue-secreted factors, effects of proteins on physiology, and receptor-directed cellular alterations by impedance spectroscopy and/or electrophysiological recording. For this purpose, e.g. spontaneously beating neonatal rat cardiomyocytes or human smooth muscle cells were cultured on substrate integrated microelectrodes for measuring activity by extracellular recording of field potential. Each single cell can be addressed and monitored on a single micro-electrode by bioelectronical and/or optical measurements. Therefore, the positive chronotropic effect or activity of angiotensin II and its binding to the AT_1 receptor resulting in an initiation of the signalling pathway can be detected in real time, online and with an extreme high sensitivity (at very low concentrations of 10^{-11}M). Furthermore, at the interface of micro-sensor system technology and cellular bioengineering a multiwell-multi-microelectrode array (MMEA) has been designed and developed for the screening of cellular ischemia or of cytostatica, phorbol-12-myristate-13-acetate (PMA) etc. induced cellular changes in single cells.

The biochip consists of cells as the biological target expressing target molecules for the detection of ligand / receptor interactions via impedance spectroscopy (reflecting the molecular and physiological alterations dependent on the frequency range up to 10^6 Hz by a change of the extracellular resistance). The status of a viable cell (e.g. stem cell, precursor cell etc.) e.g. its proliferation, differentiation, maturation, apoptosis *versus* necrosis or migration can be detected in real time using the described biophysical measurement systems. For a parallel real time monitoring in viable cells the sensor platform are adapted to a 96 well plate-MMEA for a high content screening and characterisation of viable cells depending on their proteome and *toponome*.

MICROFLUIDIC PLATFORMS FOR ANALYSIS OF LIVING CELLS AND 3D LIPID STRUCTURES

Petra Dittrich

Department of Miniaturisation
ISAS - Institute for Analytical Sciences
Bunsen-Kirchhoff-Str. 11
44139 Dortmund
Germany

Living cells stay in permanent contact with other cells, surfaces and solutions. The near environment strongly influences the behaviour of the cell – its viability, differentiation and metabolism. Studying these influences will provide detailed and deep insights into the underlying processes of the cellular life cycle. However, to address cells on an individual level requires tools of the same dimensions, which is perfectly accomplished by microstructured chips. In this presentation, an overview of different approaches and new developments towards cell manipulation and analysis by means of optical and microfluidic methods will be given.

Living cells are complex systems though. A versatile approach for many applications is the replacement of cells by biomimetic objects such as lipid vesicles, which are simple model systems for cells as well as cell organelles due to their similarities in size and membrane composition. The second part of the presentation focuses on methods to form, detect and analyse biomimetic objects in microfluidic chips. In particular, a novel formation pathway for three-dimensional (3D) lipid structures will be presented that combines top-down photolithography techniques and the self-assembling ability of polar lipids. This micro-extrusion process facilitates the formation of spherical, cylindrical, tubular and helical lipid membrane structures. Such 3D lipid structures are useful tools in analytical chemistry, pharmacy, and micro-/nanotechnology since they could serve as sensors, as containers for transport and delivery of sample, and as reaction vessels for conducting biochemical processes. Moreover, they could be employed as building blocks for novel biomimetic soft matter devces.

THE HUMAN TOPONOME PROJECT: DECIPHERING THE PROTEIN NETWORK CODE UNTIL 2020

Walter Schubert
Molecular Pattern Recognition Research Group
Institute of Medical Neurobiology
Otto-von-Guericke University
Leipziger Str. 44, 39120 Magdeburg
Germany

Understanding how proteins are temporally and spatially arranged and how this relates to function is a major post-genomic challenge. We have recently described a multi-dimensional microscopic robot technology which runs cycles of fluorescence protein tagging, imaging, and bleaching *in situ* (Schubert W *et al.* Nature Biotechnology 24, 1270-1278, 2006). It provides a major milestone in proteomics (Proteomics Research highlight "Mapping togetherness", Nature 443, Oct 12, 2006). This technology combines three advances: a technique capable of mapping hundreds of different proteins at light microscopic resolution in one tissue section or cell sample; a method for selecting the most prominent combinatorial molecular patterns by representing the resulting data as binary vectors; and a system for imaging the distribution of these protein groupings in a 'toponome map'. By analysing many cell and tissue types we have shown that this approach reveals new hierarchical properties of protein network organisation, in which the frequency distribution of different protein groupings obeys Zipf's law, and state specific lead proteins control protein network topology and function. The technology offers a rapid route to new diagnostic features and targeted therapies. A toponome initiative has been launched to establish Toponome Centres worldwide, all jointly contributing to deciphering the human toponome (the entirety of protein networks) across human tissues and cell types in health and disease within the next 12 years. The project is open for cooperation and contributions from all sides (www.Toposnomos.com).

NEW INSIGHTS INTO MOLECULAR MECHANISMS UNDERLYING AXON REGENERATION THROUGH CELL SORTING

Dietmar Fischer
Experimental Neurology, Festpunkt O25, Raum 541
Universitätsklinikum Ulm, Albert-Einstein-Allee 11
89081 Ulm, Germany

Following optic nerve injury in mature rats, retinal ganglion cells (RGCs) normally fail to regenerate their axons and soon undergo apoptosis. However, inducing inflammatory reactions in the eye, *e.g.*, by injuring the lens, enables many axotomized RGCs to regenerate lengthy axons into the optic nerve. To study the molecular changes that underlie the transformation of mature RGCs from a degenerative state to a regenerative state *in vivo* we pre-labeled RGCs by injecting the fluorescent dye 4Di-10ASP into the superior colliculus, performed optic nerve surgery with or without lens injury, and 4 days later, dissected retinas, dissociated their cells, and isolated RGCs by fluorescence-activated cell sorting. The population of isolated RGCs revealed a purity of 95%. The extracted RNA from these sorted cells was amplified and used for microarray analysis. The mRNA profiling revealed that although axotomy alone significantly alters the expression of a large cohort of genes, switching RGCs to a regenerative state involves the differential expression of a relatively small number of additional genes. A number of the regeneration-specific genes have previously been linked to development or axon regeneration in the PNS, but most were novel. Among these was also an uncharacterized ring finger protein that we named LINA. LINA upregulation in regenerating RGCs was confirmed at the RNA and protein levels. Endogenous LINA expression was also upregulated in differentiating PC12 cells and it's over expression enhanced neurite outgrowth. A point mutation in the ring finger domain decreased neurite outgrowth in a dominant negative fashion. AAV mediated over expression of LINA in RGCs improved survival of RGCs and strongly enhanced axon regeneration into the optic nerve. On the contrary, the dominant negative form of LINA compromised the beneficial effects of intraocular inflammation. This data suggest that LINA plays an important role in axon regeneration of neurons and demonstrates that purification of adult CNS neurons is a powerful tool to achieve new insights into molecular mechanisms underlying diseases or regenerative processes.

SINGLE CELL WHOLE GENOME AMPLIFICATION: UNLEASHING A WORLD WITHIN A CELL

Markus Veit
Sigma-Aldrich Chemie GmbH
Eschenstr. 5, 82024 Taufkirchen
Germany

Analyzing genomic DNA at the level of the single cell will provide the ultimate level of variation analysis. Amplifying the picogram quantities of DNA in a single cell has been difficult. The recent advent of commercial kits for whole genome amplification have provided scientists the means to amplify the information from ~3000 cells, affording amplification that is a complete and faithful representation of the original DNA. While advancing the field of DNA variation analysis, these kits still have limitations in fields such as oncology, molecular pathology, and in vitro fertilization where analyzing the DNA from a single cell is the optimal choice. The GenomePlex® Single Cell Whole Genome Amplification Kit was created to address these limitations. GenomePlex is a whole genome amplification (WGA) method that allows the researcher to generate a representative amplification of genomic DNA from a single cell.

REGULATION AND FORCE GENERATION OF THE DNA IMPORT MOTOR OF *BACILLUS SUBTILIS*

Berenike Maier
Institut für Allgemeine Zoologie und Genetik
Westfälische Wilhelms Universität
Schlossplatz 5, 48149 Münster
Germany

At high cell concentration *Bacillus subtilis* becomes competent for DNA transformation. Transformation enables the bacterium to acquire novel genetic information including antibiotic resistance traits. Intriguingly, only a fraction of a genetically identical cell population switches into the competent state, where the master regulator for competence is expressed at high level. Thus the population can be considered to divide into a fraction that is "on" (K-state) and into an "off" fraction (B-state). In the K-state cells express the proteins required for DNA import and assemble the DNA import machine.

Using a single cell assay, we have addressed the question how the fraction of competent cells is determined. We used a fluorescence reporter for the expression of the master regulator on competence genes. Our results supported the hypothesis that noise in expression of the master regulator is essential for bi-stability. Furthermore we found that cells can switch into the K-state exclusively during a very well defined "switching-window" and that by tuning the size of this window we were able to control the fraction of competent cells. Basal expression rate of the master regulator was involved in setting the size of the "switching window". Once the cells were in the K-state we characterized the dynamics of proteins that form the DNA-import machine and characterized DNA import at the single molecule level.

TWO-FOCUS FLUORESCENCE CORRELATION SPECTROSCOPY: A VERSATILE TOOL FOR PRECISE MEASUREMENTS OF MOLECULAR DIFFUSION

Jörg Enderlein
Institute of Physical and Theoretical Chemistry
Eberhard Karls University Tübingen
Auf der Morgenstelle 8, 72076 Tübingen
Germany

Thermally induced translational diffusion is one of the fundamental properties exhibited by molecules within a solution. Via the Stokes-Einstein relation it is directly coupled with the hydrodynamic radius of the molecules [1]. Any change in that radius will change the associated diffusion coefficient of the molecules. Such changes occur to most biomolecules – in particular proteins, RNA and DNA – when interacting with their environment (e.g. binding of ions or other biomolecules) or performing biologically important functions (e.g. enzymatic catalysis) or reacting to changes in environmental parameters such as pH, temperature, or chemical composition (e.g. protein unfolding). Therefore, the ability to precisely measure diffusion coefficients has a large range of potential applications, for monitoring e.g. conformational changes in proteins upon ion binding or unfolding. However, many biologically relevant conformational changes are connected with rather small changes in hydrodynamic radius on the order of Angstroms (see e.g. [2]). To monitor these small changes, it is necessary to measure the diffusion coefficient with an accuracy of better than a few percent.

An elegant technique capable of measuring diffusion coefficients of fluorescent molecules at nanomolar concentrations is Fluorescence Correlation Spectroscopy (FCS) which was originally introduced by Elson, Magde and Webb in the early seventies [3]. In its original form it was invented for measuring diffusion, concentration, and chemical/biochemical interactions/reactions of fluorescent or fluorescently labelled molecules at nanomolar concentrations in solution. However, standard FCS is prone to a wide array of optical and photophysical artefacts which make precise quantitative and absolute measurements of e.g. diffusion coefficients rather difficult [4]. The main problem of standard FCS is the absence of a reliable extrinsic length scale in the measurements, which is, however, necessary for obtaining absolute values of the diffusion coefficient.

Here, we report on our recently developed new technique of 2-focus fluorescence-correlation spectroscopy [5], allowing for measuring the hydrodynamic radius of molecules at pico- and nanomolar concentrations with sub-Angstrom precision. In 2fFCS, the problem of an extrinsic length scale is solved by generating two excitation foci with well defined distance from each other. Several applications of 2fFCS are presented, for example monitoring conformational changes of proteins upon ion binding, or monitoring protein unfolding curves upon chemical and thermal denaturation.

REFERENCES

[1] A. Einstein, *Investigations on the Theory of the Brownian Movement*, Dover, New York, 1985.

[2] A. M. Weljie et al., *Protein Science* 12 (2003) 228-236.

[3] D. Magde et al. *Phys. Rev. Lett.* 29 (1972) 705-8

[4] J. Enderlein et al. *Chem. Phys. Chem.* 6 (2005) 2324-36.

[5] T. Dertinger et al. *Chem. Phys. Chem..* 8 (2007) 433-443.

INVESTIGATION OF CONFORMATION AND DYNAMICS OF PROTEINS INSIDE LIVING CELLS BY IN-CELL NMR SPECTROSCOPY

Volker Dötsch

Institute of Biophysical Chemistry
Johann Wolfgang Goethe University
Marie Curie Str. 9, 60439 Frankfurt, Germany

The non-invasive character of NMR spectroscopy allows researchers to investigate the conformation and dynamics of biological macromolecules in their natural environment, for example the cytoplasm of cells. However, many proteins and other macromolecules interact with other cellular components, which increase their rotational correlation time, leading to broad peaks or even complete disappearance of the signals. Several small proteins that in principle should tumble fast are, therefore, invisible in in-cell NMR experiments. Since studying the interaction between the protein of interest and other cellular partners is the key focus of in-cell NMR experiments we have started to develop methods that allow us to observe proteins forming complexes as well. Two different strategies seem very promising. The first one is labeling of methyl groups, in particular of methionines as well as the δ-methyl groups of isoleucines. Their slow relaxation makes it possible to observe proteins that cannot be studied by amide proton based experiments. For even larger complexes solid state in-cell NMR experiments are the method of choice. Here the main difficulty is the labeling strategy since strong background signals are observed in these experiments.

From Mass Spectrometric Peptide Profiling of Single Neurons towards Imaging Mass Spectrometry

Connie Jimenez
OncoProteomics Laboratory
VUmc-Cancer Center Amsterdam, CCA 1-60
VU University Medical Center
De Boelelaan 1117
1081HV Amsterdam

Neuropeptide profiling is the analysis of the neuropeptides present in a single cell or tissue extract. Mass spectrometry has proven to be a great tool to detect and identify the peptides, which provides information on the processing and post-translational modifications of each peptide. We pioneered the use of MALDI-TOF mass spectrometry for detection and differential display of peptides in single cells (J. Neurochem. 1994), and for semi-quantitative pattern analysis of peptide profiles directly in tissue biopsies (PNAS 1997). In this presentation, I will give a historical overview and present our more recent work on the dissection of PTMs and functions of neuropeptides discovered by our MALDI-TOF-MS-based single cell peptide profiling studies and the use of imaging mass spectrometry at sub-cellular resolution.

Fine Scale Measurement and Mapping of Cancer Biomarkers via Laser Ablation ICP Mass Spectrometry

Jaume Seuma
Department of Chemistry
Centre for Analytical Sciences, University of Sheffield
730-740 Centennial Court, Centennial Park
WD6 3 SZ Elstree, Herfordshire, UK
J. Bunch, A. Cox, C.W. McLeod and C. Murray[1]
[1]Medical Solutions plc, Orchard Place, Nottingham Business Park,
Nottingham NG8 6PX United Kingdom

A wide range of analytical technologies have been developed for imaging applications and MS represents a unique sub-set of the 'imaging tool kit' which includes techniques such as SEM, TEM, MRI., μXRF, SR-XRF, X-ray CT, fluorescence, autoradiography etc.. Thus MS in the form of SIMS, MALDI-MS and LA-ICP-MS represents a powerful route for visualization of the spatial (2D/3D) distribution of chemical constituents in diverse sample types. Collectively the laser and ion-beam probing methodologies provide comprehensive measurement and mapping capability across the mass and length scale.

ICP-MS has been used extensively in the life sciences for trace inorganic determinations in diverse sample types. The analytical capabilities of ICP-MS are extended through the use of laser ablation (LA) which permits direct probing of element in solid samples. An opportunity for extending the scope of LA-ICP-MS, particularly in the life sciences, relates to measurement of elemental-tagged antibodies through interrogation of immunohistochemical sections. Such a strategy represents a powerful new approach for quantitative measurement and mapping of peptides and proteins in biological tissue, an activity that traditionally has been the preserve of molecular MS. The presentation will briefly review the status and scope of MS imaging and then give emphasis to bioimaging via measurement of elemental-tagged antibodies. This aspect of research will be illustrated by reference to mapping of β-amyloid peptides (and associated metal ions) in Alzheimer plaques and imaging of cancer biomarkers (HER 2, MUC 1) in breast cancer-derived immunohistochemical sections. Method development studies were designed to benchmark to optical microscopy and selection of fine laser beam (5 μm) and low line raster rate were essential for imaging at the cellular level (MUC 1) whereas wide beam diameter (>100 μm) was suited to scoring of HER 2 status. Most recent work concerning the use of SIMS for interrogation of immunohistochemical sections will also be elaborated.

MALDI SYNAPT HDMS SYSTEM: ADVANCES IN HIGH-EFFICIENCY IMS/MS

Keith Compson
Waters Corporation
Atlas Park, Simonsway
Manchester, UK

MALDI is a well established ionisation technique for the analysis of large molecules by mass spectrometry. As samples are generally introduced into the mass spectrometer in the solid state, adding additional dimensions of separation, such as liquid chromatography usually requires an intermediate step, *e.g.* fraction collection. Ion mobility separation (IMS), however, can be coupled directly with MALDI MS analysis, as it provides gas phase separation of ions post ionisation. The use of ion mobility spectrometry adds another dimension of separation to the experiment, providing separation of species by their associated drift time, a factor which is dependant on ion size, charge and shape. The Triwave device used for performing the ion mobility separation contains three T-wave regions arranged in series. The Trap T-wave is used for ion storage prior to IMS, the second T-wave, maintained at a higher pressure (typically 0.5 mbar), is used to perform the mobility separation and the Transfer T-wave is used to transmit the separated packets of ions into the TOF mass analyser. Fragmentation can be induced in the Trap T-wave and Transfer T-wave, making it possible to perform a number of novel fragmentation experiments that assist in the characterisation of analytic.

We will give a detailed description of this novel MALDI mass spectrometer and present IMS MS data. For example, using this technique it is possible to isolate low abundance multiply charged peptide ions generated in a MALDI peptide mass fingerprinting experiment and generate fragment ion spectra from co-resident isobaric species.

New Biotechnology Method for Localised Analysis of Single Cells

Tanja Ninković, Peter Lanigan, Oscar Ces, Mark Neil and David Klug
Single Cell Proteomics Group,
Chemical Biology Centre,
Imperial College London

Cell polarisation and the localised distribution of specific proteins are hallmarks of cell migration, which drives processes such as organogenesis and cancer metastasis. We have developed a microfluidic-based system for a locally targeted manipulation of the cell using optical tweezers. The microfluidic device includes a chamber for cell growth and channels for the flow of analytes. This system can deliver chemicals to a selected area of the cell surface. Furthermore, the chemicals can include detergents digesting the defined part of the cell membrane, allowing discrimination between the leading and trailing edges of a migrating cell. Such controlled delivery of substances to confined cellular regions has previously not been possible. A targeted membrane digestion also reduces contamination and enables real-time tracking of membrane changes during migration. In future studies we plan to focus on the analysis of migrational mechanisms during cancer cell metastasis, which may reveal the spatial distribution of proteins, lipids and nucleic acids involved in these processes.

Stem Cells for Therapy of Inherited Liver Diseases

Andree Zibert, Vanessa Sauer, Jörg Haberland and Hartmut H.-J. Schmidt*
Universitätsklinikum Münster, Transplantationshepatologie,
Domagkstr. 3a, 48149 Münster, Germany

Mesenchymal stem cells (MSC) derived from bone marrow possess enormous potential for cell-based therapy of liver disease. *In vitro* and *in vivo* studies have shown that MSC differentiate into hepatocyte-like cells although efficiency of differentiation was generally low. To study the effect of MSC after transplantation into animal models of liver disease, we have isolated, expanded and cryopreserved MSC from human bone marrow. Cells were characterized by flow-cytometry and expressed typical surface markers ($CD45^-$ $CD34^-$, $CD14^-$, $CD29^+$, $CD90^+$, $CD44^+$, $CD166^+$, $CD105^+$). By RT-qPCR anlysis, MSC showed low expression of marker genes associated with pluripotency (e. g. Oct4 and Nanog). However, upon hepatogenic stimulation *in vitro*, MSC differentiated almost completely into hepatocyte-like cells. Cell morphology changed from a spindle-like to a polygonal-shape typically observed in adult hepatocytes. Relative gene expression of early (AFP, CK19, CK7) and late hepatocyte markers (CYP, CX32, CK18, albumin) were significantly up-regulated by 1-5

* hepar@ukmuenster.de

orders of relative gene expression. Since gene transfer may further enhance the therapeutic effectivity of these cells, e. g. by over-expression of therapeutic transgenes or correction of inherent mutations, we also established retroviral gene transfer protocols. Cells will be transplanted into a rat model of Wilson disease (WD) which leads to copper overload of liver followed by acute and chronic liver disease. Our results suggest that the *in vitro* generated hepatocyte-like cells carry many functions of adult human hepatocytes, which is highly significant for clinical applications.

ARE THE GENOMICALLY PREDICTED ADIPOKINETIC PEPTIDES OF *TRIBOLIUM CASTANEUM* PRESENT IN THE CORPORA CARDIACA?

***Gerd Gäde*[*1], *Heather G. Marco*[1], *Petr Šimek*[2], and *Robert J. Weaver*[3]**

[1]Zoology Department, University of Cape Town, Private Bag, ZA-7701 Rondebosch, Republic of South Africa

[2] Laboratory of Analytical Biochemistry, Biology Centre, Academy of Sciences of the Czech Republic, Branišovská 31/1160, CZ-37005, České Budějovice, Czech Republic

[3] Environmental Biology Group, Central Science Laboratory, Sand Hutton, York YO41 1LZ, U.K.

With the ever-expanding range of genomic sequences produced over the past years, several whole insect genomes are known. This genomic information can be "mined" to annotate potentially expressed gene products. We were interested to see how close such predictions come to reality. In our case study, we use the red flour beetle, *Tribolium castaneum*, to analyse the most abundant neuropeptides that are synthesised in the corpora cardiaca, viz. the adipokinetic peptides. Such peptides belong to a large family which is characterised by blocked termini (pGlu and carboxyamide) and the aromatic amino acids Phe and Trp at position 4 and 8. Two octapeptides that fulfil these structural criteria are predicted to be encoded in the genome of *T. castaneum*. Although the adult beetle is minute in size and about 100-fold smaller than an adult yellow mealworm beetle, *Tenebrio molitor*, we were able to dissect the corpora cardiaca. One to three of the glands were put on a MALDI plate, rinsed very briefly with ice-cold distilled water and covered with a droplet of matrix in 60% methanol plus acetonitrile. Four distinct mass peaks at *m/z* 1014.2/1030.2 and 1023.3/1039.3 were observed. This shows the typical AKH mass signature; the mass difference of 16 units represents $[M+Na]^+$ and $[M+K]^+$. In a second set of experiments, a methanolic extract of 30 glands was separated on a C-18 column by liquid chromatography and detected by an LTQXL mass spectrometer (Thermo Fisher). MS/MS sequencing of the peaks of interest revealed two octapeptides: one with $[M+H]^+$ at *m/z* 1001.4 is assigned as pGlu-Leu-Asn-Phe-Thr-Pro-Asn-Trp amide and the other with $[M+H]^+$ at *m/z* 992.4 is assigned as pGlu-Leu-Asn-Phe-Ser-Thr-Asp-Trp amide. These sequences are consistent with the processed peptides predicted in the genome project.

[*] E-mail: gerd.gade@uct.ac.za

ACKNOWLEDGEMENT

Funding by the Royal Society-National Research Foundation (SA-UK Networking Grant) is gratefully acknowledged.

SATELLITE EVENTS

BEING FAST AND EFFICIENT – NEW TECHNOLOGIES FOR RELIABLE SAMPLE PREPARATION

Noushin Delmdahl
Lab Technologies
Sartorius Stedim Biotech
Weender Landstrasse 94-108
37075 Göttingen
Germany

Sample preparation is always a crucial step when working with biological samples, especially proteins. It is mandatory to have easy to use and highly reproducible methods in order to achieve scientifically reliable results in down stream applications. In this presentation, we will introduce fast ultra-filtration devices for sample preparation along with passivation methods for increasing recoveries when working with diluted protein samples. For sample fractionation, e.g. prior to 2D-PAGE, an innovative method using ion exchange spin columns based on membrane adsorbers as a matrix will be introduced as a reliable and affordable technology.

BIOMARK MICRO FLUIDIC ARRAYS: A PROVEN METHOD TO QUANTIFY GENE EXPRESSION WITHIN A SINGLE CELL

Martin Pieprzyk
Fluidigm Corporation
7000 Shoreline Court, Suite 100
94080 South San Francisco, USA

Historically, single-cell gene expression experiments have been difficult and expensive to perform. Fluidigm has introduced a new single-cell gene expression technique that, when used with the BioMark System, produces inexpensive, easily reproducible, gene expression results from single-cell samples. The BioMark system which utilizes micro fluidic technology is ideally suited to single cell work by minimizing reaction size and allowing up to 48 individual reactions to be created from as little as 1 ul of sample.

The method is ideally suited for high-throughput cell-line studies to determine individual cell behavior in what has been believed to be homozygous populations. The method is also ideally suited to determine single gene cell expression levels in everything from circulating tumor cells (CTCs) to stem cells. The data we will present includes data from single human cells from 8-cell stage embryos which were collected and analyzed for expression of 46 developmental genes.

GENE EXPRESSION PROFILING OF LASER CAPTURE MICRODISSECTED SAMPLES USING SUPERAMP™ AMPLIFICATION METHOD

Kurt Sieberns
Miltenyi Biotec GmbH
Friedrich-Ebert-Straße 68
51429 Bergisch Gladbach,Germany

The analysis of single or a few cells is of increasing importance for the identification of cell-specific marker genes. For example, tumors represent a heterogeneous mixture of neoplastic and non-neoplastic cells. Cells of interest might be represented by a minority of cells and thus, their gene expression profiles might be under the detection limit of conventional methods. The isolation of a certain cell population using laser-assisted microdissection, fluorescent activated cell sorting, or magnetic cell sorting is often a prerequisite to gain insight into cell type-specific gene expression patterns.

The SuperAmp™ Service was developed to enable the microarray-based gene expression analysis of low cell numbers, starting with a minimum amount of one cell. The Laser Microdissection and Pressure Catapulting technology (P.A.L.M. Microlaser Technologies) represents a method for e.g. the isolation of single cells from tissue sections. The aim of the present study was to validate the suitability of the SuperAmp™ Technology for laser-capture microdissected cells derived from cryo-fixed and paraffin-embedded melanoma tissue sections. A total of 36 samples were isolated from sections of cryo-fixed or formalin-fixed paraffin-embedded tissue sections. Sample sizes ranged from 1-1000 cells; the sections were excised from the central regions of the melanoma. Gene expression profiling was performed using two-color PIQOR™ Skin Microarrays.

The SuperAmp™ Technology offers a robust and reliable method for the analysis of gene expression even of small cell numbers such as cells isolated by laser capture microdissection. The high correlation values support the high reproducibility of the modified global PCR strategy. To our knowledge no comparable method for the microarray analysis of low cell counts is available.

LASER-MICRODISSECTION OF SINGLE LIVING CELLS OPENS NEW FIELDS OF APPLICATIONS

Monika Stich
P.A.L.M. Microlaser Technologies GmbH
Carl Zeiss MicroImaging GmbH
Am Neuland 9 + 12, 82347 Bernried, Germany

Laser Microdissection and Pressure Catapulting (LMPC) is a non-contact method to separate cell areas, single cells or sub cellular compartments out of a fixed tissue section. In recent investigations this technique was modified and optimized for the work with living cells.

The isolation of individual, e.g. fluorescently marked, cells out of a mixed culture and the subsequent cultivation of these captured cells was established on different cell lines in our lab. After capturing selected cells grow out to new cell colonies, not only preserving their viability and proliferation characteristics, but also keeping their genetic information unaltered. This was proved using LMPC on the carcinoma cell line HCT116. After re-cultivation the cells were analysed by CGH (comparative genomic hybridization) and M-FISH (multiplex-fluorescence in-situ hybridization).

Further experiments were done with different stem cell lines to present a solution for the challenge of receiving homogenous cell clones. Embryonic stem cells differentiate into various kinds of cells and it is difficult to obtain a specific differentiated cell type from this assay. To overcome this problem, laser-microdissection techniques are the most convenient way to work only with defined stem cells, especially concerning therapy approaches. LMPC offers a method to receive homogenous cell cultures out of one single isolated cell (clonal expansion). In our experiments it was shown that the cells keep their stem cell character, i.e. by detection of different stem cell markers. For example the expression of CD34 and OCT-4 was unchanged after LMPC and re-cultivation. Additionally these results were proved by FACS (Fluorescent Activated Cell Sorting).

The technology of LMPC represents a progress in the contamination-free isolation and analysis of living cells. It opens a wide field of interesting applications in cell and developmental biology and even drug development research, where homogeneous cell clusters or defined clones are needed.

SEX HORMONES AND CARDIOVASCULAR DISEASE; NOVEL LIGANDS, NEW HOPES AND CLINICAL IMPLICATIONS

Theo Pelzer
AG Molekulare Kardiologie, Medizinische Klinik I - Raum 3.455
Josef-Schneider Str. 2, 97078 Würzburg, Germany

Advances in gender based medicine are inherently linked to our improved understanding of sex hormone effects on the cardiovascular system. The talk will update clinicians and basic scientists in the field of cardiology, gynaecology and endocrinology with recent results on the function of newly developed oestrogen- and progestin in heart failure, hypertension and cardiac hypertrophy, which have previously been challenged by the negative outcome of clinical trials on the prevention of cardiovascular disease by conventional hormone replacement therapy. The presentation will assist the audience to make their own judgements whether "New ligands" may translate into "New hopes".

INDEX

A

B

C

D

E

H

J

K

L

M

N

O

P

T

U

V